Remote Sensing of Cryosphere and Related Processes

Remote Sensing of Cryosphere and Related Processes

Guest Editors

Andrey Abramov
Stefano Ponti

MDPI Basel • Beijing • Wuhan • Barcelona • Belgrade • Novi Sad • Cluj • Manchester

Guest Editors

Andrey Abramov
Institute of Physicochemical and Biological Problems of Soil Science
Russian Academy of Sciences
Pushchino
Russia

Stefano Ponti
Theoretical and Applied Sciences
Università degli Studi dell'Insubria
Varese
Italy

Editorial Office

MDPI AG
Grosspeteranlage 5
4052 Basel, Switzerland

This is a reprint of the Special Issue, published open access by the journal *Remote Sensing* (ISSN 2072-4292), freely accessible at: https://www.mdpi.com/journal/remotesensing/special_issues/Photo_Glacier_Permafrost.

For citation purposes, cite each article independently as indicated on the article page online and as indicated below:

Lastname, A.A.; Lastname, B.B. Article Title. *Journal Name* **Year**, *Volume Number*, Page Range.

ISBN 978-3-7258-4317-6 (Hbk)
ISBN 978-3-7258-4318-3 (PDF)
https://doi.org/10.3390/books978-3-7258-4318-3

Cover image courtesy of Andrey Abramov

Contents

About the Editors

Andrey Abramov

Andrey Abramov, Head of Soil Cryology Department, Institute of Physicochemical and Biological Problems of Soil Science, Russian Academy of Sciences, is interested in permafrost, glaciology, climate change, physical geography, remote sensing, quaternary geology, sedimentology, and astrobiology. Finished PhD at the Geocryology department of Lomonosov Moscow State University, followed by numerous expeditions to the Arctic, Antarctic, and mountain areas.

Stefano Ponti

Stefano Ponti, a researcher at University of Insubria, Theoretical and Applied Sciences department. Interested in spatial analysis and statistics, geomorphological and vegetation mapping, geostatistical analysis, geo-processing, photogrammetry, satellite image processing, thermal analysis. Finished PhD at the University of Insubria, followed by numerous expeditions to the Arctic and Antarctic areas.

remote sensing

MDPI

Article

Spatial Distribution and Variation in Debris Cover and Flow Velocities of Glaciers during 1989–2022 in Tomur Peak Region, Tianshan Mountains

Weiyong Zhou [1,2], Min Xu [1,3,4,*] and Haidong Han [1,3]

1 Key Laboratory of Cryospheric Science and Frozen Soil Engineering, Northwest Institute of Eco-Environment and Resources, Chinese Academy of Sciences, Lanzhou 730070, China; 11210912@stu.lzjtu.edu.cn (W.Z.); hhd@lzb.ac.cn (H.H.)
2 Faculty of Geomatics, Lanzhou Jiaotong University, Lanzhou 730070, China
3 University of Chinese Academy of Sciences, Beijing 100049, China
4 National Cryosphere Desert Data Center, Northwest Institute of Eco-Environment and Resources, Chinese Academy of Sciences, Lanzhou 730070, China
* Correspondence: xumin@lzb.ac.cn

Abstract: In this study, we utilized a feature optimization method combining texture and topographical factors with the random forest (RF) approach to identify changes in the extent of the debris cover around the Tianshan Tomur Peak between 1989 and 2022. Based on Sentinel-1 image data, we extracted glacier flow velocities using an offset tracking method and conducted a long-term analysis of flow velocities in combination with existing datasets. The debris identification results for 2022 showed that the debris-covered area in the study region was 409.2 km^2, constituting 22.8% of the total glacier area. Over 34 years, the area of debris cover expanded by 69.4 km^2, reflecting a growth rate of 20.0%. Analysis revealed that glaciers in the Tomur Peak area have been decelerating at an overall rate of −4.0% per decade, with the complexity of the glacier bed environment and the instability of the glacier's internal structure contributing to significant seasonal and interannual variability in the movement speeds of individual glaciers.

Keywords: debris-covered glaciers; supraglacial debris identification; feature optimization; glacier velocity; offset-tracking

1. Introduction

According to the Sixth Assessment Report by the Intergovernmental Panel on Climate Change (IPCC), there is an increasing global rate of mass loss from mountain glaciers, with human-induced climate warming identified as the primary cause of glacier mass depletion [1]. Local glacial climate (temperature, precipitation, snowfall, etc.) along with non-climatic factors (topography, size, glacial ice temperature, etc.) result in significant spatial variability in glacier changes [1–3]. Located primarily in the ablation zones of glaciers, debris cover is another significant non-climatic factor influencing glacier dynamics. Thin layers of debris reduce the surface albedo of glaciers, increasing radiation absorption and thereby accelerating the melting of the underlying ice; conversely, thick debris layers act as an insulating barrier, inhibiting the melting of the ice below [4–7].

With the intensification of changes to glaciers, many debris-covered glaciers are experiencing varying degrees of expansion in their debris-covered areas [8–13], gradually highlighting the role of debris in glacier melting and movement processes. The spectral reflectance characteristics of debris, similar to those of surrounding rocks and moraine material, present a "different object with same spectrum" phenomenon, leading to the ineffectiveness of conventional glacier information extraction methods in identifying debris cover. Addressing this challenge involves considering various factors (topography, texture, surface temperature, etc.) and integrating multi-source data, becoming the primary means of enhancing the accuracy of debris cover identification [14–17]. Remote sensing spectral information of mountain glaciers exhibits high variability temporally and spatially, and debris cover characteristics like the distribution slope and surface temperature range vary as well. Machine learning methods significantly enhance work efficiency by learning target patterns and features from large datasets and automatically selecting appropriate thresholds for image classification. Neural networks, random forests, and support vector machines are machine learning methods widely applied in debris cover identification studies [11,16–19].

Glacier flow velocity serves as a key indicator reflecting glacier mass redistribution. Primarily controlled by glacier thickness, a reduction in thickness due to mass loss leads to decreased glacier velocity [20,21]. The presence of debris cover slows glacier retreat and thinning, substantially preserving terminus activity. Debris moves with the underlying ice, with its displacement speed essentially equivalent to the glacier's surface speed. Debris movement on the ice surface and local thickness changes are mainly governed by the glacier's horizontal velocity field [22–24]. The accumulation of moraine material and the development of supraglacial ponds and terminal moraine lakes make debris-covered glaciers more prone to disasters like glacial debris-flows and moraine-dammed lake outbursts [25–28]. Glacier movement is closely related to glacial disasters, and its monitoring provides crucial evidence for studying and predicting such disasters [29–31].

Glacier flow velocity extraction methods are classified into optical and radar remote sensing techniques. In optical remote sensing, cloud cover severely impacts image quality. For instance, in mid-latitude regions such as the Tianshan Mountains, frequent cloud cover above glaciers often results in significant data gaps [32]. The Synthetic Aperture Radar (SAR) imaging system demonstrates advantages in all-weather operation and cloud penetration [33]. Glacier velocity estimation methods using SAR data are divided into differential interferometric synthetic aperture radar (DInSAR) and offset-tracking [34]. Although DInSAR provides considerable accuracy in estimating velocities, decorrelation phenomena impede its further development and application. Conversely, offset-tracking does not face decorrelation issues and is minimally impacted by terrain undulations with short spatial baselines, suiting the mountainous terrain of debris-covered glaciers. Satellite-borne SAR systems operate in varying bands, with L, C, and X bands suitable for estimating debris-covered glacier flow velocities using offset-tracking [32,35–37].

The Central Tianshan Tomur Peak Region hosts China's largest concentration of debris-covered glaciers [38]. It is relatively complicated to correctly identify debris-covered glaciers based on remote sensing data, due to the "different object with same spectrum" issue. This limitation may be addressed by employing a feature optimization method that combines texture and topographical factors with the random forest approach. To achieve this, we used Landsat satellite data to identify changes in the extent of debris cover on glaciers around the Tomur Peak from 1989 to 2022. The study estimated the surface flow velocities of Tomur-type glaciers within the study area for 2020–2022, focusing on identified debris-covered areas, using the offset-tracking method. This analysis was augmented by combining glacier flow velocity data from 1989–2020 from the ITS_LIVE dataset and the Sentinel-1 glacier motion velocity dataset [39], to conduct a long-term analysis of surface flow velocities of debris-covered glaciers within the study area.

2. Study Area

Located at the trijunction of Kazakhstan, Kyrgyzstan, and China (41°10′N–42°40′N, 79°20′E–80°55′E), the Tomur-Khan Tengri Mountain Range is depicted in Figure 1. Around 60% of this region's land exceeds 4000 m in elevation, and its highest point, Tomur Peak, ascends to 7443.8 m. The glaciers in the Tomur Peak region are primarily dendritic large valley glaciers, characterized by their long, narrow tongues, extensive supraglacial debris cover, and well-developed thermokarst features. These glaciers can be referred to as "Tomur-type" glaciers. This study focused on Tomur-type glaciers in the Tomur Peak region, specifically those with an area greater than 40 km^2.

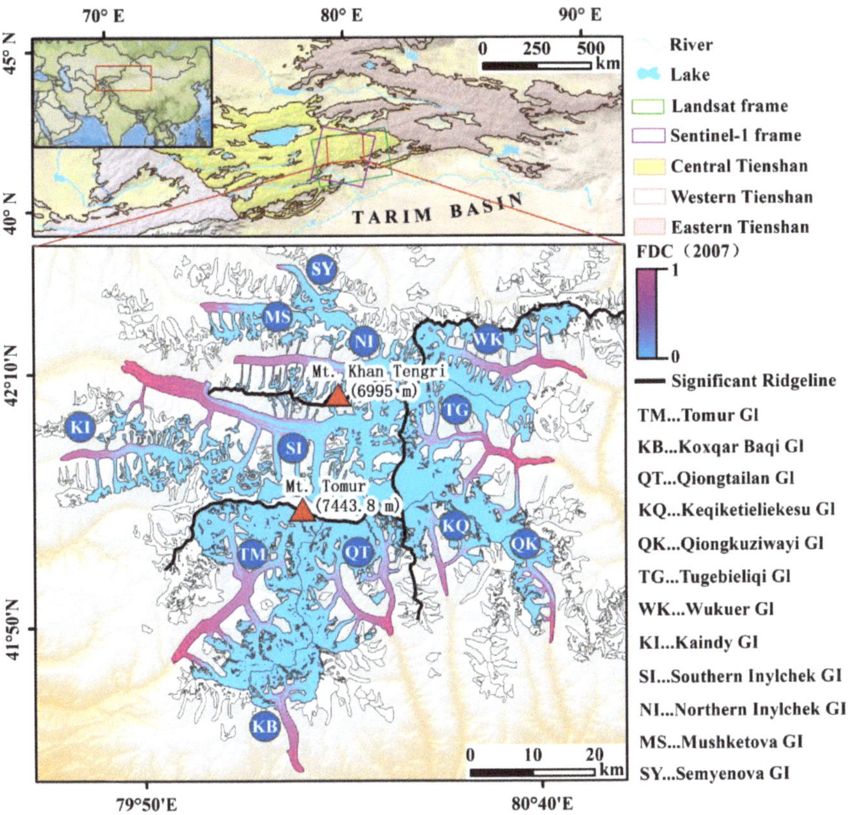

Figure 1. The geographical location and glacier distribution in the study area. Glaciers exceeding 40 km^2 are emphasized with FDC (fractional debris cover from linear spectral unmixing) color scales.

The Tomur Peak Region features a temperate continental climate, influenced by prevailing westerlies, with abundant precipitation mainly concentrated in the glacier melt season, accounting for 70% of the annual total [40,41]. The extensive elevation range of the glaciers and abundant material supply in the study area contribute to relatively high glacier velocities. Glacial processes, including erosion, transport, and deposition, are pronounced, fostering conditions for extensive development of debris cover. The melt season of glaciers in the Tomur Peak Region is from May to September. Debris cover serves as an "insulating barrier", mitigating glacial mass loss and curtailing glacier retreat and thickness reduction, thereby preserving glacier terminus activity to some extent.

3. Data and Methods

The experimental workflow of this study is shown in Figure 2. We utilized a feature optimization random forest method, combining texture and topographical factors, to identify the extent of debris cover around Tomur Peak in Central Tianshan. The method involved extracting and optimizing spectral, topographical, and texture features. Given the significant impact of cloud cover on optical image quality, we conducted flow velocity estimation from 2021 to 2022 using the SAR-based offset-tracking method. Debris distribution enhances glacier surface features, offering favorable conditions for remote sensing estimation of glacier velocities [42]. Glacier velocity estimations within debris-covered areas tend to be more stable and accurate [43]. To ensure the reliability of our conclusions, the analysis of glacier velocity was limited to the debris-covered areas identified in 2007.

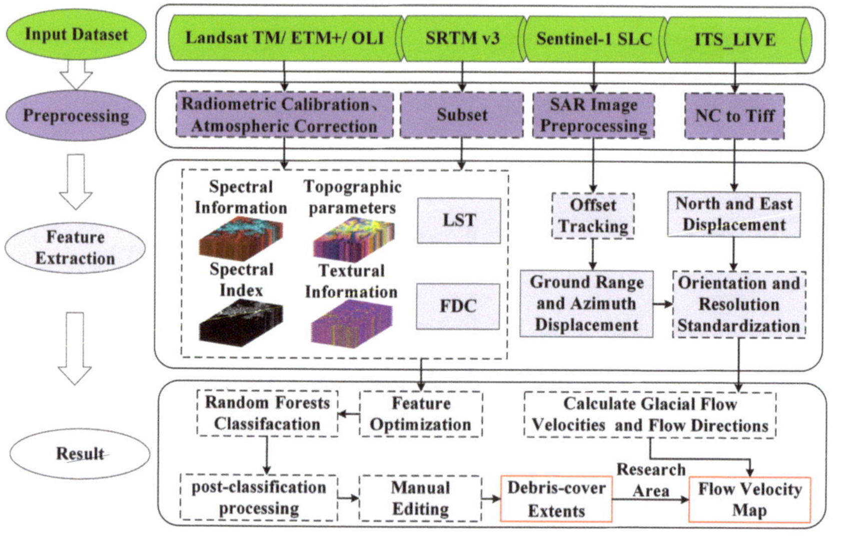

Figure 2. Workflow schematic for the semi-automatic delineation of the extent of debris cover and methodology for mapping glacier flow velocities. NC: network common data format; LST: land surface temperature.

3.1. Data Sources

3.1.1. Satellite Data

Landsat image data were obtained from the USGS Earth Explorer website (http://earthexplorer.usgs.gov/, accessed on 1 July 2024), as detailed in Table 1. Cloud-covered images were excluded, and selections were made from the latter part of the glacier melt season to reduce the impact of seasonal snow cover. For 2022, two images were combined to address cloud cover. Atmospheric correction was applied to all Landsat images, converting the original brightness values (digital numbers, DNs) to surface reflectance, to mitigate sensor differences and variations in observational conditions.

Table 1. List of Landsat scenes used to map supra-glacial debris.

Path-Row	Date	LANDSAT_SCENE_ID	Sensor	Cloud Cover (%)
147-031	22 August 1989	LT51470311989234ISP00	TM	3
147-031	10 September 1996	LT51470311996254ISP00	TM	3
147-031	24 August 2007	LT51470312007236IKR00	TM	10
147-031	1 September 2016	LC81470312016245LGN01	OLI	6.79
147-031	13 July 2022	LE71470312022194NPA00	ETM+	33
147-031	24 July 2022	LC91470312022205LGN01	OLI	26.48

We acquired Sentinel-1A data from the National Aeronautics and Space Administration (NASA)'s Alaska Satellite Facility. For glacier flow velocity estimation, this study utilized 49 single look complex (SLC) images, collected by Sentinel-1A from 2019 to 2022. The images' orbit and frame numbers were 158 and 132, respectively, with temporal baselines of 24 or 36 days. Table 2 provides a list of the Sentinel-1A images.

Table 2. Sentienl-1A SAR offset-tracking data pairs used in this study.

No.	Date-1_Date-2	No.	Date-1_Date-2
1	31 December 2020_5 February 2021	13	26 December 2021_31 January 2022
2	5 February 2021_1 March 2021	14	31 January 2022_24 February 2022
3	1 March 2021_25 March 2021	15	24 February 2022_1 April 2022
4	25 March 2021_30 April 2021	16	1 April 2022_7 May 2022
5	30 April 2021_5 June 2021	17	7 May 2022_31 May 2022
6	5 June 2021_29 June 2021	18	31 May 2022_24 June 2022
7	29 June 2021_4 August 2021	19	24 June 2022_30 July 2022
8	4 August 2021_28 August 2021	20	30 July 2022_4 September 2022
9	28 August 2021_3 October 2021	21	4 September 2022_28 September 2022
10	3 October 2021_27 October 2021	22	28 September 2022_3 November 2022
11	27 October 2021_2 December 2021	23	3 November 2022_9 December 2022
12	2 December 2021_26 December 2021	24	9 December 2022_2 January 2023

3.1.2. Reference Data

This study utilized the Shuttle Radar Topography Mission (SRTM) version 3 DEM, with a 30 m resolution, to provide the necessary topographical information. Land surface temperature (LST) data were derived from the Landsat Level-2 Surface Temperature Science Product. Both datasets were downloaded from the USGS Earth Explorer. Temperature and precipitation data were sourced from the European Centre for Medium-Range Weather Forecasts (ECMWF)'s ERA5 dataset.

The Randolph Glacier Inventory version 6.0 (RGI 6.0), obtained from Global Land Ice Measurements from Space (GLIMS) (http://www.glims.org/RGI/, accessed on 1 July 2024), served as a reference boundary for manual editing of the extent of debris cover. Google Earth™ provided crucial high-resolution imagery and topographical information as a reference for manual editing. Furthermore, Google Earth™ enhanced the identification of landform types for training sample selection, improving the sample quality and accuracy. Herreid et al. [44] conducted image threshold segmentation within RGI 6.0 glacier boundaries, considering the remaining area after excluding snow and bare ice as debris cover, yielding the global debris cover extent product, referred to as the H2020 dataset in this text. The H2020 dataset was used to verify the accuracy of the debris cover identification.

The Inter-mission Time Series of Land Ice Velocity and Elevation (ITS_LIVE) dataset, from NASA's MEaSUREs project is accessible at https://nsidc.org/apps/itslive/ (accessed on 1 July 2024). The glacier surface velocities derived from Sentinel-1 data (hereafter referred to as the S1 dataset) were obtained from the glacier portal of the University of Erlangen-Nuremberg (http://retreat.geographie.uni-erlangen.de/search, accessed on 22 April 2024). This study analyzed long-term changes in flow velocities of debris-covered glaciers by selecting ITS_LIVE data for 1989–2018 and S1 dataset data for 2019–2020.

Solar radiation data were obtained from the National Centers for Environmental Prediction and the National Center for Atmospheric Research (NCEP/NCAR) reanalysis datasets, with acquisition times matching those of the Landsat imagery. The MCD43A3 product is part of the MODIS global albedo product series, providing surface albedo values. In this study, it was used to calculate net radiation.

3.2. Debris Cover Extent Identification

3.2.1. Feature Extraction

Thirty feature variables were chosen for further refinement, as indicated in Table 3. Spectral analysis utilized six Landsat imagery bands: blue, green, red, near infrared (NIR), shortwave infrared 1 (SWIR1), and shortwave infrared 2 (SWIR2). Using these bands, remote sensing indices such as the normalized difference vegetation index (NDVI), normalized difference water index (NDWI), normalized difference snow index (NDSI), and

RATIO (red to shortwave infrared band ratio) were constructed. The calculation formulas for these remote sensing indices are as follows:

$$\mathrm{NDVI} = \frac{\rho_{NIR} - \rho_{red}}{\rho_{NIR} + \rho_{red}} \tag{1}$$

$$\mathrm{NDWI} = \frac{\rho_{green} - \rho_{NIR}}{\rho_{green} + \rho_{NIR}} \tag{2}$$

$$\mathrm{NDSI} = \frac{\rho_{green} - \rho_{SWIR1}}{\rho_{green} + \rho_{SWIR1}} \tag{3}$$

$$\mathrm{RATIO} = \frac{\rho_{red}}{\rho_{SWIR1}} \tag{4}$$

where ρ_{NIR}, ρ_{red}, ρ_{green}, and ρ_{SWIR1} correspond to the reflectance values of the NIR, red, green, and SWIR1 bands, respectively.

Table 3. List of feature variables before optimization.

Category	No.	Feature Variable Name	Category	No.	Feature Variable Name
Spectral information	1	Blue band	Textual information	16	Entropy
	2	Green band		17	Second Moment
	3	Red band		18	Correlation
	4	NIR	Topographic parameters	19	Elevation
	5	SWIR1		20	Slope
	6	SWIR2		21	Aspect
Spectral index	7	NDVI		22	Shaded relief
	8	NDWI		23	profile convexity
	9	NDSI		24	Plan Convexity
	10	RATIO		25	Longitudinal Convexity
Textual information	11	Mean		26	Cross-sectional Convexity
	12	Variance		27	Minimum Curvature
	13	Homogeneity		28	Maximum Curvature
	14	Contrast	Others	29	LST
	15	Dissimilarity		30	FDC

Debris, resulting from glacier movement, feature unique textures and spatial distributions. Identifying debris cover with slope constraints and multispectral imagery yielded favorable outcomes [45]; adding texture information as a classification basis further improved the accuracy [14]. Following principal component analysis, texture information was derived via gray-level co-occurrence matrix (GLCM) operations with a 3×3 sliding window [46]. This texture information comprised mean, variance, homogeneity, contrast, dissimilarity, entropy, second moment, and correlation, with formulas detailed in the literature [47]. SRTM DEM provided topographical information, processed into ten features (Table 3) with ENVI 5.6 [18]. The single-period DEM did not match the image dates; however, as topographic features mainly distinguish debris from mountainous rock, this discrepancy minimally affected the classification outcomes. Changes at the glacier terminus received special attention during post-classification.

The lower temperature of debris cover compared to the surrounding terrain, due to underlying ice, facilitated debris-covered glacier boundary identification via thermal infrared remote sensing [48,49]. This study's LST data came from the Landsat Level-2 Land Surface Temperature Science Product. Within remote sensing imagery, spectral signals of debris and bare ice frequently blend within a single pixel. To distinguish them, a linear spectral unmixing approach with debris and bare ice as endmembers generated the FDC (fractional debris cover) index as a classification feature. Debris moves along with the glacier motion, and some studies explored using glacier flow velocity for debris identification [15,50]. However, in the study area experiments, incorporating coarse-resolution flow velocity products (ITS_LIVE) as classification features did not enhance the debris identification accuracy, resulting in jagged misclassifications.

3.2.2. Feature Optimization and RF Classification

The H2020 dataset utilized the Landsat image LT51470312007236IKR00 for information gathering around Tomur Peak. This study's sample point selection and random forest model training were based on this image. After roughly determining the proportion of various land cover types within the image frames using the H2020 dataset, we selected sample points using the stratified random sampling method. Land cover type uncertainties were verified using Google Earth™. Ultimately, 1726 sample points covering four debris categories—bare ice/snow, rock/bare land, and vegetation—were selected. In classifying multi-feature data with the random forest method, each feature's importance is assessed. Within a classification and regression tree (CART), the Gini impurity is used to evaluate a feature's classification efficacy, assigning normalized importance scores to features throughout the random forest [51].

Inherently, decision trees can select features and thresholds stepwise, but multiple local optima do not ensure a globally optimal outcome. Inspired by wrapper-based feature selection methods, this study sifted through 30 feature variables. Upon obtaining feature variable importance scores, they were ranked accordingly, and the random forest model was trained by incrementally adding features based on this ranking. The F1-score and Kappa coefficient evaluated the debris cover identification accuracy, with the F1-score as a weighted average of precision and recall [52]. The debris cover identification accuracy varied with feature count, peaking at 23 features, as depicted in Figure 3. Adhering to the highest-accuracy criterion for debris cover identification, the top 23 feature vectors were chosen as the optimal feature combination to train the random forest model. This model then classified the imagery data with the same feature combination across different periods, identifying the extent of debris cover.

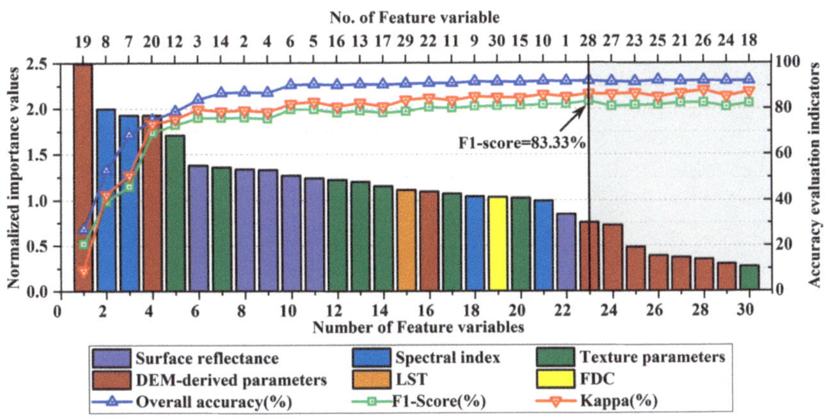

Figure 3. Normalized importance ranking for 30 feature variables, with No. corresponding to Table 3. This curve illustrates how the accuracy metrics varied as the number of feature variables increased.

Interference from moraines, seasonal snow, and shadows necessitated manual visual interpretation and correction of the automatically identified debris cover extents using remote sensing, for uses like mapping and monitoring changes in debris cover area [11,53–55]. Prior to manual editing, the classification results underwent median filtering to diminish "salt-and-pepper noise", with the kernel size set to 5 × 5. For manual editing of the extent of debris cover, ArcGIS 10.8 and Google Earth™ were employed, with visual interpretation markers including debris color texture, supraglacial ponds, glacial terminus hydrology, and topography. Debris cover identification, when combined with topographic data, effectively differentiated debris from mountainous rock, with manual editing mainly focusing on the glacier tongue terminus.

3.3. Glacier Flow Velocity Extraction

Located in a mid-latitude high mountain area, Tomur Peak Region's severe cloud cover limited the optical image availability for the study, prompting a focus on radar remote sensing methods for extracting glacier flow velocities. While the DInSAR method provides considerable precision in velocity estimates, it requires high image coherence, due to its differential interferometric technique. Mountain glaciers' relatively rapid flow can lead to decorrelation between images, resulting in the loss of deformation information. Conversely, offset-tracking does not encounter decorrelation issues and is minimally impacted by terrain undulations, better suiting the mountainous terrain of the study area.

Prior to cross-correlation matching, Sentinel-1A SLC images underwent processing to eliminate thermal noise, implementing multilooking and ground range projection, thereby generating backscatter intensity images. These images were subsequently registered using orbit information. For offset estimation, this study employed a frequency domain-based image cross-correlation algorithm. GCPs were uniformly placed in the master image, surrounded by 60 × 60 pixel search windows. Search windows slide across the slave image's search area, performing cross-correlation calculations at specified step sizes. Equation (5) illustrates the frequency domain cross-correlation algorithm:

$$CC(i,j) = \text{IFFT}\left(\frac{F_o(u,v)G_o^*(u,v)}{|F_o(u,v)G_o^*(u,v)|}\right) \tag{5}$$

In this equation, *(i,j)* represents the search window's center, F_0 *(u,v)* and G_0 *(u,v)* are the fast-Fourier-transformed master and slave images, "*" indicates complex conjugation, and IFFT is the inverse Fourier transform. The correlation matrix *CC* measures image similarity, with its peak coordinates matching the GCP's position in the slave image. The offset was calculated using homologous point coordinates in both master and slave images. A higher *CC* peak value signifies a more reliable matching result [56]. Flow velocities at GCPs exceeding 5 m/d were deemed outliers [57,58] and replaced with local weighted averages. Orbital shift errors in the total offset, arising from satellite orbit differences or attitude changes across passages, were computed and removed using a second-degree polynomial fit for global deformation. Ultimately, converting azimuthal and range displacement into north and east directions, based on radar incidence angle and satellite orientation, and performing vector composition yielded the glacier's planar flow velocity.

4. Result

4.1. Spatial Distribution of Debris Cover

The identification results of the extent of debris cover in 2022 were analyzed, showing that the area of debris cover within the study region (409.2 km^2) accounted for 22.8% of the total glacier area. The area–elevation distribution of the debris cover, as illustrated in Figure 4, shows that the debris was distributed across an elevation range of 2500 m to 4700 m. A total of 75% of the debris area was concentrated between 3300 m and 3900 m, while glaciers from 2500 m to 3100 m elevation were entirely debris-covered.

Spatial distribution patterns of debris cover across different regions are detailed in Figure 5. The elevation's normal distribution for debris cover predominantly fell within the interquartile range of 3250 m to 3950 m, with its slope distribution similarly within 3° to 10°. Given debris primarily forms on glaciers' gentler downstream slopes, its elevation and slope distributions are notably focused, starkly contrasting with clean ice surfaces. This rationale underscores the importance of elevation and slope as key features in the debris classification process, as depicted in Figure 3. In diverse regions, slight variances in the debris distribution's elevation and slope predominantly stem from the glaciers' inherent topographic features. Glaciers of greater length and scale show a narrower range in their debris slope distribution. On Tomur Peak's northwest, larger glaciers such as North Inylchek, South Inylchek, and Kaindy exhibited a concentrated slope distribution of 1.9° to 5.4° (Figure 5b), markedly gentler and more focused than those in other regions. The aspect of the debris distribution mirrored the glaciers' orientation.

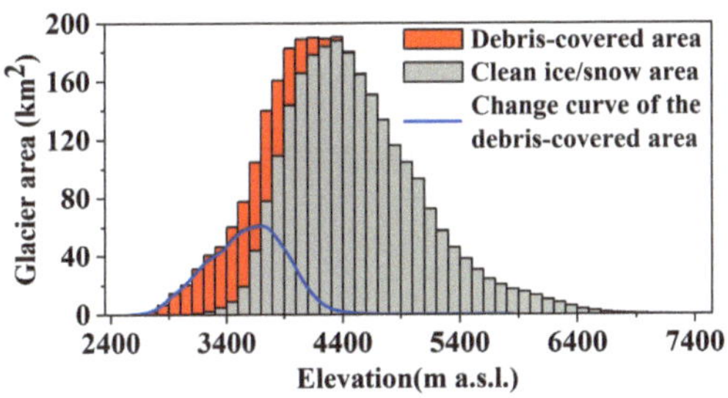

Figure 4. Area–altitude distributions of the debris-covered and clean ice/snow in the study region.

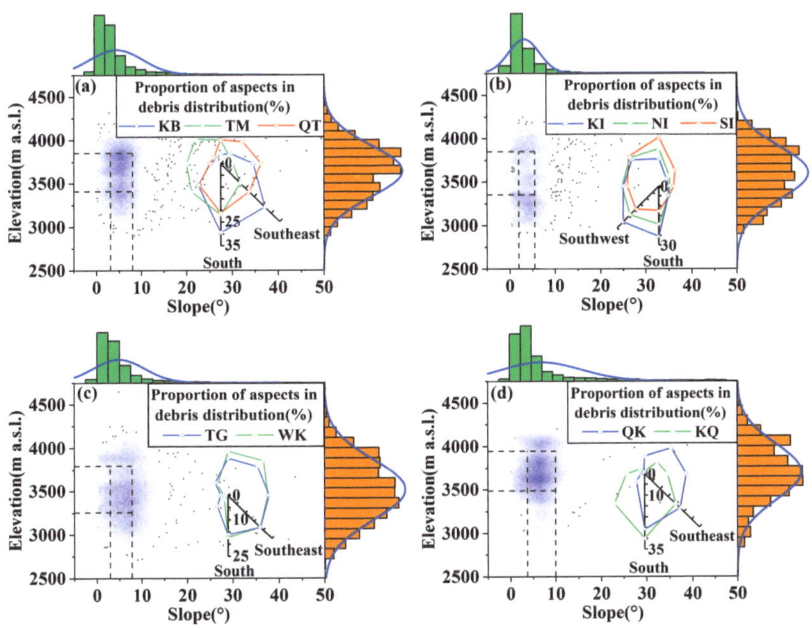

Figure 5. Histogram showing the normal distribution of the elevation and slope for the debris-covered area, with dashed lines marking the upper and lower quartiles. (**a**) Southwest region (KB, TM, QT in Figure 1); (**b**) northwest region (KI, NI, SI in Figure 1); (**c**) northeast region (TG, WK in Figure 1); (**d**) southeast region (QK, KQ in Figure 1).

4.2. Changes in the Extent of Debris Cover

Table 4 and Figure 6 show the changes in the extent of debris cover on glaciers within the Tomur Peak Region from 1989 to 2022. Over 34 years, Tomur-type glaciers saw a 69.4 km^2 expansion in debris cover, marking a 20.0% increase. The significant expansion of debris-covered areas on glaciers may have influenced the overall melting state in the Tomur Peak Region. From 1989 to 1996, an unusual reduction in debris cover area was observed across glaciers, with a rate of $-1.09\%\cdot a^{-1}$. Imagery observations indicated the extension of bare ice areas downward with glacier motion. In other time frames, all glaciers' debris cover expanded, with the highest annual growth rate of 1.96% between 2016 and 2022, and a subsequent rate of 1.02% from 1996 to 2007.

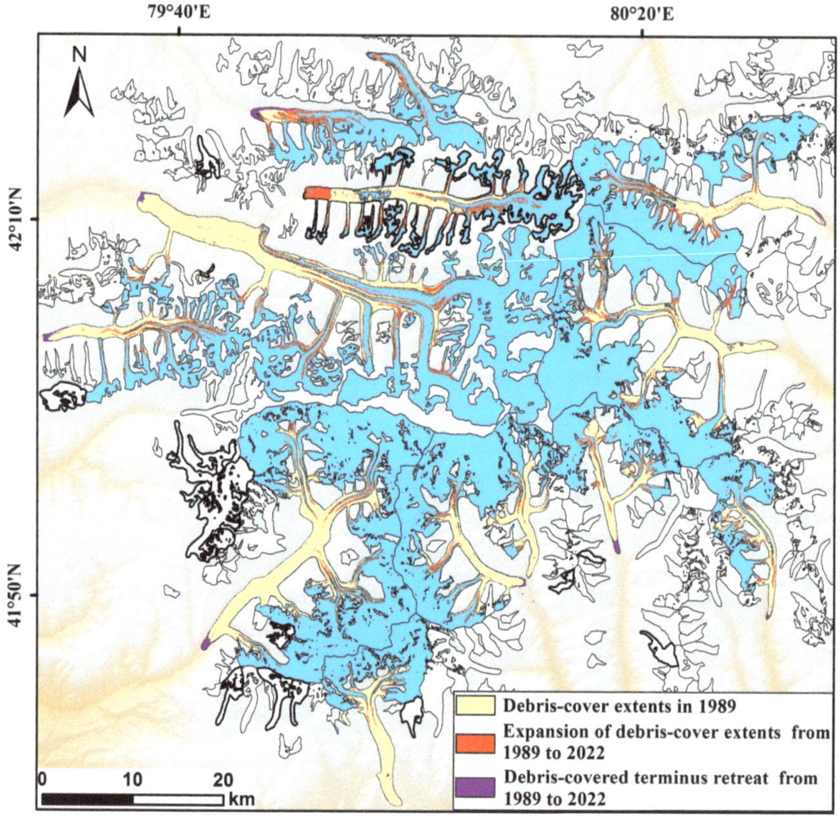

Figure 6. Changes in the extent of debris-cover on glaciers (larger than 40 km^2) in Tomur Peak Region from 1989 to 2022.

Upward evolution, delineated by the spread of debris cover at the interface between debris-covered and clean ice areas, represented the predominant form of debris expansion in the study area. Debris expansion at glacier termini, triggered by glacier advance and surging, was more common for smaller glaciers. Larger glaciers typically possess more stable internal structures. During the study period, only the North Inylchek Glacier exhibited surge phenomena [59], resulting in the expansion of its terminal debris. Tomur Glacier, Koxqar Glacier, and Qiongtailan Glacier, located on the southwestern side of the Tomur-Khan Tengri range, showed relatively little expansion in supraglacial debris cover. Conversely, glaciers at Tomur Peak's northern base saw a significantly higher debris area growth rate than those on the southern slope. These differences in the change in debris cover area between the northern and southern slopes can be attributed to (1) the northern slope's glaciers having numerous narrow tributaries and pronounced fracture faces, aiding in the debris' upward evolution; (2) the relatively isolated and smaller accumulation areas on the northern slope, which might have experience intensified mass loss due to climate warming, leading to the rapid exposure and subsequent formation of debris from moraine material at the glacier tongue; and (3) variances in sunlight exposure and climate change impacts between the slopes potentially influencing debris expansion rates.

Table 4. Changes in the extent of debris cover on glaciers in the study region from 1989 to 2022. The abbreviations of glacier names are detailed in Figure 1. ΔDebrisCov, debris coverage change; Abs. rate, absolute rate; Ann. rate, mean annual rate; Cum. area, cumulative area.

Glacier or Region	ΔDebrisCov (1989–1996)		ΔDebrisCov (1996–2007)		ΔDebrisCov (2007–2016)		ΔDebrisCov (2016–2022)		ΔDebrisCov (1989–2022)	
	Abs. Rate	Ann. Rate	Abs. Rate	Ann. Rate	Abs. Rate	Ann. Rate	Abs. Rate	Ann. Rate	Cum. Area	Cum. Rate
	$km^2 \cdot a^{-1}$	$\% \cdot a^{-1}$	$km^2 \cdot a^{-1}$	$\% \cdot a^{-1}$	$km^2 \cdot a^{-1}$	$\% \cdot a^{-1}$	$km^2 \cdot a^{-1}$	$\% \cdot a^{-1}$	km^2	%
KB	−0.10	−0.43	0.05	0.24	0.01	0.05	0.05	0.24	0.24	1.06
TM	−0.46	−0.72	0.30	0.48	−0.09	−0.13	1.11	1.75	5.22	8.04
QT	−0.29	−0.72	0.19	0.49	0.02	0.04	0.45	1.11	2.51	6.14
KI	−0.09	−0.52	0.01	0.07	0.17	1.02	0.62	3.45	4.91	28.99
NI	−0.41	−1.69	0.51	2.40	0.57	2.17	0.58	1.83	10.89	44.70
SI	−0.87	−1.12	0.71	1.00	0.56	0.73	1.79	2.15	16.69	21.60
TG	−0.39	−1.00	0.24	0.67	0.53	1.39	0.67	1.55	8.78	22.74
WK	−0.37	−1.50	0.35	1.56	0.44	1.73	0.42	1.42	7.40	29.84
QK	−0.37	−2.73	0.26	2.32	0.03	0.23	0.30	2.17	1.79	13.05
KQ	−0.47	−2.72	0.31	2.23	−0.06	−0.33	0.57	3.52	2.39	13.77
SY	0.00	0.20	0.11	9.80	0.02	0.75	0.31	13.95	3.02	276.79
MS	0.04	0.92	0.23	4.40	0.07	0.99	0.42	5.36	5.59	115.71
Total	−3.78	−1.09	3.27	1.02	2.26	0.65	7.31	1.96	69.41	20.00

4.3. Glacier Flow Velocity Extraction Results

4.3.1. Error Analysis

Deformation in non-glacial areas (such as bare land) is theoretically static, serving as reliable control points for validating velocity extraction accuracy [60]. Numerous control points, well-distanced from the glaciers, were selected in the flow velocity extraction images, with their displacement values illustrated in Figure 7. Displacement values in static areas, both east-west and north–south, exhibited a roughly normal distribution, indicative of the random error distribution within the velocity extraction outcomes. The east–west directional mean error stood at 1.79 m with a standard deviation of 2.69 m, whereas the north–south errors slightly exceeded the east–west, with a mean error of 2.39 m and a standard deviation of 4.40 m.

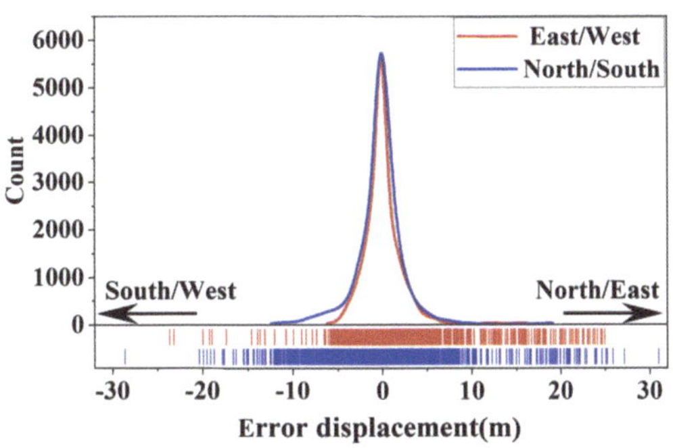

Figure 7. An example frequency distribution of surface displacement over stable terrain areas.

Errors in extracting glacier velocities chiefly arise from image registration inaccuracies, errors induced by terrain, and geocoding discrepancies [61]. The primary errors occur during the image cross-correlation phase; image resolution and algorithmic limitations restrict the accuracy of velocity results. Additionally, complex surface changes on glaciers, such as precipitation, snowfall, and melting, may lead to inaccurate image matching. This study opted for shorter temporal baselines and focused on stable debris-covered areas to reduce the introduction of errors from external factors, with interpolation addressing "void values" or abrupt velocity changes due to misregistration. Given the expansive riverbeds of Tomur-type glaciers and the gentle debris-covered terrain, the terrain's impact on the flow velocity extraction was minimal. The SRTM v3's average absolute error in glacier regions is approximately ±8 m [62], ensuring its reliability for geocoding flow velocity extractions.

4.3.2. Seasonal Velocity Variation

Flow velocity variations along the central flow lines of glaciers are spatially and temporally stable, offering a representative overview of overall glacier flow changes. The SAR offset-tracking method was employed to extract monthly flow velocities of Tomur-type glaciers for the years 2021 and 2022. Figure 8 illustrates the variation in mean flow velocity along the glaciers' central flow lines.

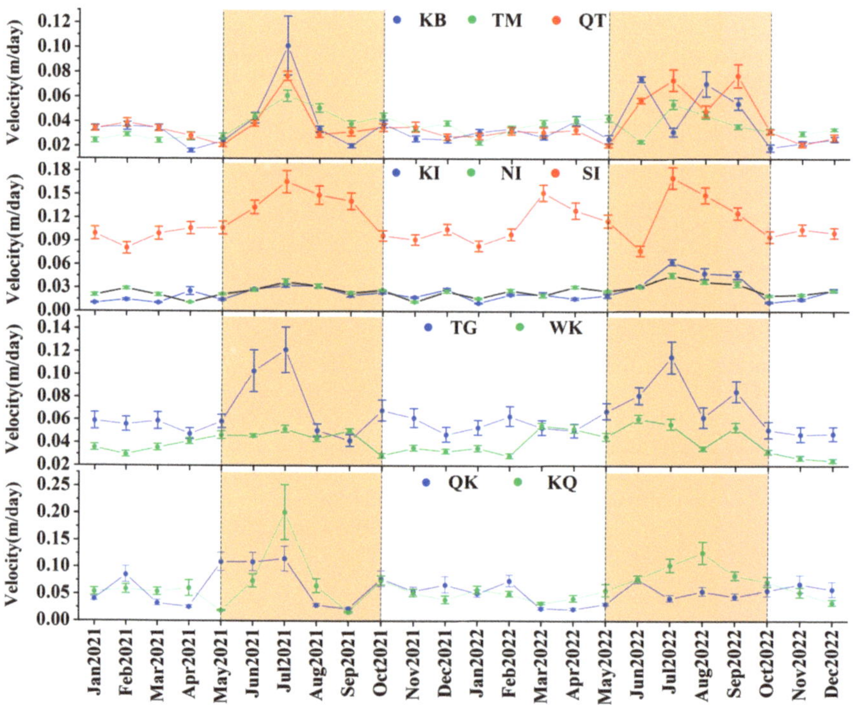

Figure 8. Seasonal variations in the average flow velocity along the central flow line from January 2021 to December 2022. The orange region covers the melt season of glaciers, and interval symbols indicate the 95% confidence interval of the mean flow velocity. The abbreviations of glacier names are shown in Figure 1.

Glacier flow velocity fluctuates throughout the year, with stronger fluctuations during the melt season compared to the non-melt season. Moreover, mean velocities are generally higher in the melt season than in the non-melt season, with peak velocities frequently occurring during this period, particularly in July. Increased glacial meltwater and precipitation enhance drainage at the glacier's base during the melt season, promoting basal motion and thereby accelerating glacier flow [42,63,64]. Furthermore, rising ice temperatures can intensify the plastic deformation of glaciers, and short-term material replenishment may heighten gravitational stress, albeit these elements have a secondary influence on glacier flow velocity.

4.4. Spatiotemporal Velocity Variations

We integrated the ITS_LIVE dataset, the S1 dataset, and the glacier velocity extraction results from this study to construct a time series from 1989 to 2022. The three datasets were generated using consistent principles, namely image feature tracking. The analyzed values were averaged to prevent any single value from having an excessive error. Consequently, the time series constructed from these values reliably reflected the trends in glacier velocity.

4.4.1. Northwest Region

A time series analysis showcasing flow velocities for three glaciers in the northwest region of the Tomur-Khan Tengri Mountain Range is presented in Figure 9, with the changes in mean velocity along the central flow lines over 34 years depicted in Figure 9e.

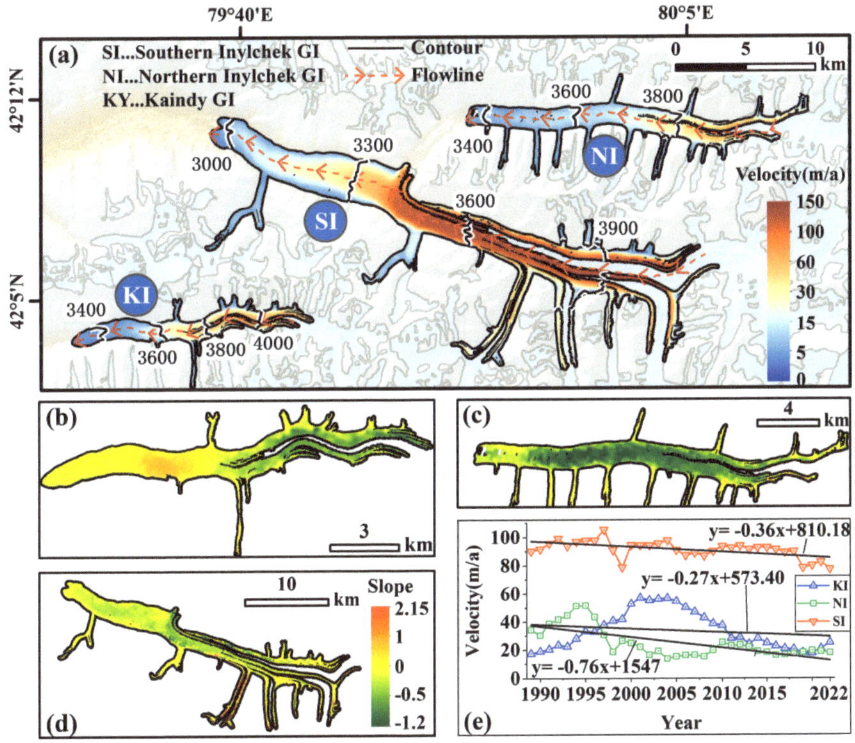

Figure 9. (**a**) Spatial distribution of average glacier flow velocities in the northwest region from 1989 to 2022; (**b–d**) spatial distributions of the flow velocity change trend of the Kaindy Glacier, Northern Inylchek Glacier, and Southern Inylchek Glacier. Slope, slope of linear regression; (**e**) changes in average flow velocities along the central flow line over the past 34 years.

As demonstrated in Figure 9a, the flow velocities exhibited a downward trend from the glaciers' upper reaches to their lower extents, with velocities along the central flow lines surpassing those at the glacier edges. Owing to its large size, the average peak velocity of the South Inylchek Glacier reached 149.7 m/a. Nearly three-quarters of the main glacier stream (from 3400 m a.s.l to 4200 m a.s.l) exhibited exceptionally high movement speeds. In the last 50 years, Central Tianshan's glaciers have experienced notable mass reductions and ongoing thinning [65,66], culminating in a widespread slowdown of glacier movement speeds. Kaindy Glacier was an exception, showing a notable uptick in flow velocity between 3450 m and 3680 m. Li et al. [66]'s research indicated that while the glacier's forefront underwent substantial thinning between 2000 and 2012, the segment from 3500 m to 3700 m slightly thickened. A marked discrepancy in glacier thickness alterations on either side of the 3500 m contour line induced an elevation in local slope, concurrently amplifying the gravitational driving stress and thereby accelerating the glacier flow velocities.

4.4.2. Southwest Region

Time-series analyses of glacier flow velocities in the southwest sector of the Tomur-Khan Tengri Mountain Range are illustrated in Figure 10.

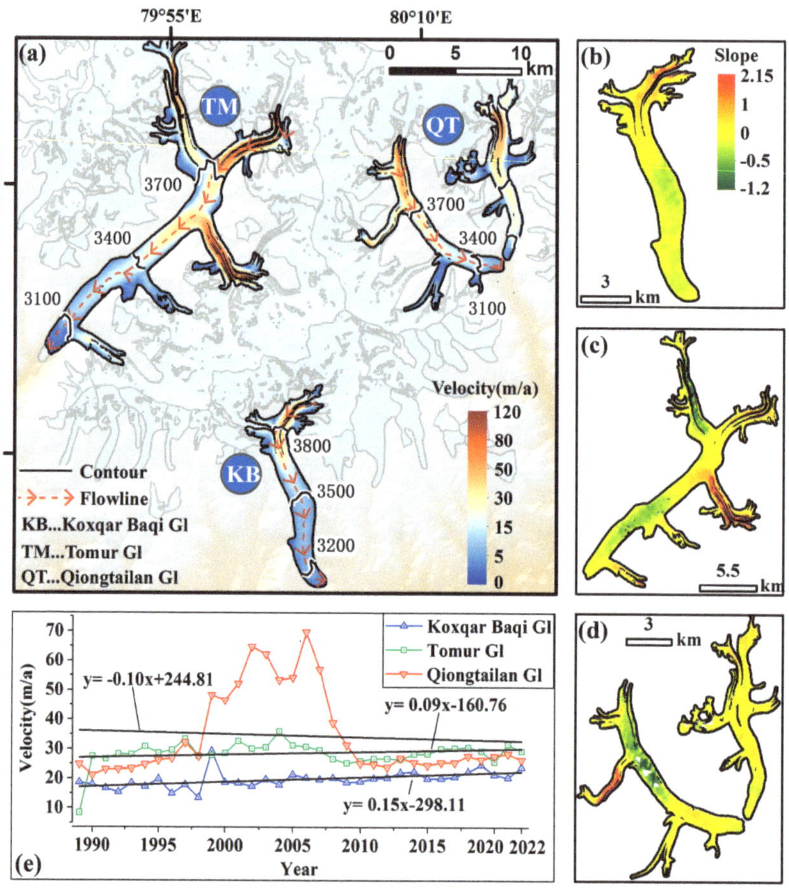

Figure 10. (**a**) Same as Figure 9a, but for the southwest region; (**b–d**) spatial distributions of the flow velocity change trend of the Koxqar Baqi Glacier, Tomur Glacier, and Qiongtailan Glacier. Slope, slope of linear regression; (**e**) same as Figure 9e, but for the southwest region.

The large size of the tributaries of the Tomur Glacier significantly impacted the gradient distribution of the glacier velocity (Figure 10a). In the ablation zones of the Tomur and Koxqar Glaciers, certain areas were covered by thick debris layers exceeding 2 m in depth, which significantly slowed down the thinning of these glaciers. Despite a reduction in flow velocities downstream (Figure 10b,c), these glaciers' overall velocity trend was upward (Figure 10e). The West Qiongtailan Glacier saw an unusual increase in flow velocity between 1999 and 2009 (Figure 10e), with the mean velocity at the flow line reaching 69.7 m/a in 2006, nearly 2.8 times that of the pre-1999 period. This was accompanied by a notable thickening in the glacier's accumulation area (3430 m to 3700 m) [66]. The acceleration in the West Qiongtailan Glacier's western branch from 1999 to 2009 might have been primarily due to thermodynamic changes at the glacier base. An increase in ice temperature likely facilitated the deformation of basal sediments and elevated the pore water pressure, thereby intensifying basal sliding or potentially causing detachment [67].

4.4.3. East Region

Time-series analyses of glacier flow velocities in the eastern sector of the Tomur-Khan Tengri Mountain Range, as depicted in Figure 11.

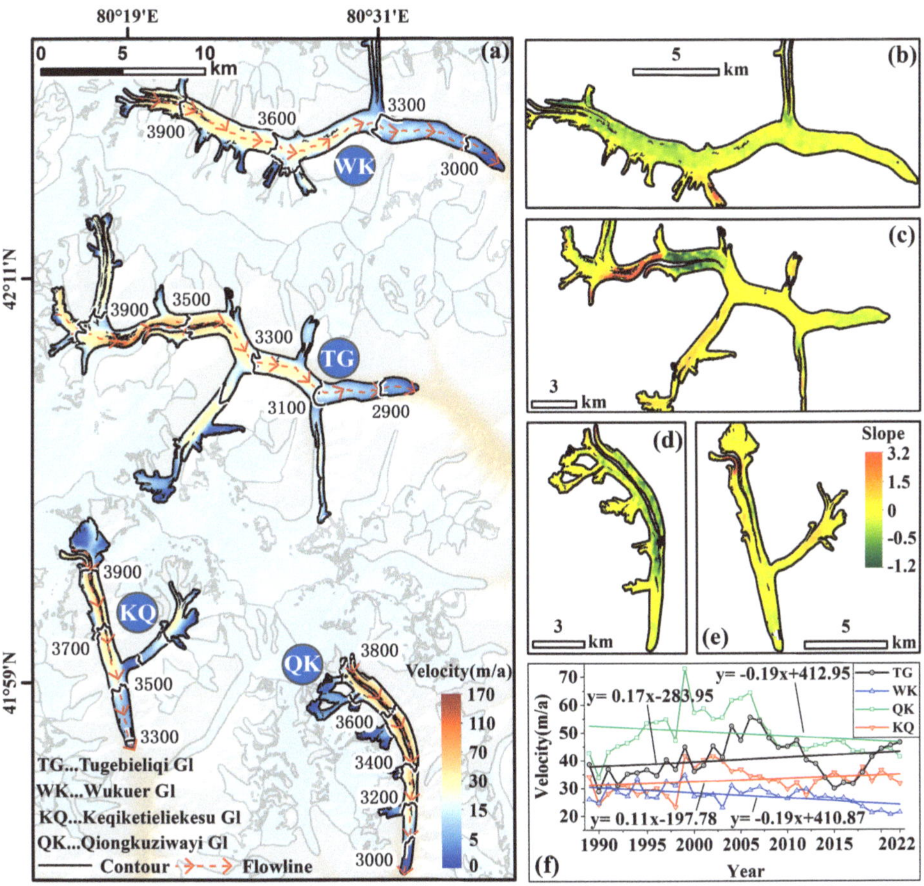

Figure 11. (**a**) Same as Figure 9a, but for the east region; (**b–e**) spatial distributions of the flow velocity change trend of the Wukuer Glacier, Tugebieliqi Glacier, Qiongkuziwayi Glacier, and Keqiketieliekesu Glacier. Slope, slope of linear regression; (**f**) same as Figure 9e, but for the east region.

Among the four glaciers, the Qiongkuziwayi Glacier maintained the highest overall flow velocity. Influenced by the confluence of four western tributaries, average velocities at elevations above 3200 m exceeded 30 m/a (Figure 11a). In comparison with the Tugebieliqi and Wukuer glaciers, despite the Tugebieliqi Glacier extending only 3.7 km more, its ice volume amounted to 2.2 times that of Wukuer's, guaranteeing higher flow velocities due to the ample material supply. Over a 34-year period, the overall velocities of Wukuer and Qiongkuziwayi Glaciers exhibited a decreasing trend (Figure 11b,d,f), while those of Tugebieliqi and Keqiketieliekesu Glaciers demonstrated an increasing trend (Figure 11c,e,f).

Synthesizing data from Figures 9–11, it is evident that increases in glacier flow velocity generally occurred at tributaries and their confluence points, including the southern tributaries of the Tomur Glacier and the segment between 3580 m and 3830 m of the Tugebieliqi Glacier. Areas of velocity increase are typically situated near the glacier accumulation zones, where elevations are high, melt rates are low, and material supply is abundant, resulting in increased gravitational stress between these areas and the continually thinning main glacier tongues, leading to an increase in flow velocities.

5. Discussion

5.1. Accuracy Assessment of Debris Cover Identification

Employing the optimal feature combination, the random forest classifier attained an overall classification accuracy of 92.23%, a debris cover identification accuracy of 83.33%, and a Kappa coefficient of 0.87 (Figure 3), falling within an acceptable range, with a Kappa coefficient greater than 0.8 deemed an acceptable accuracy range [68,69]. To examine the differences in performance of various machine learning methods in debris cover identification and the impact of study area size on this accuracy, the glaciers in the Tomur Peak Region and the Koxqar Glacier were classified with various machine learning methods, with their accuracies subsequently validated, as illustrated in Table 5.

Table 5. Accuracy assessment of various machine learning methods for debris-cover identification in the Tumur Peak Region and Koxqar Baqi Glacier.

Feature Composite	Classifier	Tomur Peak Region		Koxqar Baqi Gl	
		F1-Score	Kappa	F1-Score	Kappa
Spectral information	Random Forest	56.0%	0.53	67.8%	0.65
	Maximum Likelihood	42.3%	0.32	45.8%	0.35
	Support Vector Machine	27.9%	0.22	29.5%	0.25
	Artificial Neural Network	53.3%	0.49	61.4%	0.48
All features	Random Forest	82.4%	0.76	85.4%	0.84
	Support Vector Machine	73.2%	0.65	76.0%	0.69
Selected features	Random Forest	83.3%	0.87	88.6%	0.86

The accuracy of classification methods relying solely on spectral information for debris cover identification was insufficient, failing to meet the accuracy requirements for remote sensing classification. The random forest method, when considering all feature vectors, demonstrated high accuracy in debris cover identification, a metric that could be further enhanced through feature optimization. Overall, the feature optimized random forest method demonstrated significant advantages in debris cover identification. Remote sensing spectral information of mountain glaciers exhibits high spatial variability, with the characteristics of debris cover, including the distribution slope and surface temperature range, also showing considerable variation. The more limited the research scope, the more concentrated the distribution interval of debris cover features, leading to an enhanced debris identification effect. Irrespective of the classification method employed, the accuracy of debris cover identification within the Koxqar Glacier range surpassed that of the experiments conducted across the entire Tomur Peak Region, underscoring the influence of study area size on the accuracy of debris cover identification.

Although visual interpretation markers in debris-covered areas of Tomur-type glaciers are abundant and explicit, the manual editing process may still introduce subjectivity. Given the complex surface conditions in high mountain areas, the accuracy of digital elevation models (DEM) and surface temperature products may be compromised [64,70], introducing errors for which there are currently no effective solutions. Another source of uncertainty is that temporal mismatches in topographical data may result in the misclassification of features at glacier termini, necessitating adjustments during the manual editing process.

5.2. Determinants of Debris Cover Upward Evolution

Temperature and precipitation respectively govern the melting and accumulation processes of glaciers, thereby influencing their development and morphological evolution. This study obtained surface temperature and cumulative precipitation data for the melt season in the Tomur Peak region from 1989 to 2022, sourced from the ERA5 dataset, and conducted trend analyses, as depicted in Figure 12. The results of the linear regression analyses revealed a gradual upward trend in both surface temperature and cumulative precipitation within the study area over the preceding 34 years (Figure 12a,b).

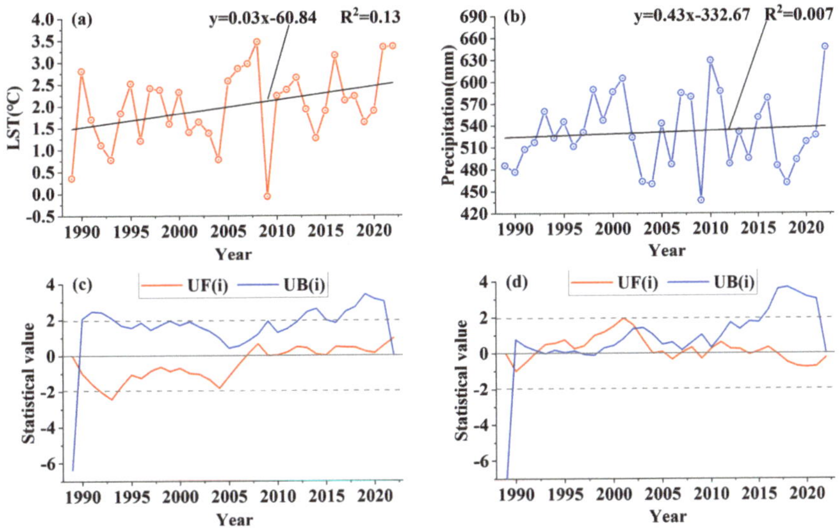

Figure 12. (**a**,**b**) Temporal variation and linear regression of regional average land surface temperature and cumulative precipitation; (**c**,**d**) Mann–Kendall test for regional average land surface temperature and cumulative precipitation.

Glacier length and area may expand due to glacier advancement or surging, consequently leading to the expansion of the extent of debris cover. Glacier advancement and surging are driven by complex factors, among which changes in temperature and precipitation play a significant role in affecting a glaciers' internal stability and material accumulation upstream, often triggering surging. Upon examining the spatial distribution of surging glaciers in the Tianshan region, it was noted that these glaciers are predominantly situated in the central and western parts of the Tianshan mountains, regions where the temperature incrementally increases and precipitation gradually rises annually, with a notable concentration of surging glaciers in the Tomur Peak area [59]. The advancing and surging glaciers around Tomur Peak are predominantly small in size and not universally debris-covered; thus, the debris cover expansion attributed to glacier advancement and surging represents only a minor portion. The predominant mechanism behind debris cover expansion is its upward progression, specifically, the expansion of debris at the interface between debris-laden and clean glacier ice. In the context of global warming, similar trends in debris upward have been documented in debris-covered glaciers in diverse locations, such as the Zmutt Glacier in the Alps [71] and glaciers in the Greater Caucasus region [12]. In regions maintaining a stable glacier mass balance, the area covered by debris has not exhibited a significant increase [72].

Debris cover formation predominantly occurs through two processes: the exposure of englacial debris by glacier movement and melting, and the accumulation of weathered rock and fallen debris on the glacier surface, forming supraglacial debris. A minor increase in precipitation impacts glaciers significantly less than the effects of temperature elevation.

Glaciers in the Tomur area are undergoing continuous mass loss. As englacial debris is exposed through melting, the area covered by debris continues its upward evolution. Considering the Mushketova Glacier as an example, which has exhibited significant debris cover expansion, despite a modest retreat at the glacier terminus, the debris cover area on this glacier has still expanded by nearly 1.2 times over 34 years (Table 4, Figure 13).

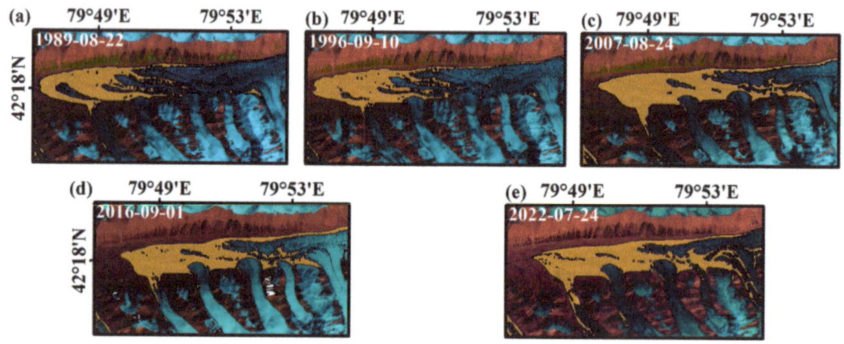

Figure 13. (**a**–**e**) Upward evolution in debris cover and debris-covered terminus retreat on Mushketova Glacier from 1989 to 2022.

The Mann–Kendall test results for precipitation and temperature changes in the study area indicated a downward temperature trend and an upward precipitation trend between 1989 and 2006 (Figure 12c,d), contributing to a slower glacier mass loss. Certain studies verified a reduction in the glacier mass loss rate in the Tomur area during this period [73]. In addition to climatic factors, elements like debris cover effects, low ice temperatures, and high altitudes contribute to the glaciers in the Tomur area being more stable than those in other central Tianshan Mountain regions [40]. This may related to the minor decrease in the debris cover area in the study area from 1989 to 1996. Considering the source of debris material, glacier movement erodes the underlying and surrounding rocks, while accumulated moraine material is redistributed with glacier movement, continually supplying materials for debris cover formation. A warming climate and thinning glaciers may lead to permafrost melting and high-altitude mountain slope instability [74], triggering increased rockfalls, ice–rock avalanches, and snow–rock avalanches, which transport significant debris to glaciers [75,76], thereby facilitating the upward evolution of debris cover.

5.3. *Control Factors of Glacier Flow Velocity*

Gravity triggers internal deformation and basal sliding within glaciers, with the intensity of gravitational driving stress being dictated by the glacier's thickness and slope. Therefore, glacier flow velocity predominantly relies on the glacier's thickness and slope [3,20]. Additionally, a glacier's size, shape, length, and topographical attributes (including elevation, altitude, and aspect) also exert an influence on its flow velocity, though to a lesser extent. Typically, the glacier's upper and middle sections are thicker than its lower parts, with a steeper slope, resulting in a general trend of decreasing glacier flow velocities from upper to lower reaches. This pattern of velocity distribution has been documented in numerous glaciers throughout the high-mountain regions of Asia [43,77–79].

In the context of global climate warming, glaciers generally exhibit a thinning trend, resulting in decreased velocities for most glaciers in the high-mountain areas of Asia. Notably, glaciers in the Nyainqentanglha region have experienced the most significant deceleration, with a decrease of (37.2 ± 1.1)% per decade [20,80]. Conversely, glaciers located in the western part of the Kunlun Mountains and the Karakoram region, where the glacier mass is nearly balanced or slightly increasing, have seen a slight acceleration in flow velocities [81,82]. The annual mean velocity for all glaciers in the study area was analyzed statistically, revealing a gradual deceleration of overall glacier velocities in the Tomur region at a rate of −4.0% per decade. However, when examining individual glaciers, significant variability in flow velocities emerged, marked by considerable seasonal differences. Specifically, certain glaciers registered an increase in flow speeds over the 34-year period, a phenomenon not solely explainable by changes in glacier thickness. Basal sliding constitutes the greatest uncertainty in determining glacier flow velocities. Numerous studies agreed that variations in temperature and precipitation in the study area control the discharge at the glacier bed. Changes in subglacial water pressure influence the intensity of basal sliding, thereby causing fluctuations in glacier flow velocity [83,84]. During the mid-study period, the Kaindy and Qiongtailan Glaciers displayed marked increases in flow velocity, notwithstanding stable glacier terminus lengths, indicative of rapid material transfer and accumulation—a phenomenon characteristic of swift glacier advance [66]. More specifically, ongoing accumulation in the upper zones and alterations in thermal conditions at the glacier's base may facilitate enhanced basal sliding or even detachment, resulting in glacier material moving from the accumulation zones to the ablation areas, thus culminating in periods of rapid movement [67].

5.4. Relationship between Glacier Movement and Debris Distribution

The interaction and feedback mechanisms among climate, supraglacial debris, hydrological conditions, and glacier dynamics collectively control the ablation and morphological evolution of debris-covered glaciers, as illustrated in Figure 14. Identifying the extent of debris cover is a fundamental task in the study of debris-covered glaciers. Glacier dynamic processes are the primary driving forces behind the formation, transfer, and accumulation of supraglacial debris. The glacier's horizontal velocity field governs the transportation of debris on the ice surface and the variations in local debris thickness.

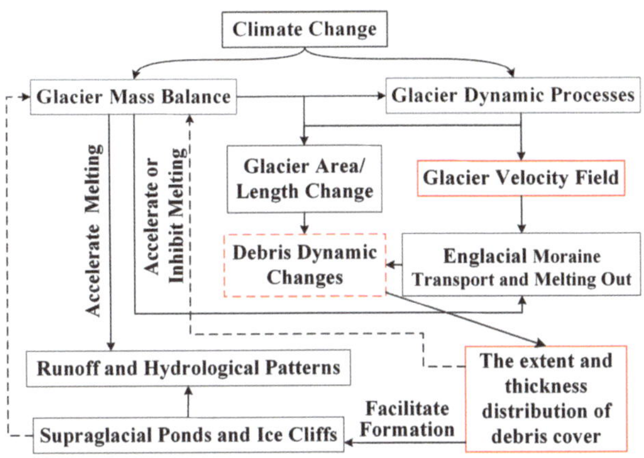

Figure 14. Diagram of climate–debris–glacier system interactions and feedback mechanisms.

The method of estimating the distribution of debris thickness on glaciers in the Tomur Peak region was based on remote sensing-derived land surface temperature combined with the energy balance equation, using the debris thermal resistance as a proxy indicator for debris thickness. The results are shown in Figure 15. Similar methods have been applied to study typical debris-covered glaciers in the Himalayas [85,86], Gongga Mountain [87], the Caucasus Mountains, and the China-Pakistan Economic Corridor [88]. These studies demonstrated a common characteristic: the debris thickness at the glacier terminus was greater than that in the upper middle part of the ablation zone, with the thickness increasing closer to the terminus. This characteristic is in agreement with observations and simulations of the glacier debris thickness distribution in many regions of Asian high mountains and can be regarded as a widespread phenomenon [89–91].

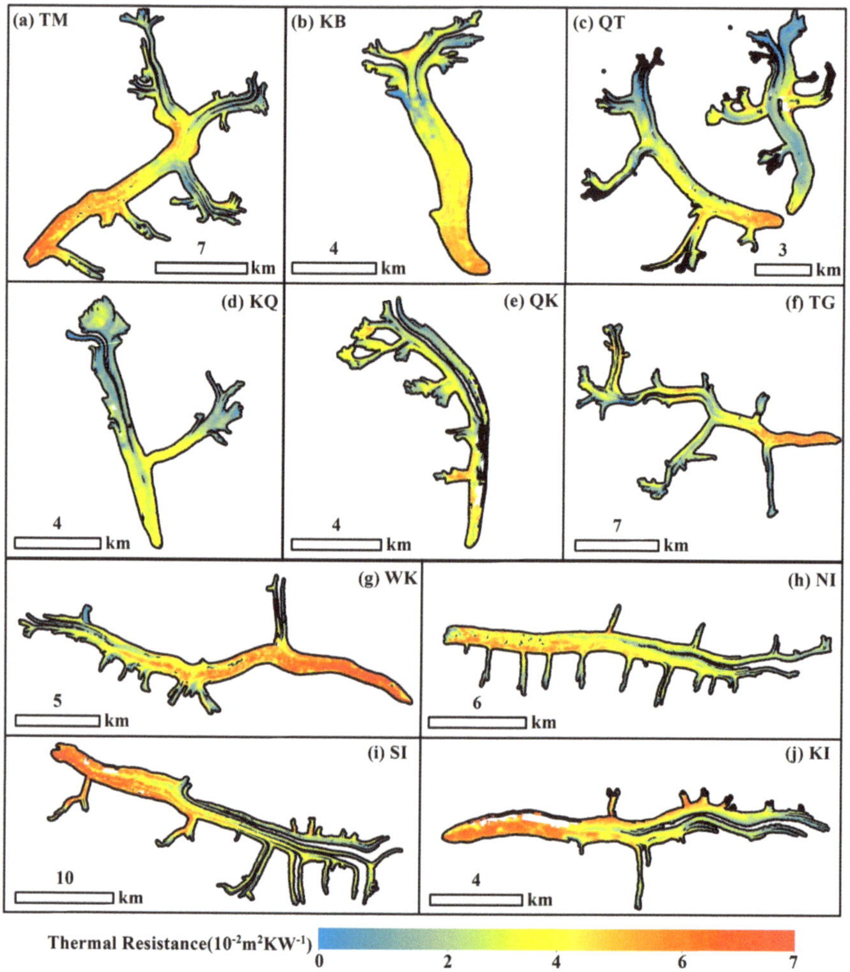

Figure 15. Spatial distribution of thermal resistances of debris layers in the study area. The abbreviations of glacier names are detailed in Figure 1.

As seen in the cross-sectional profile of the glaciers, the debris thickness on both sides of the glacier generally exceeded that observed along the central flow line. A special case occurs when one side of the glacier receives a substantial input from tributaries, resulting in greater debris thickness on that side compared to the central flow line and the opposite side, as observed on the south side of the North Inylchek Glacier tongue (Figure 15h) and the north side of the Kaindy Glacier tongue (Figure 15j).

During the formation of supraglacial debris, glacier movement acts like a "conveyor belt": debris are continuously transported to the glacier terminus, and existing debris typically remain on the glacier surface. As the glacier velocity gradually decreases to zero, the position of the debris stabilizes, and the continuous accumulation of debris leads to an increase in debris thickness. This process results in debris being primarily distributed in the middle to lower parts of the glacier, with the thickness increasing closer to the glacier terminus.

Due to the lack of extensive measured parameters (e.g., debris porosity, debris density, and glacier ablation rates), we adopted the analytical model proposed by Anderson et al. [22] instead of a numerical model to analyze the formation of the debris thickness distribution pattern. This approach simplifies the complex physical processes inherent in glaciers, only considering the instantaneous state of the glacier system. In the ablation zone of a glacier, the debris thickness can be expressed as

$$d(x) = \frac{1}{U_s} \int_{x_e}^{x} \frac{Cb}{(1-\phi)\rho} dx \tag{6}$$

C represents the concentration of moraine within the ice, b represents the glacier ablation rate under the debris cover, φ represents the porosity of the debris, ρ represents the density of the debris, U_s represents the glacier surface velocity, d represents the debris thickness, x represents the distance from the glacier's origin, and x_e denotes the position where debris first appears. Assuming C, φ, and ρ are constants throughout the glacier, the debris thickness d is controlled by the glacier surface ablation rate b and the glacier surface velocity U_s. Using the Koxqar Glacier as a case study to analyze the trend of debris thickness variation, the relationship between debris thickness and glacier elevation is shown in Figure 16.

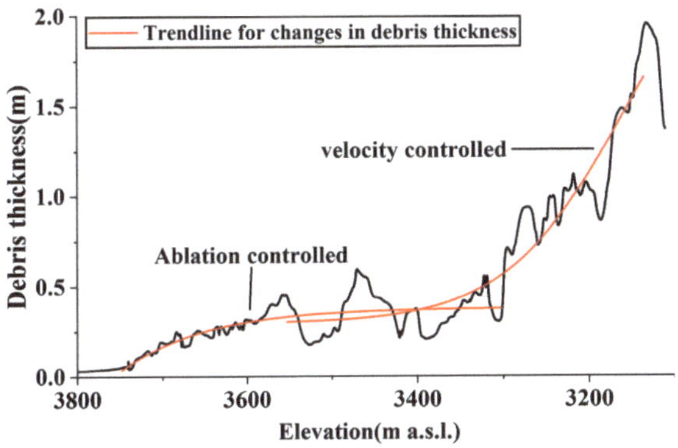

Figure 16. The relationship between debris thickness and elevation along the central flowline of the Koxkar Glacier, with debris thickness estimates derived from Rounce et al. [89] is accessible at https://nsidc.org/data/hma_dte/versions/1 (accessed on 1 July 2024).

In the elevation range of 3500 m to 3880 m, the debris thickness variation is mainly controlled by the ablation rate. Assuming a constant glacier surface velocity U_s, as the debris thickness increases, the glacier ablation rate b will continuously decrease, leading to a reduction in the debris emergence rate and a slower increase in debris thickness. This explains the trend observed in the 3500 m to 3880 m elevation range. As the debris thickness continues to increase, the glacier ablation rate b approaches zero, and the debris emergence rate also approaches zero. At this point, the glacier ablation rate b and the concentration of debris C no longer influence the debris thickness growth rate. Below 3500 m, the glacier flow velocity becomes the primary factor controlling the variation in debris thickness, with the debris thickness increasing rapidly as the inverse of the glacier surface velocity (1/Us) increases.

In the study area, the debris thickness on both sides of many glaciers is greater than that along the central flow line. A one-dimensional debris thickness analysis model can be simply applied to a two-dimensional plane, where the glacier velocity on both sides is lower than that along the central flow line. Ideally, the debris thickness would symmetrically increase from the central flow line toward both sides of the glacier. The replenishment of loose rock material from both sides of the glacier may also be a significant reason for the thicker lateral supraglacial debris.

6. Conclusions

A feature optimization random forest method, which integrated texture and topographical factors, was utilized to identify changes in the extent of debris cover on glaciers around the central Tianshan Tomur Peak from 1989 to 2022. For 2022, the study region's debris cover area was recorded at 409.2 km^2, constituting 22.8% of the total glacier area, with a predominant distribution on the gentler slopes of the glaciers' downstream areas. Over the span of 34 years, the debris cover area witnessed an expansion of 69.4 km^2, marking an increase of 20.0%. Influenced by various factors including the number of tributaries, glacier scale, sunlight exposure, and climate change, the rate of debris cover expansion at glaciers' northern base on Tomur Peak significantly surpassed that on the southern slopes.

Over a 34-year period, the analysis of the changes in glacier velocity indicated that the South Inylchek Glacier exhibited the highest glacier velocity in the study area, with a peak average velocity of 149.7 m/a. Nearly three-quarters of the main glacier stream (from 3400 m a.s.l to 4200 m a.s.l) maintained exceptionally high movement speeds. As the glacier thickness decreased, the overall glacier movement in the study area gradually decelerated, at a rate of 4.0% per decade. When examining individual glaciers, there was significant variability in their velocity change trends, characterized by notable fluctuations in speed. We hypothesized that possible causes might include changes in basal water pressure, thermodynamic conditions, and internal glacier structure, which alter the intensity of basal sliding.

Author Contributions: M.X. designed the overall research plan of this study and revised the manuscript. W.Z. performed the experiments and drafted the first manuscript. H.H. reviewed and revised the manuscript. All authors have read and agreed to the published version of the manuscript.

Funding: This research was funded by the National Natural Science Foundation of China (42371145; 41971094); China–Pakistan joint program of the Chinese Academy of Sciences (046GJHZ2023069MI), project of State Key Laboratory of Cryospheric Science (SKLCS-ZZ-2024) and Open Fund Project of the National Cryosphere Desert Data Center (E01Z790201).

Data Availability Statement: The raw data supporting the conclusions of this article will be made available by the authors on request.

Acknowledgments: The authors wish to acknowledge the European Space Agency (ESA), United States Geological Survey (USUG), National Snow and Ice Data Center (NSIDC) and the National Aeronautics and Space Administration (NASA) for offering free access to satellite imagery and reference data.

Conflicts of Interest: The authors declare no conflicts of interest.

References

1. Hock, R.; Rasul, G.; Adler, C.; Cáceres, B.; Gruber, S.; Hirabayashi, Y.; Jackson, M.; Kääb, A.; Kang, S.; Kutuzov, S.; et al. High Mountain Areas. In *IPCC Special Report on the Ocean and Cryosphere in a Changing Climate*; Masson-Delmotte, V., Zhai, P., Pirani, A., Connors, S.L., Péan, C., Berger, S., Caud, N., Chen, Y., Goldfarb, L., Gomis, M.I., et al., Eds.; IPCC: Geneva, Switzerland, 2019; pp. 131–202.
2. Johnson, E.; Rupper, S. An examination of physical processes that trigger the albedo-feedback on glacier surfaces and implications for regional glacier mass balance across high mountain asia. *Front. Earth Sci.* **2020**, *8*, 129. [CrossRef]
3. Paterson, W. *The Physics of Glaciers*, 3rd ed.; Pergamon Press: Oxford, UK, 1994.
4. Østrem, G. Ice melting under a thin layer of moraine, and the existence of ice cores in moraine ridges. *Geogr. Ann.* **1959**, *41*, 228–230. [CrossRef]
5. Mattson, L.E.; Gardner, J.S.; Young, G.J. Ablation on debris covered glaciers: An example from the rakhiot glacier, punjab, himalaya. *Int. Assoc. Hydrol. Sci. Publ.* **1993**, 289–296.
6. Khan, M.I. Ablation on Barpu Glacier, Karakoram Himalaya, Pakistan a Study of Melt Processes on a Faceted, Debris-Covered Ice Surface. Master's Thesis, Wilfrid Laurier University, Waterloo, ON, Canada, 1989.
7. Takeuchi, Y.; Kayastha, R.B.; Nakawo, M. Characteristics of ablation and heat balance in debris-free and debris-covered areas on khumbu glacier, nepal himalayas, in the pre-monsoon season. *Int. Assoc. Hydrol. Sci.* **2000**, *264*, 53–62.
8. Kellerer-Pirklbauer, A.; Lieb, G.; Gspurning, J. The response of partially debris-covered valley glaciers to climate change: The Example of the Pasterze Glacier (Austria) in the period 1964 to 2006. *Geogr. Ann. Ser. A Phys. Geogr.* **2008**, *90*, 1–17. [CrossRef]

9. Thakuri, S.; Salerno, F.; Smiraglia, C.; Bolch, T.; D'Agata, C.; Viviano, G.; Tartari, G. Tracing glacier changes since the 1960s on the south slope of mt. Everest (central southern himalaya) using optical satellite imagery. *Cryosphere* **2014**, *8*, 1297–1315. [CrossRef]
10. Glasser, N.F.; Holt, T.O.; Evans, Z.D.; Davies, B.J.; Pelto, M.; Harrison, S. Recent spatial and temporal variations in debris cover on patagonian glaciers. *Geomorphology* **2016**, *273*, 202–216. [CrossRef]
11. Xie, F.; Liu, S.; Wu, K.; Zhu, Y.; Gao, Y.; Qi, M.; Duan, S.; Saifullah, M.; Tahir, A.A. Upward expansion of supra-glacial debris cover in the hunza valley, karakoram, during 1990~2019. *Front. Earth Sci.* **2020**, *8*, 308. [CrossRef]
12. Tielidze, L.G.; Bolch, T.; Wheate, R.D.; Kutuzov, S.S.; Lavrentiev, I.I.; Zemp, M. Supra-glacial debris cover changes in the greater caucasus from 1986 to 2014. *Cryosphere* **2020**, *14*, 585–598. [CrossRef]
13. Fleischer, F.; Otto, J.C.; Junker, R.R.; Hölbling, D. Evolution of debris cover on glaciers of the eastern alps, austria, between 1996 and 2015. *Earth Surf. Process. Landf.* **2021**, *46*, 1673–1691. [CrossRef]
14. Racoviteanu, A.; Williams, M.W. Decision tree and texture analysis for mapping debris-covered glaciers in the kangchenjunga area, eastern himalaya. *Remote Sens.* **2012**, *4*, 3078–3109. [CrossRef]
15. Smith, T.; Bookhagen, B.; Cannon, F. Improving semi-automated glacier mapping with a multi-method approach: Applications in central asia. *Cryosphere* **2015**, *9*, 1747–1759. [CrossRef]
16. Alifu, H.; Vuillaume, J.; Johnson, B.A.; Hirabayashi, Y. Machine-learning classification of debris-covered glaciers using a combination of sentinel-1/-2 (sar/optical), landsat 8 (thermal) and digital elevation data. *Geomorphology* **2020**, *369*, 107365. [CrossRef]
17. Lu, Y.; Zhang, Z.; Kong, Y.; Hu, K. Integration of optical, sar and dem data for automated detection of debris-covered glaciers over the western nyainqentanglha using a random forest classifier. *Cold Reg. Sci. Tech.* **2022**, *193*, 103421. [CrossRef]
18. Zhang, J.; Jia, L.; Menenti, M.; Hu, G. Glacier facies mapping using a machine-learning algorithm: The parlung zangbo basin case study. *Remote Sens.* **2019**, *11*, 452. [CrossRef]
19. Khan, A.A.; Jamil, A.; Hussain, D.; Taj, M.; Jabeen, G.; Malik, M.K. Machine-learning algorithms for mapping debris-covered glaciers: The hunza basin case study. *IEEE Access* **2020**, *8*, 12725–12734. [CrossRef]
20. Dehecq, A.; Gourmelen, N.; Gardner, A.S.; Brun, F.; Goldberg, D.; Nienow, P.W.; Berthier, E.; Vincent, C.; Wagnon, P.; Trouvé, E. Twenty-first century glacier slowdown driven by mass loss in high mountain asia. *Nat. Geosci.* **2019**, *12*, 22–27. [CrossRef]
21. Millan, R.; Mouginot, J.; Rabatel, A.; Morlighem, M. Ice velocity and thickness of the world's glaciers. *Nat. Geosci.* **2022**, *15*, 124–129. [CrossRef]
22. Anderson, L.S.; Anderson, R.S. Modeling debris-covered glaciers: Response to steady debris deposition. *Cryosphere* **2016**, *10*, 1105–1124. [CrossRef]
23. Benn, D.; Evans, D.J. *Glaciers and Glaciation*; Routledge: London, UK, 2014.
24. Wirbel, A.; Jarosch, A.H.; Nicholson, L. Modelling debris transport within glaciers by advection in a full-stokes ice flow model. *Cryosphere* **2018**, *12*, 189–204. [CrossRef]
25. Watanabe, T.; Ives, J.D.; Hammond, J.E. Rapid growth of a glacial lake in khumbu himal, himalaya: Prospects for a catastrophic flood. *Mt. Res. Dev.* **1994**, *14*, 329–340. [CrossRef]
26. Benn, D.I.; Bolch, T.; Hands, K.; Gulley, J.; Luckman, A.; Nicholson, L.I.; Quincey, D.; Thompson, S.; Toumi, R.; Wiseman, S. Response of debris-covered glaciers in the mount everest region to recent warming, and implications for outburst flood hazards. *Earth-Sci. Rev.* **2012**, *114*, 156–174. [CrossRef]
27. Miles, E.S.; Watson, C.S.; Brun, F.; Berthier, E.; Esteves, M.; Quincey, D.J.; Miles, K.E.; Hubbard, B.; Wagnon, P. Glacial and geomorphic effects of a supraglacial lake drainage and outburst event, everest region, nepal himalaya. *Cryosphere* **2018**, *12*, 3891–3905. [CrossRef]
28. Zaginaev, V.; Petrakov, D.; Erokhin, S.; Meleshko, A.; Stoffel, M.; Ballesteros-Canovas, J.A. Geomorphic control on regional glacier lake outburst flood and debris flow activity over northern tien shan. *Glob. Planet. Chang.* **2019**, *176*, 50–59. [CrossRef]
29. Shangguan, D.; Liu, S.; Ding, Y.; Guo, W.; Xu, B.; Xu, J.; Jiang, Z. Characterizing the may 2015 karayaylak glacier surge in the eastern pamir plateau using remote sensing. *J. Glaciol.* **2016**, *62*, 944–953. [CrossRef]
30. Zhang, M.; Chen, F.; Tian, B.; Liang, D.; Yang, A. Characterization of kyagar glacier and lake outburst floods in 2018 based on time-series sentinel-1a data. *Water* **2020**, *12*, 184. [CrossRef]
31. Racoviteanu, A.E.; Nicholson, L.; Glasser, N.F.; Miles, E.; Harrison, S.; Reynolds, J.M. Debris-covered glacier systems and associated glacial lake outburst flood hazards: Challenges and prospects. *J. Geol. Soc.* **2022**, *179*, jgs2021-084. [CrossRef]
32. Li, J.; Li, Z.; Zhu, J.; Ding, X.; Wang, C.; Chen, J. Deriving surface motion of mountain glaciers in the tuomuer-khan tengri mountain ranges from palsar images. *Glob. Planet. Chang.* **2013**, *101*, 61–71. [CrossRef]
33. Huang, L.; Li, Z. Comparison of sar and optical data in deriving glacier velocity with feature tracking. *Int. J. Remote Sens.* **2011**, *32*, 2681–2698. [CrossRef]
34. Luckman, A.; Quincey, D.; Bevan, S.; Tedesco, M. The potential of satellite radar interferometry and feature tracking for monitoring flow rates of himalayan glaciers. *Remote Sens. Environ.* **2007**, *111*, 172–181. [CrossRef]
35. Wang, Q.; Fan, J.; Zhou, W.; Tong, L.; Guo, Z.; Liu, G.; Yuan, W.; Sousa, J.J.; Perski, Z. 3d surface velocity retrieval of mountain glacier using an offset tracking technique applied to ascending and descending sar constellation data: A case study of the yiga glacier. *Int. J. Digit. Earth* **2019**, *12*, 614–624. [CrossRef]
36. Yasuda, T.; Furuya, M. Dynamics of surge-type glaciers in west kunlun shan, northwestern tibet. *J. Geophys. Res. Earth Surf.* **2015**, *120*, 2393–2405. [CrossRef]

37. Neckel, N.; Loibl, D.; Rankl, M. Recent slowdown and thinning of debris-covered glaciers in south-eastern Tibet. *Earth Planet. Sci. Lett.* **2017**, *464*, 95–102. [CrossRef]
38. Guo, W.; Liu, S.; Xu, J.; Wu, L.; Shangguan, D.; Yao, X.; Wei, J.; Bao, W.; Yu, P.; Liu, Q.; et al. The second chinese glacier inventory: Data, methods and results. *J. Glaciol.* **2015**, *61*, 357–372. [CrossRef]
39. Friedl, P.; Seehaus, T.; Braun, M. Global time series and temporal mosaics of glacier surface velocities derived from sentinel-1 data. *Earth Syst. Sci. Data* **2021**, *13*, 4653–4675. [CrossRef]
40. Li, Z.W.; Li, J.; Ding, X.L.; Wu, L.X.; Ke, L.H.; Hu, J.; Xu, B.; Peng, F. Anomalous glacier changes in the southeast of tuomuer-khan tengri mountain ranges, central tianshan. *J. Geophys. Res. Atmos.* **2018**, *123*, 6840–6863. [CrossRef]
41. Ma, Q.; Li, Z.; Chen, Z.; Su, T.; Wu, Y.; Feng, G. Moisture changes with increasing summer precipitation in qilian and tienshan mountainous areas. *Atmos. Sci. Lett.* **2023**, *24*, e1154. [CrossRef]
42. Kääb, A. Combination of srtm3 and repeat aster data for deriving alpine glacier flow velocities in the bhutan himalaya. *Remote Sens. Environ.* **2005**, *94*, 463–474. [CrossRef]
43. Das, S.; Sharma, M.C.; Miles, K.E. Flow velocities of the debris-covered miyar glacier, western himalaya, india. *Geogr. Ann. Ser. A Phys. Geogr.* **2022**, *104*, 11–34. [CrossRef]
44. Herreid, S.; Pellicciotti, F. The state of rock debris covering earth's glaciers. *Nat. Geosci.* **2020**, *13*, 621–627. [CrossRef]
45. Paul, F.; Huggel, C.; Kääb, A. Combining satellite multispectral image data and a digital elevation model for mapping debris-covered glaciers. *Remote Sens. Environ.* **2004**, *89*, 510–518. [CrossRef]
46. Mohanaiah, P.; Sathyanarayana, P.; Gurukumar, L. Image texture feature extraction using glcm approach. *Int. J. Sci. Res. Publ.* **2013**, *3*, 1–5.
47. Haralick, R.M.; Shanmugam, K.; Dinstein, I.H. Textural features for image classification. *IEEE Trans. Syst. Man Cybern.* **1973**, *6*, 610–621. [CrossRef]
48. Bolch, T.; Buchroithner, M.; Kunert, A.; Kamp, U. Automated delineation of debris-covered glaciers based on ASTER data. In *GeoInformation in Europe*; Gomarasca, M.A., Ed.; Millpress: Rotterdam, The Netherlands, 2007; pp. 403–410.
49. Shukla, A.; Arora, M.K.; Gupta, R.P. Synergistic approach for mapping debris-covered glaciers using optical–thermal remote sensing data with inputs from geomorphometric parameters. *Remote Sens. Environ.* **2010**, *114*, 1378–1387. [CrossRef]
50. Lu, Y.; Zhang, Z.; Huang, D. Glacier mapping based on random forest algorithm: A case study over the eastern pamir. *Water* **2020**, *12*, 3231. [CrossRef]
51. Disha, R.A.; Waheed, S. Performance analysis of machine learning models for intrusion detection system using gini impurity-based weighted random forest (giwrf) feature selection technique. *Cybersecurity* **2022**, *5*, 1. [CrossRef]
52. Yacouby, R.; Axman, D. Probabilistic extension of precision, recall, and f1 score for more thorough evaluation of classification models. In Proceedings of the First Workshop on Evaluation and Comparison of NLP Systems, Punta Cana, Dominican Republic, 16 March 2020; pp. 79–91.
53. Paul, F.; Bolch, T.; Kääb, A.; Nagler, T.; Nuth, C.; Scharrer, K.; Shepherd, A.; Strozzi, T.; Ticconi, F.; Bhambri, R.; et al. The glaciers climate change initiative: Methods for creating glacier area, elevation change and velocity products. *Remote Sens. Environ.* **2015**, *162*, 408–426. [CrossRef]
54. Robson, B.A.; Nuth, C.; Dahl, S.O.; Hölbling, D.; Strozzi, T.; Nielsen, P.R. Automated classification of debris-covered glaciers combining optical, sar and topographic data in an object-based environment. *Remote Sens. Environ.* **2015**, *170*, 372–387. [CrossRef]
55. Haireti, A.; Tateishi, R.; Alsaaideh, B.; Gharechelou, S. Multi-criteria technique for mapping of debris-covered and clean-ice glaciers in the shaksgam valley using landsat tm and aster gdem. *J. Mt. Sci.* **2016**, *13*, 703–714. [CrossRef]
56. Amitrano, D.; Guida, R.; Di Martino, G.; Iodice, A. Glacier monitoring using frequency domain offset tracking applied to sentinel-1 images: A product performance comparison. *Remote Sens.* **2019**, *11*, 1322. [CrossRef]
57. James, M.R.; How, P.; Wynn, P.M. Pointcatcher software: Analysis of glacial time-lapse photography and integration with multitemporal digital elevation models. *J. Glaciol.* **2016**, *62*, 159–169. [CrossRef]
58. Gómez, D.; Salvador, P.; Sanz, J.; Urbazaev, M.; Casanova, J.L. Analyzing ice dynamics using sentinel-1 data at the solheimajokull glacier, iceland. *GISci. Remote Sens.* **2020**, *57*, 813–829. [CrossRef]
59. Zhou, S.; Yao, X.; Zhang, D.; Zhang, Y.; Liu, S.; Min, Y. Remote sensing monitoring of advancing and surging glaciers in the tien shan, 1990–2019. *Remote Sens.* **2021**, *13*, 1973. [CrossRef]
60. Strozzi, T.; Luckman, A.; Murray, T.; Wegmuller, U.; Werner, C.L. Glacier motion estimation using sar offset-tracking procedures. *IEEE Trans. Geosci. Remote Sens.* **2002**, *40*, 2384–2391. [CrossRef]
61. Kumar, V.; Venkataraman, G.; Høgda, K.A.; Larsen, Y. Estimation and validation of glacier surface motion in the northwestern himalayas using high-resolution sar intensity tracking. *Int. J. Remote Sens.* **2013**, *34*, 5518–5529. [CrossRef]
62. Liu, K.; Song, C.; Ke, L.; Jiang, L.; Pan, Y.; Ma, R. Global open-access dem performances in earth's most rugged region high mountain asia: A multi-level assessment. *Geomorphology* **2019**, *338*, 16–26. [CrossRef]
63. Benn, D.I.; Thompson, S.; Gulley, J.; Mertes, J.; Luckman, A.; Nicholson, L. Structure and evolution of the drainage system of a himalayan debris-covered glacier, and its relationship with patterns of mass loss. *Cryosphere* **2017**, *11*, 2247–2264. [CrossRef]
64. Kraaijenbrink, P.; Meijer, S.W.; Shea, J.M.; Pellicciotti, F.; De Jong, S.M.; Immerzeel, W.W. Seasonal surface velocities of a himalayan glacier derived by automated correlation of unmanned aerial vehicle imagery. *Ann. Glaciol.* **2016**, *57*, 103–113. [CrossRef]
65. Farinotti, D.; Longuevergne, L.; Moholdt, G.; Duethmann, D.; Mölg, T.; Bolch, T.; Vorogushyn, S.; Güntner, A. Substantial glacier mass loss in the tien shan over the past 50 years. *Nat. Geosci.* **2015**, *8*, 716–722. [CrossRef]

66. Li, J.; Li, Z.; Zhu, J.; Li, X.; Xu, B.; Wang, Q.; Huang, C.; Hu, J. Early 21st century glacier thickness changes in the central tien shan. *Remote Sens. Environ.* **2017**, *192*, 12–29. [CrossRef]
67. Clarke, G.K.; Collins, S.G.; Thompson, D.E. Flow, thermal structure, and subglacial conditions of a surge-type glacier. *Can. J. Earth Sci.* **1984**, *21*, 232–240. [CrossRef]
68. Thenkabail, P.S. *Remotely Sensed Data Characterization, Classification, and Accuracies*; CRC Press: Boca Raton, FL, USA, 2015; ISBN 9781482217865.
69. Congalton, R.G.; Green, K. *Assessing the Accuracy of Remotely Sensed Data: Principles and Practices*; CRC Press: Boca Raton, FL, USA, 2019.
70. Kolecka, N.; Kozak, J. Assessment of the accuracy of srtm c-and x-band high mountain elevation data: A case study of the polish tatra mountains. *Pure Appl. Geophys.* **2014**, *171*, 897–912. [CrossRef]
71. Mölg, N.; Bolch, T.; Walter, A.; Vieli, A. Unravelling the evolution of zmuttgletscher and its debris cover since the end of the little ice age. *Cryosphere* **2019**, *13*, 1889–1909. [CrossRef]
72. Herreid, S.; Pellicciotti, F.; Ayala, A.; Chesnokova, A.; Kienholz, C.; Shea, J.; Shrestha, A. Satellite observations show no net change in the percentage of supraglacial debris-covered area in northern pakistan from 1977 to 2014. *J. Glaciol.* **2015**, *61*, 524–536. [CrossRef]
73. Pieczonka, T.; Bolch, T.; Junfeng, W.; Shiyin, L. Heterogeneous mass loss of glaciers in the aksu-tarim catchment (central tien shan) revealed by 1976 kh-9 hexagon and 2009 spot-5 stereo imagery. *Remote Sens. Environ.* **2013**, *130*, 233–244. [CrossRef]
74. Deline, P.; Gruber, S.; Delaloye, R.; Fischer, L.; Geertsema, M.; Giardino, M.; Hasler, A.; Kirkbride, M.; Krautblatter, M.; Magnin, F.; et al. Ice Loss and Slope Stability in High-Mountain Regions. In *Snow and Ice-Related Hazards, Risks, and Disasters*; Academic Press: Cambridge, MA, USA, 2014; pp. 521–561. ISBN 9780123964731.
75. Deline, P. Interactions between rock avalanches and glaciers in the mont blanc massif during the late holocene. *Quat. Sci. Rev.* **2009**, *28*, 1070–1083. [CrossRef]
76. Reznichenko, N.V.; Davies, T.R.; Alexander, D.J. Effects of rock avalanches on glacier behaviour and moraine formation. *Geomorphology* **2011**, *132*, 327–338. [CrossRef]
77. Satyabala, S.P. Spatiotemporal variations in surface velocity of the gangotri glacier, garhwal himalaya, india: Study using synthetic aperture radar data. *Remote Sens. Environ.* **2016**, *181*, 151–161. [CrossRef]
78. Shukla, A.; Garg, P.K. Spatio-temporal trends in the surface ice velocities of the central himalayan glaciers, india. *Glob. Planet. Chang.* **2020**, *190*, 103187. [CrossRef]
79. Das, S.; Sharma, M.C. Glacier surface velocities in the jankar chhu watershed, western himalaya, india: Study using landsat time series data (1992–2020). *Remote Sens. Appl. Soc. Environ.* **2021**, *24*, 100615. [CrossRef]
80. Heid, T.; Kääb, A. Repeat optical satellite images reveal widespread and long term decrease in land-terminating glacier speeds. *Cryosphere* **2012**, *6*, 467–478. [CrossRef]
81. Yasuda, T.; Furuya, M. Short-term glacier velocity changes at west kunlun shan, northwest tibet, detected by synthetic aperture radar data. *Remote Sens. Environ.* **2013**, *128*, 87–106. [CrossRef]
82. Dehecq, A.; Gourmelen, N.; Trouve, E. Deriving large-scale glacier velocities from a complete satellite archive: Application to the pamir–karakoram–himalaya. *Remote Sens. Environ.* **2015**, *162*, 55–66. [CrossRef]
83. Nienow, P.W.; Hubbard, A.L.; Hubbard, B.P.; Chandler, D.M.; Mair, D.W.F.; Sharp, M.J.; Willis, I.C. Hydrological controls on diurnal ice flow variability in valley glaciers. *J. Geophys. Res. Earth Surf.* **2005**, *110*, F04002. [CrossRef]
84. Vincent, C.; Moreau, L. Sliding velocity fluctuations and subglacial hydrology over the last two decades on argentière glacier, mont blanc area. *J. Glaciol.* **2016**, *62*, 805–815. [CrossRef]
85. Rounce, D.R.; Mckinney, D.C. Debris thickness of glaciers in the everest area (nepal himalaya) derived from satellite imagery using a nonlinear energy balance model. *Cryosphere* **2014**, *8*, 1317–1329. [CrossRef]
86. Fujita, K.; Sakai, A. Modelling runoff from a himalayan debris-covered glacier. *Hydrol. Earth Syst. Sci.* **2014**, *18*, 2679–2694. [CrossRef]
87. Zhang, Y.; Hirabayashi, Y.; Fujita, K.; Liu, S.; Liu, Q. Heterogeneity in supraglacial debris thickness and its role in glacier mass changes of the mount gongga. *Sci. China Earth Sci.* **2016**, *59*, 170–184. [CrossRef]
88. Wang, L.; Li, Z.; Wang, F. Spatial distribution of the debris layer on glaciers of the tuomuer peak, western tian shan. *J. Earth Sci.* **2011**, *22*, 528–538. [CrossRef]
89. Rounce, D.R.; Hock, R.; Mcnabb, R.W.; Millan, R.; Sommer, C.; Braun, M.H.; Malz, P.; Maussion, F.; Mouginot, J.; Seehaus, T.C.; et al. Distributed global debris thickness estimates reveal debris significantly impacts glacier mass balance. *Geophys. Res. Lett.* **2021**, *48*, e2020GL091311. [CrossRef] [PubMed]
90. Patel, L.K.; Sharma, P.; Thamban, M.; Singh, A.; Ravindra, R. Debris control on glacier thinning—A case study of the batal glacier, chandra basin, western himalaya. *Arab. J. Geosci.* **2016**, *9*, 309. [CrossRef]
91. Shah, S.S.; Banerjee, A.; Nainwal, H.C.; Shankar, R. Estimation of the total sub-debris ablation from point-scale ablation data on a debris-covered glacier. *J. Glaciol.* **2019**, *65*, 759–769. [CrossRef]

remote sensing

MDPI

Article

A Long-Duration Glacier Change Analysis for the Urumqi River Valley, a Representative Region of Central Asia

Lin Wang [1,2], Shujing Yang [1,2,3,*], Kangning Chen [4], Shuangshuang Liu [1,2,3], Xiang Jin [1,2,3] and Yida Xie [1,2,3]

1 Northwest Institute of Eco-Environment and Resources, Chinese Academy of Sciences, Lanzhou 730000, China
2 Key Laboratory of Cryospheric Science and Frozen Soil Engineering, Northwest Institute of Eco-Environment and Resources, Chinese Academy of Sciences, Lanzhou 730000, China
3 University of Chinese Academy of Sciences, Beijing 100049, China
4 School of Geography, Nanjing Normal University, Nanjing 210023, China
* Correspondence: yangshujing21@mails.ucas.ac.cn

Abstract: The increasing global warming trend has resulted in the mass loss of most glaciers. The Urumqi Vally, located in the dry and cold zone of China, and its widely dispersed glaciers are significant to the regional ecological environment, oasis economic development, and industrial and agricultural production. This is representative of glaciers in Middle Asia and represents one of the world's longest observed time series of glaciers, beginning in 1959. The Urumqi Headwater Glacier No. 1 (UHG-1) has a dominant presence in the World Glacier Monitoring Service (WGMS). This paper supplies a comprehensive analysis of past studies and future modeling of glacier changes in the Urumqi Valley. It has received insufficient attention in the past, and the mass balance of UHG-1 was used to verify that the geodetic results and the OGGM model simulation results are convincing. The main conclusions are: The area of 48.68 ± 4.59 km^2 delineated by 150 glaciers in 1958 decreased to 21.61 ± 0.27 km^2 delineated by 108 glaciers in 2022, with a reduction of 0.47 ± 0.04 km^2· a^{-1} (0.96% a^{-1} in 1958–2022). The glacier mass balance by geodesy is −0.69 ± 0.11 m w.e.a^{-1} in 2000–2022, which is just deviating from the measured result (−0.66 m w.e.a^{-1}), but the geodetic result in this paper can be enough to reflect the glacier changes (−0.65 ± 0.11 m w.e.a^{-1}) of the URB in 2000–2022. The future loss rate of area and volume will undergo a rapid and then decelerating process, with the fastest and slowest inflection points occurring around 2035 and 2070, respectively. High temperatures and large precipitation in summer accelerate glacier loss, and the corresponding lag period of glacier change to climate is about 2–3 years.

Keywords: Urumqi Valley; glacier change; climate change; OGGM model

1. Introduction

As the largest reservoirs of freshwater resources, glaciers are critical to the hydrological cycle and ecosystems because of their sensitive and synergistic relationship with the global climate [1,2]. Glaciers store solid water during cold temperatures and melt during the ablation period to supply water to rivers and downstream populations, especially in cold and dry regions. [3,4]. However, the global sea level has been rising at a rate of 0.2–0.4 mm per year during the 20th century due to the accelerated shrinkage of the cryosphere; this is as a result of man-made climate change [5,6]. A proposal to designate 2025 as the International Year of Glacier Protection, initiated by the Republic of Tajikistan, has been implemented, and a further plan to establish an international fund for glacier protection emphasizes the importance and urgency of glacier change [7].

There are two aspects to the study of systematic refinement of glacier change in a particular region: past systematic analyses and future change trends. The past period of glacier shrinkage is manifested in many aspects, such as reduction in number, area, and volume, while the mass balance is one of the most direct quantitative reflections of glacier accumulation and mass loss [8–10]. The traditional glaciological method of acquiring mass balance results using flower pole and snow pit observations is the most accurate method [11,12]. Moreover, a series of techniques, such as aerial photogrammetry, gravity satellite, SAR interferometry, and laser measurement, have been well used to study glacier elevation, area, and volume changes [13,14]. With the development of geodesy and the increase in DEM resolution, the results have become more accurate [15,16]. Currently, the most widely popular research method is a combination of geodetic and traditional glaciological methods, which ensures the credibility of research on glacial change, and is made more accessible by remote sensing technology [14,17]. Previous research has focused on trends in glacier change over an historical period, which is of much less reference value in the context of global warming trends [18,19]. There is a developed global scale, modular, open source numerical model called OGGM Model, which can simulate the dynamics of glaciers for both past and future change [16]. Glacial borders and DEM data are the underlying input data, and meteorological data provided by Coupled Model Intercomparison Project 6 (CMIP6) are used to drive the OGGM model [20]. Future glacier trends under different climate scenarios have been approximated by selecting glacier area changes, and the conclusions have been shown to be reliable and convincing [21,22].

The Urumqi Valley in Tianshan is an important region in the northwest dry region of China. There are many glaciers in Middle Asia and one glacier, UGH-1, has been observed for the past 60 years [23]. Most of the previous research has focused on the single glacier, UHG-1 [24–26]. The study shows that mass balance of UHG-1 is -0.46 ± 0.14 m w.e.a^{-1}, with much more area reduction and a faster rate of loss than in Tianshan [27,28]. The average glacier flow rate is 0.07 m a^{-1}, which may be related to glacier retreat and thinning [23]. The modeling of the mass balance in the basin is mostly focused on UHG-1 [29,30] and extends at most to a few surrounding glaciers [31]. Glaciers within such an important and typical basin have yet to be studied as a whole, let alone as part of future glacial changes in the basin.

This paper attempts to fill glacier information gaps in the Urumqi Valley and initial predictions of glacier ablation dynamics under different scenarios. Multi-source imagery, including topographic maps, high-resolution images, and multi-source DEM collections, were used to study a variety of indicators that can evaluate glacier changes, especially the mass balance by DEM variations. Moreover, the DEMs generated spatial resolution by stereo image pairs (Ziyuan No. 3) is up to 2.5 m, which significantly improves the accuracy of the study. In addition to discussing climate and past glacier change in detail, this paper uses the OGGM model to predict glacier change within the basin to provide a basic reference for domestic water and Industrial and agricultural development in Urumqi, a city of 3,505,800 people.

2. Study Region

The Urumqi Valley is located on the north side of Tengger Peak in middle Tianshan, China. It consists of the Urumqi River on the west and the East Mountain River system on the east. The region (43°00′N~44°07′N, 86°45′ E~87°56′E) covers 4684 km^2 area with a length of 214.30 km and the glacial region studied can be seen in Figure 1. According to the China Second Glacier Inventory, there are 129 glaciers with an area of 29.26 km^2 [32]. UHG-1, located in this basin, is the most well-recorded glacier in China, and has been studied for the longest time; the earliest recorded information began in 1958. The climate of the region can be summarized as low temperature, low humidity, high evaporation, and monsoonal character [33]. Land usage consists mainly of alpine meadows and wastelands at altitude distribution from 3391 m to 4459 m a.s.l. According to meteorological data gathered at Daxigou station between 1959 and 2018, the average temperature is −4.9 °C, with 7–8 months below 0 °C. The temperature distribution ranges from 5 °C in July to −1.5 °C in January. The precipitation reaches 466 mm, and summer contributes the most precipitation.

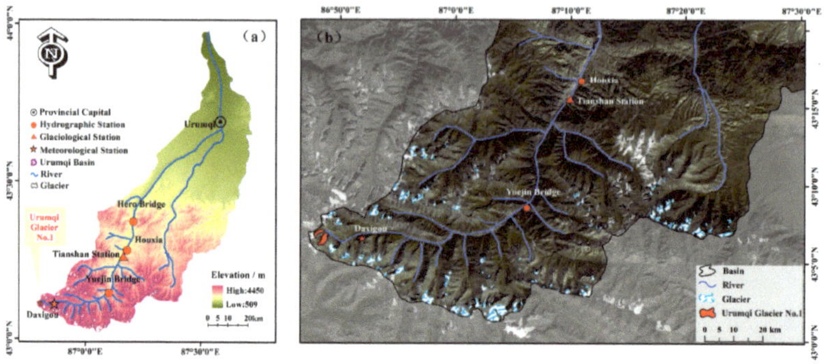

Figure 1. Location of the Urumqi Valley and glacier distribution. (**a**) is a diagram of the entire Urumqi River basin; (**b**) is a diagram of all the glaciers involved in the study area.

3. Data and Method

3.1. Data and Pre-Processing

The earliest glacial boundary extraction was based on six topographic maps involving this basin that were scanned, aligned, spliced, and corrected based on ENVI 5.3 to ensure that the corrected mean-variance was less than one cell before delineation. All topographic maps have a scale of 1:50,000. Landsat imagery has been widely used to study glacial change [34]; given the difficulty of obtaining early high-resolution imagery, we chose Landsat TM imagery from 1989. The Landsat images were downloaded from the website abbreviated as USGS and have been pre-processed, including radiometric and control point correction. (https://glovis.usgs.gov/). This paper studies glacier changes based on higher-resolution multi-source remote sensing images, such as the Spot 5 in 2005 and Ziyuan No. 3 stereo pairs in 2022. These images had <10% clouds, occurred within the ablation period, and were used to generate glacial contours. The data used for past glacier changes in the Urumqi Valley are presented in Table 1. Data are available free of charge.

Table 1. Attributes and applications of the data used in the past glacier change.

Data	Image ID/Attribution	Date	Resolution	Application
Topographical maps	——	August 1964	1:50,000	Boundary Extraction
Landsat4 TM	LT41430301989230XXX01	18 August 1989	30 m	Boundary Extraction
Spot 5	5217-262/805/09/0704:57:242A	7 September 2005	10 m/5 m	Boundary Extraction
ZY3 02 TMS	_L1A0001048489	14 September 2022	2.5 m	Boundary Extraction and Generating DEMs from Stereo Images
	_L1A0001048441	14 September 2022	2.5 m	
	_L1A0001130875	22 July 2022	2.5 m	
	_L1A0001130874	22 July 2022	2.5 m	
landsat9 OLI_TIRS	LC91430302022257LGN01	14 September 2022	30 m/15 m	Calibration Control Points
SRTM DEM	——	February 2000	30 m	Glacial Elevation Changes
ALOS PALSAR DEM	ALPSRP253320850	26 October 2010	12.5 m	Glacial Elevation Changes
ASTER GDEM V3	——	5 August 2019	30 m	Basin Extraction and Comparison Aids
ZY3-DEM	——	July–September 2022	2.5 m	Glacial Elevation Changes
The Second Glacier Inventory of China		2006–2011	30 m/15 m	Reference to Basic Glacier Properties

SRTM DEM is readily available as open-source data and, being within 1 month of 2000, has a unique advantage for studying glacier elevation change with the 30 m resolution (https://earthexplorer.usgs.gov/). The ALOS satellite from Japan is dedicated to earth observation. The PALSAR satellite has the advantage of round-the-clock, all-weather land observation, and was selected for the elevation difference analysis (https://search.asf.alaska.edu/). To measure glacier mass balance changes, this paper takes elevation changes since the 21st century at 10-year intervals. The China Daxigou Meteorological Station, at the source of the Urumqi River, provides meteorological data that can reveal synergistic changes in glaciers and climate. Data are available free of charge.

3.2. Method

3.2.1. Delineation of Glaciers and Uncertainty Assessment

Delineation of accurate glacial boundaries is important when glaciers are used to represent glacial changes in a region. Automated edge delineation is determined by many factors such as snow cover, clouds, shadows, and the resolution of images [35]. In this study, glacial margins in different periods were delineated to meet the recognized standards [36]. In addition to combining the experience of the experts in the field, Google Earth pro 2023 software also contributed [37]. Clustered glaciers are separated by ridgelines and referred to glacier cataloging to retain subsequent accuracy.

The error in delineating the glacier can be regarded as caused only by the resolution of the remotely sensed images after the glacial boundaries have been captured by the visual interpretation method, which is the most accurate method available. The uncertainty of delineation EA can be evaluated as follows:

$$E_A = \frac{N \times \lambda^2}{2} \quad (1)$$

where N is the number of pixels and λ is the resolution of different remote sensing images. The Urumqi glaciers uncertainties in 1964, 1989, 2005, and 2022 were ±4.59 km^2, ±3.98 km^2, ±0.96 km^2, and ±0.27 km^2, respectively, accounting for 9.43%, 9.81%, 3.38%, and 1.20% of total glacierized areas, respectively.

The uncertainty of changed area (E_B) is expressed as follows [38]:

$$E_B = \sqrt{E_1{}^2 + E_2{}^2} \quad (2)$$

where E_1 and E_2 are uncertainties in glacier area at the beginning and ending years, respectively.

3.2.2. Ziyuan No.3 DEMs

Ziyuan No.3 was launched in January 2012 as a forerunner of China's civilian high-resolution stereo mapping satellites, with three panchromatic cameras and a multispectral camera, which can acquire stereo image pairs of the same area from three different observation angles. The resolutions are 5.8 m multispectral, 2.1 m panchromatic, and 3.5 m front and rear view. Based on this feature, which can provide rich 3D geometric information, it can be used to generate DEMs, and Resource 3-02 has a higher front- and rear-view resolution of 2.5 m, which can meet the requirements of higher precision stereo mapping. Considering the seasonal snow and cloud cover, we chose four pairs of images from July and September 2022 to extract the DEM. The process of extracting the DEM using stereo image pairs was carried out in the DEM extraction model of the ENVI 5.3 software. After comparing the extraction results, we found that the use of frontal-view and rear-view image pairs to extract the DEM provided better results. The extraction process was carried out by selecting 20 ground control points and 100 correlation points in relatively stable non-glaciated areas, using Landsat 9 OLI images of the same period as reference positioning. All DEMs are georeferenced in the WGS84 coordinate system and stitched together with a spatial resolution of 2.5 m.

3.2.3. DEMs Co-Registration and Error Analyses

In the geodetic method, the elevation difference between DEMs is the value that characterizes the variation of glacier surface elevation; this is fundamental for our assessment of glacier mass balance in the Urumqi Valley. The ALOS DEM in the intermediate periods was selected as the reference DEM to assess the SRTM DEM and ZiyuanNo.3 DEM. The 1964 glacier margin was chosen to differentiate non-glacier topography, as the glacier area shrank the most in 2022. The rationale for the alignment is that the topography of the non-glaciated regions remains essentially unchanged, which can be used to test and iteratively adjust for errors in the multi-source DEMs. The relationship between vertical deviation and slope and aspect will tend to sine or cosine if the DEMs in non-glaciated areas are not perfectly aligned with each other [39]. The offset difference equation proposed by Nuth is as follows:

$$\frac{\mathbf{dh}}{\mathbf{tan}(\alpha)} = \mathbf{a} \times \mathbf{cos}(\mathbf{b} - \varphi) + \mathbf{c} \tag{3}$$

In order to display the results more intuitively and conveniently, we linearize Equation (3) and then solve Equation (4) to obtain the value of the offset in each direction, thus fulfilling the DEM alignment.

$$\frac{\mathbf{dh}}{\mathbf{tan}(\alpha)} = \mathbf{xsin}(\varphi) + \mathbf{ycos}(\varphi) + \frac{\mathbf{z}}{\mathbf{tan}(\overline{\alpha})} \tag{4}$$

$$\mathrm{x} = \mathrm{a} \times \sin(\mathrm{b}) \tag{5}$$

$$\mathrm{y} = \mathrm{a} \times \cos(\mathrm{b}) \tag{6}$$

$$\mathrm{z} = \mathrm{c}\ \times\ \tan(\overline{\boldsymbol{\alpha}}) \tag{7}$$

In Equations (3)–(7), dh represents the height difference between DEMs of different periods due to offset displacement: a is the magnitude of vertical movement, b is the direction of the movement vector, α is the slope of terrain, φ is the vertical direction of the terrain, and c represents the mean deviation between the different DEMs divided by the mean slope tangent, which can be calculated by $\overline{\mathrm{dh}}/\tan(\overline{\alpha})$. After linearisation, x represents the offset in EW direction, y represents the offset in the NS direction, and z is the offset in the vertical direction. After aligning the multiphase DEM with the topographic information of the non-glacier area, if elevation residual error satisfies the normal distribution, then the elevation standard deviation residual in the non-glaciated stable area approves of the estimate of the accuracy of the surface elevation change in the glacier area [40].

3.2.4. Geodetic Mass Balance and Uncertainty Assessment

Density should be taken into account when volume changes are converted to glacier mass balance [15,41]. This paper applied 850 ± 60 kg·m^{-3} to provide a parameter for taking the snow and pure ice into account to assess the mass changes by water equivalent (w.e.) [42]:

$$M = \frac{\rho}{S}\sum_{i=1}^{n} \Delta \mathrm{h_i} \times \mathrm{S_i} \tag{8}$$

where ρ is ice density transition; S is consequential glacier area; n is pixel numbers; Δhi is the single pixel height variation; and Si is single pixel area. The uncertainty was calculated as follows:

$$E = \sqrt{\left(\frac{\Delta \mathrm{h_i}}{t} \times \frac{\Delta \rho}{\rho_{\mathrm{w}}}\right)^2 + \left(\frac{\sigma}{t} \times \frac{\rho_1}{\rho_{\mathrm{w}}}\right)^2} \tag{9}$$

where Δh is the mean height variation of glacier areas; t is studied period; $\Delta\rho$ is the error of ice density, which value is taken as 60 kg·m^{-3}; ρ_{w} is water density, which value is taken as 1000 kg·m^{-3}; σ is the errors of height variation; and ρ_1 is ice density transition, which value is taken as 850 kg·m^{-3}.

3.2.5. Glacial Mass Balance and Uncertainty

In 1959, we began to measure the glacier mass balance during ablation (May~August) by using flower sticks to confirm the height variation and snow pits to assess the density [43,44]. As illustrated in Figure 2, the stake network spans the whole UHG-1 in order to provide the result of mass balance. For UHG-1, we manually interpolated the measured data between adjacent contours by manually drawing lines of equal mass balance [45].

$$Bn = \frac{\sum_{i=1}^{n} B_i \cdot S_i}{S} \tag{10}$$

where *Si* is the pixel area, *Bi* is the obtained mass balance of the corresponding pixel glacier, and *S* is the total area of UHG-1.

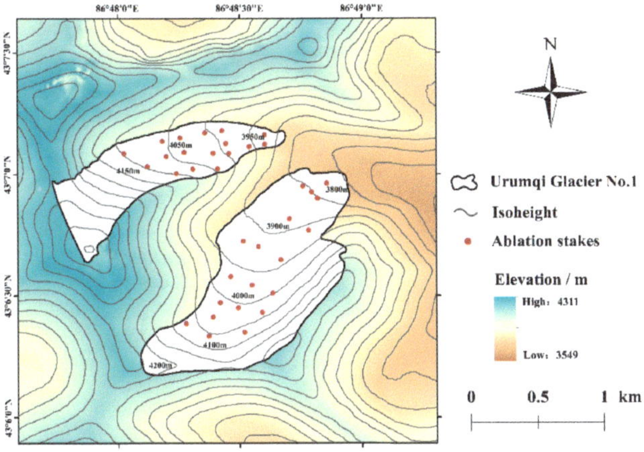

Figure 2. Schematic diagram of flower pole deployment under traditional glaciological methods.

3.2.6. OGGM Model

Open Global Glacier Model gives full consideration to glacial geometry and consists of an explicit ice dynamics module and a calving parametrization. The modular OGGM supports being redefined, remixed, and repeated, with reliance on publicly available data for calibration and validation. In the model, glacial boundaries and DEM data are used as model base input data. It is driven by climate scenarios to forecast future area, volume changes, etc. The model is globally consistent in predicting ice thickness and glacier mass loss. To maintain the purity of the results, the observed glacier mass balance is the best choice for checking the accuracy of OGGM. Note that model accuracy is verified by predicting glacier mass balance using an extended temperature index model [21]

$$M(h) = P_{cf} \cdot P(h)_{solid} - \mu \cdot max\{(T(h) - T_{melt}, 0)\} + \varepsilon \tag{11}$$

where characters with (h) qualify the elevation, M is monthly glacier mas balance, P_{cf} is precipitation correction factor, P_{solid} is monthly solid precipitation, and μ is updated calibration and correction factor, which represents the glacier sensitivity parameter and where a particular glacier can be set to agree with observations [20]. T is monthly temperature, Tmelt represents monthly mean air temperature, and ε is deviation correction.

3.2.7. Meteorological Data

In this study, three SSPs scenarios were selected to drive the OGGM model. In order to reduce the error of the results, for each SSPs pathway, the results of 13 climate models were selected, which were BBC, CAMS, CESM2-WACCM, CESM, EC-Earth3, EC-Earth3-Veg, FGOALS, GFDL, INM-CM4-8, INM-CM5-0, MPI-ESM1-2-HR, MPI-ESM2-0, and NorESM2-MM. The final results of glacial area and volume in the URB are obtained by averaging the standard errors of the 13 climate models; ensembles are taken as the errors of the area and volume simulations results [31]. Given the assumptions that climate change arises only over large areas and the associations of climate variables remain consistent in the base period, interpolating or inserting data where none exists with thin-plate spline eliminates errors.

4. Result

4.1. Delienation and Uncertainty of Glaciers

In 2022, 108 glaciers were mapped in the entire URB, a decrease of nearly one third compared to 150 in 1964. The number of glaciers throughout the basin has been trending downward over time. Moreover, the whole basin can be divided into four sub-basins, coded as 5Y730A, 5Y730B, 5Y730C, and 5Y730D. We discerned and statistically calculated the contours of the glacier in detail and combined the number of pixels occupied by the glacier and the image resolution to obtain the uncertainty in the delineated and vanished glacier area. Specific information is presented in Tables 2 and 3. Because the region belongs to multiple field observations, the error of manual visual interpretation is ignored in this paper. Higher resolution images are the basis and direction for the development of glacier remote sensing for the future. The areas of glaciers delineated in different years of this paper and the errors are shown below:

Table 2. Glacier delineation and uncertainty in each sub-basin of the Urumqi River.

District	1964 Area	1989 Area	2005 Area	2022 Area	Area Change Rate (1964–2022)
5Y730A	6.16 ± 0.58 km^2	4.52 ± 0.44 km^2	3.19 ± 0.11 km^2	1.99 ± 0.04 km^2	32.31%·a^{-1}
5Y730B	21.62 ± 2.04 km^2	18.41 ± 1.81 km^2	12.81 ± 0.43 km^2	9.73 ± 0.12 km^2	45.00%·a^{-1}
5Y730C	9.17 ± 0.86 km^2	8.32 ± 0.82 km^2	5.13 ± 0.17 km^2	4.10 ± 0.04 km^2	44.71%·a^{-1}
5Y730D	11.73 ± 1.11 km^2	9.32 ± 0.91 km^2	7.31 ± 0.25 km^2	5.79 ± 0.07 km^2	49.36%·a^{-1}
Total	48.68 ± 4.59 km^2	40.57 ± 3.98 km^2	28.44 ± 0.96 km^2	21.61 ± 0.27 km^2	44.39%·a^{-1}

Table 3. Metrics of glacier change and uncertainty in different periods.

Period	Overall Losses km^2	Overall Error of Variation	Annual Area Change	Area Change Rate
1964–1989	8.11 km^2	1.10 km^2	0.32 ± 0.04 km^2·a^{-1}	0.67%·a^{-1}
1989–2005	12.14 km^2	1.26 km^2	0.76 ± 0.08 km^2·a^{-1}	1.87%·a^{-1}
2005–2022	6.82 km^2	0.24 km^2	0.40 ± 0.01 km^2·a^{-1}	1.41%·a^{-1}
1964–2020	27.07 km^2	2.57 km^2	0.47 ± 0.04 km^2·a^{-1}	0.96%·a^{-1}

4.2. Area Changes and Analysis in Glaciers

The glacier area in the basin decayed rapidly, from 48.68 ± 4.59 km^2 in 1964 to 21.61 ± 0.26 km^2, a shrinkage rate of 0.47 ± 0.04 km^2·a^{-1} and an overall relative rate of change in glacier area of 55.6% (0.96%·a^{-1}). Because the glacier scale in this region is inherently not very large, it is believed that the glacier will be completely ablated. In this paper, the rate of glacier area change is divided into three periods, and the ablation rate is -0.31 km^2·a^{-1} (0.64%·a^{-1}), -0.81 km^2·a^{-1} (1.99%·a^{-1}), and -0.40 km^2·a^{-1} (1.41%·a^{-1}) in the periods of 1964–1990, 1990–2005, and 2005–2022, respectively. The years from 1990–2005 had the fastest rate of glacial ablation, and the ablation rate for the latest period has remained above the average rate for the last 58 years. Moreover, four sub-basins, coded as 5Y730A, 5Y730B, 5Y730C, and 5Y730D, involved 67.75%, 54.96%, 55.32%, and 50.65% of the ablation area from 1964 to 2022, respectively. The detailed ablation shape is shown in Figure 3.

The Urumqi Valley is on northern Tianshan, so most glacier numbers (106) and glacier area (20.79 ± 0.24 km^2) belong to the N, NE, and NW facings. The glacier area oriented towards N had the most ablation, accounting for 58% of the total. The NE and NW oriented glacier areas are about equally ablated, accounting for about 54%. The two glaciers in other orientations had an the area of only 0.81 ± 0.02 km^2, and were much smaller in 2022. (Figure 4). The mean elevation for the glaciers inventoried ranged from 3667 to 4272 m, with the mean being 3948 m in 2000; they ranged from 3671 m to 4270 m with the mean 3926 m in 2022. About 80% of the glacier area was distributed within 3800–4000 m. Over 50% of the glacier area was between 3900–4000 m. The area of glaciers is roughly normally distributed at different elevation scales. The most serious ablation of glacier area is in the altitude zone below than 3700 m, accounting for 76.55%, followed by 4000–4100 m, accounting for 68.47%, while the smallest rate of ablation is in the most aggregated altitude zone of 3900–4000 m, accounting for 52.5%, which reflects the intense ablation of glacier area in the region.

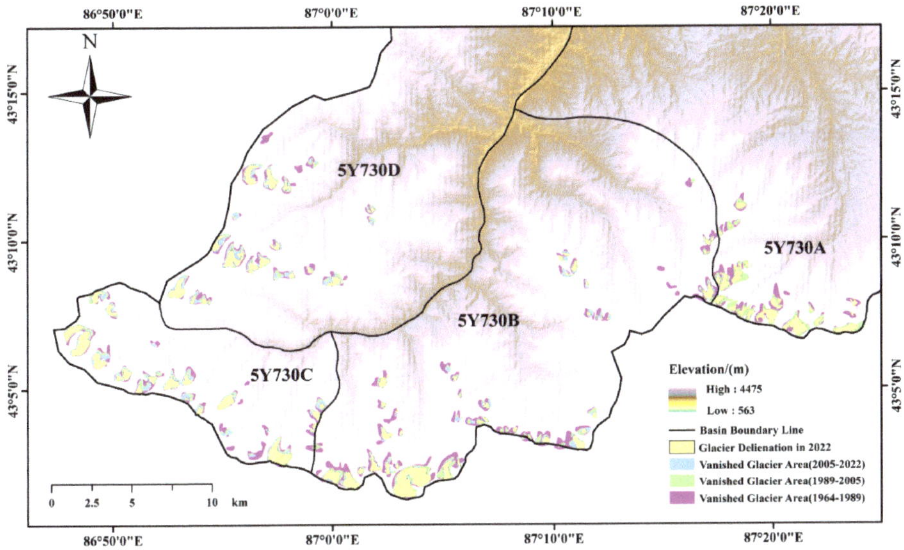

Figure 3. Distribution and delineation of glaciers in different periods of the Urumqi River basin.

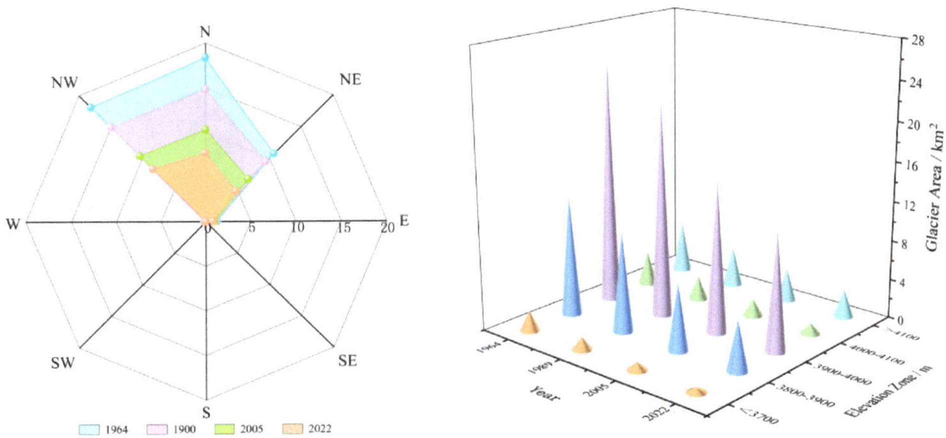

Figure 4. Schematic illustration of the distribution and variation of glacier outlines over time. The left figure shows glacier orientation over time. The right figure shows the changes in glacier area at different altitudes.

4.3. Geodetic Uncertainty Analysis

After correcting the offset bias among SRTM, ALOS, and ZiyuanNo.3 DEM data with a trigonometric correction (Figures 5 and 6), the elevation difference residuals in the non-glaciated areas can be used to evaluate the errors among the DEMs and to calculate the accuracy of the estimation results for the glacier volume change and the mass balance. Because the spatial resolution of the DEM data used is not uniform, the spatial autocorrelation distances were chosen to be 600 m, 250 m, and 50 m for the SRTM, ALOS, and ZiyuanNo.3 resolutions of 30 m, 12.5 m, and 2.5 m, respectively (Bolch T. et al., 2011). After error correction, Mean Elevation Difference (MED) tends to be close to 0; the relative error between the DEM data is significantly less than 1 m (Table 4).

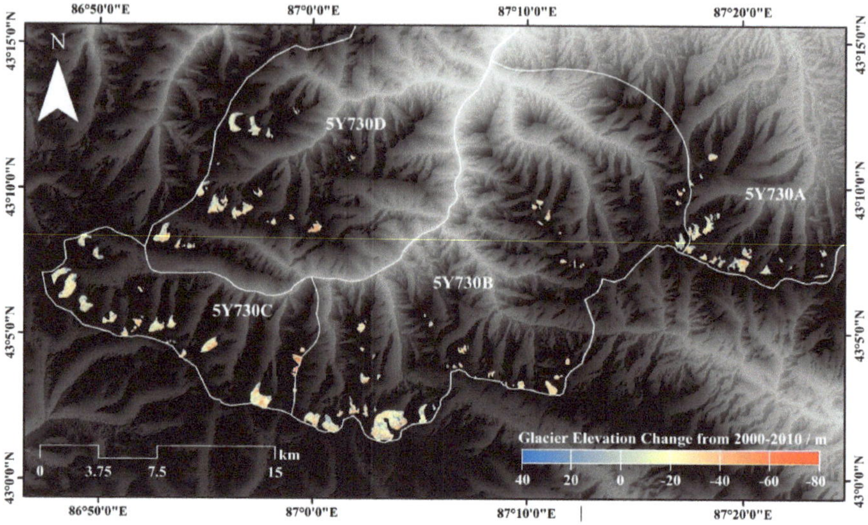

Figure 5. Glacier height variations in Urumqi Valley, 2000–2010.

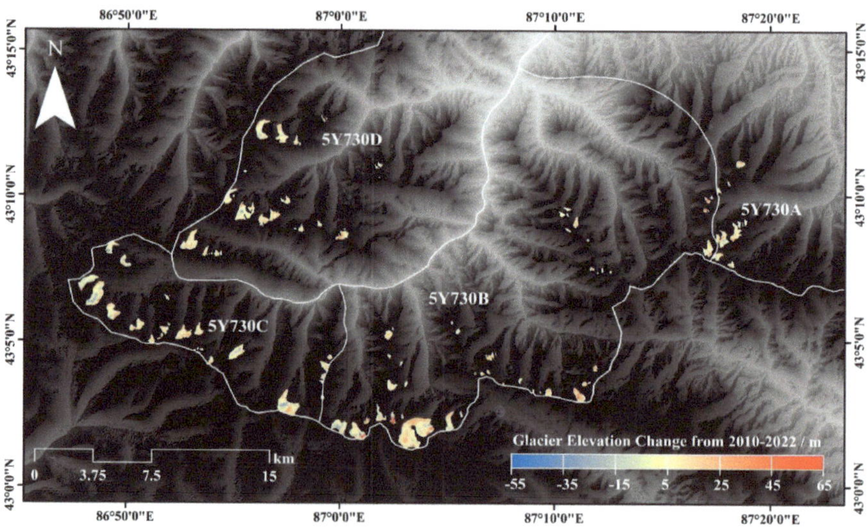

Figure 6. Glacier height variations in Urumqi Valley, 2010–2022.

Table 4. Original and adjusted errors between muti-DEMs.

Period	Item	Original (m)		Adjusted (m)		N	SE (m)	σ (m)
		MED	STDV	MED	STDV			
2000–2010	SRTM_ALOS	17.84	29.47	0.14	32.85	3872	0.53	0.53
2010–2022	ALOS_ZiyuanNo.3	2.00	12.59	−0.13	9.74	10,363	0.10	0.16

4.4. Geodetic Glacier Mass Loss

Theoretically, the difference embodied in each raster of the DEMs can respond to a single point of glacier mass balance. Given the corrected results, the glacier elevation in the URB declined significantly in various periods, with an accelerated trend of glacier retreat. The following figure is a visualization of the difference between the DEMs of the two periods processed by the Arcmap 10.3.

After discounting obviously erroneous extremes from the results of the subtraction of the DEM, most of the variables are tightly clustered around the average value, with only a few at the extremes (Figure 7). The glacier elevation in the URB declined by 44.64 $\pm$ 0.69 m during 2000–2022, of which the glacier elevation declined by 23.07 $\pm$ 0.53 m on average during 2000–2010, with an annual mean decrease of 2.31 $\pm$ 0.05 m·a^{-1}, and by 21.57 $\pm$ 0.16 m during 2010–2022, with an annual mean decrease of 1.80 $\pm$ 0.01 m·a^{-1}.

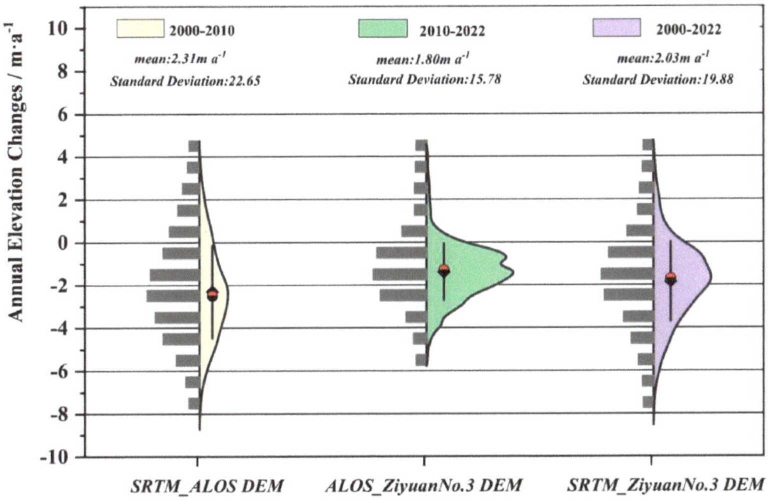

Figure 7. Box plots of annual height variations for multi-period DEMs.

According to the results of the glacier height variations converted to glacier loss, the volume of glaciers in the Urumqi Valley has decreased by 0.24 km^3 over the past 22 years, with an annual mean decrease of 0.01 km^3·a^{-1}. The annual glacier mass balance of Urumqi Valley was -0.65 ± 0.11 m w.e.a^{-1} during 2000–2022. Of this, the mean glacier mass balance was -0.67 ± 0.12 m w.e.a^{-1} before 2010 and was -0.63 ± 0.11 m w.e.a^{-1} after 2010. As a special mention, we also calculated the individual glacier mass balance separately for the key monitoring glacier (the UHG-1). The average annual glacier mass balance of the UHG-1 was -0.69 ± 0.12 m w.e.a^{-1} during 2000–2022. The average glacier mass balance was -0.71 ± 0.12 m w.e.a^{-1} in 2000–2010 and was -0.67 ± 0.12 m w.e.a^{-1} in 2010–2022.

4.5. Predicted Future Glacier Mass Loss in Urumqi River Basin

The simulation and analysis of glacier area and reserves in the Urumqi River basin under different climate scenarios of CMIP6, including SSP1-2.6, SSP2-4.5, and SSP5-8.5, found that the trends of glacier area and reserves changes under different emission scenarios coincide with each other. In 2020–2080, the glacier area and reserves in the Urumqi River Basin of the Tianshan Mountains still show an overall trend of retreat and decrease (Figure 8). Reserves as a whole still show a trend of retreat and decrease, and both basically disappear completely by the end of this century. Under the three climate scenarios, the initial and final values of SSP1-2.6, SSP2-4.5, and SSP5-8.5 are approximately the same, but the intermediate trends are very different. For glacier area, the SSP1-2.6 scenario has the slowest rate of glacier retreat, with an average annual decrease in change of 0.351 km^2·a^{-1}

and an average annual decrease in glacier volume/volume of 5.21 × 10^6 $m^3 \cdot a^{-1}$. Next is the SSP2-4.5 scenario, with an average annual decrease in change of 0.352 $km^2 \cdot a^{-1}$ and an average annual decrease in glacier volume/volume of 5. 23 × 10^6 $m^3 \cdot a^{-1}$, and the SSP5-8.5 scenario, with a decrease in glacier area/volume of 0.352 km^2 a^{-1} and an average annual decrease in glacier volume/volume of 5.23 × 10^6 $m^3 \cdot a^{-1}$. SSP5-8.5 scenario had the largest glacier area and reserve melt. The glacier area change decreases by 0.353 $km^2 \cdot a^{-1}$ and the glacier reserve/volume decreases by 5.29 × 10^6 m^3 a^{-1}.

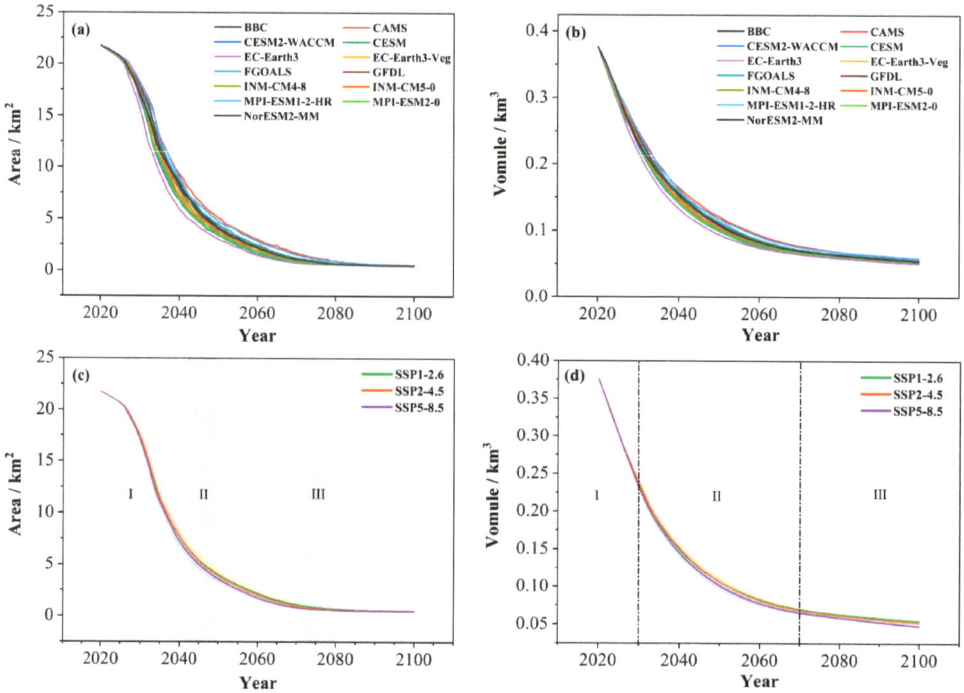

Figure 8. The area and volume change of glaciers in the Urumqi River Basin from 2020 to 2100. Where (**a**,**b**) shows the area and volume trends of the 13 climate models under different climate models, respectively; (**c**,**d**) presents the mean values of the simulation results of the 13 climate models after processing under the scenarios of SSP1-2.6, SSP2-4.5, and SSP5-8.5. The 13 model patterns described in the legend and whose simulation standard errors are displayed as shaded areas in (**c**,**d**). I, II, III represent the different stages of the trend change.

Taking the mid-century (2050) as the cut-off point, the rate of glacier area decrease in the first half of this century (2020–2050) is higher than that in the second half of this century (2050–2080) under the emission scenarios of SSP1-2.6, SSP2-4.5, and SSP5-8.5 (0.482 $km^2 \cdot$ a^{-1}, 0.485 $km^2 \cdot a^{-1}$, and 0.505 $km^2 a^{-1}$, respectively); the rate of decrease in glacier area decrease is 0.353 $km^2 \cdot a^{-1}$ and the rate of decrease in glacier storage/volume decrease is 5.29 × 10^6 $m^3 \cdot a^{-1}$. The rate of decrease of glacier reserves in the first half of this century (2020–2050) is higher than that in the second half of this century (2050–2080) by 0.226 $km^3 \cdot a^{-1}$, 0.224 $km^3 \cdot a^{-1}$, 0.232 $km^3 \cdot a^{-1}$, and the percentage of decrease is higher than that in the second half of this century (2050–2080) by 31%, 30%, and 31%, respectively. Comparing the glacier area and glacier reserve/volume, the percentage reduction of glacier area is about 10% higher than the percentage reduction of glacier reserve in the first part of this century (2020–2050), and the percentage reduction of glacier area is about 40% higher than the percentage reduction of glacier reserve in the second part of this century (2050–2080), indicating that the glaciers in the Urumqi River Basin are mainly characterized by glacier area retreat, followed by thickness reduction. This pattern is even greater in the middle- and late-century. The reason for this is that the glaciers in the Urumqi River Basin are inherently thinner than the glacier properties in other regions.

5. Discussion

5.1. Geodetic Glacier Mass Balance Validation

The mass balance obtained from the UHG-1 based on traditional glaciological measurements is shown in Figure 9. It is obvious that the period of maximum positive accumulation of mass balance is between 1960 and 1980. Since 2000, the positive values of glacier mass balance have almost disappeared and the negative accumulation trend of glacier mass balance has intensified, with the highest negative accumulation value in 2010.The annual average values of glacier mass balance of UG-1 for the periods of 2000–2010, 2010–2022, and 2000–2022 are 664 mm w.e.a^{-1}, 666 mm w.e.a^{-1}, and 662 mm w.e.a^{-1}, respectively. Overall, the annual mean mass balance for the period 2000–2022 is stable and shows little fluctuation; it is still suitable as a reference for the same period of time for obtaining the mass balance based on remote sensing. Compared with the annual average mass balance of a single glacier obtained by geodetic methods during the same period of this study, the values are -0.67 m w.e.a^{-1}, -0.71 m w.e.a^{-1}, and -0.69 m w.e.a^{-1}, respectively, and it can be concluded that the mass balances of geodetic and glaciological measurements are in agreement with each other. Taking the glaciological mass balance values as benchmarks, the relative differences between 2000–2010, 2010–2022, and 2000–2022 are only 1.5%, 7.5%, and 4.5%, respectively. Moreover, a linear fit to the mass balance trends obtained by traditional glaciological methods after 2000 showed a stronger negative accumulation trend. The interval of fitted values is from -0.83 m w.e.a^{-1} to -0.67 m w.e.a^{-1} and the results obtained in this study fall perfectly in that interval.

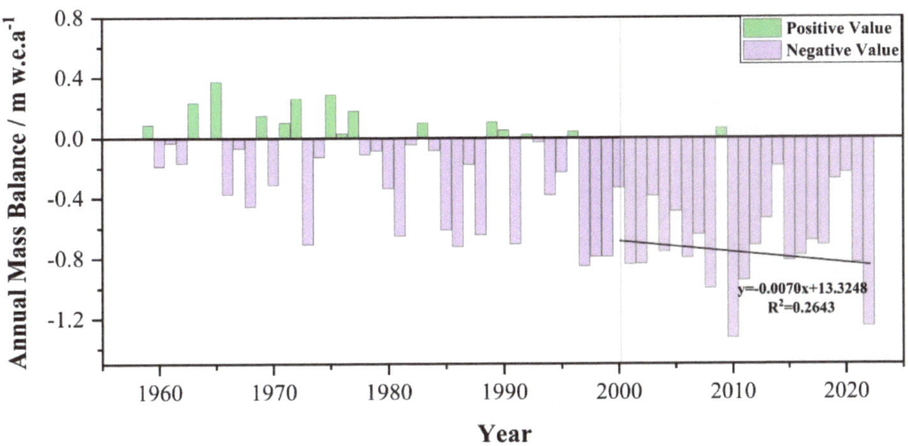

Figure 9. Annual mass balance of UHG-1 based on traditional glaciological measurements, 1958–2022.

5.2. OGGM Model Validation and Forecasting

To prove the accuracy of the future changes in glacier area and volume, we used the OGGM model to acquire the glacier mass balance from 2000 to 2020, which is displayed in Figure 10. The OGGM model and the observed values show a near-identical trend, with a correlation coefficient of 0.87, which has a strong significance. In addition, the difference between the linear fit lines of OGGM simulated and the observed values have a tendency to widen; the error range of 52–124 mm w.e.a^{-1}. is acceptable. After confirming that the results of glacier area and volume were credible, we calculated the glacier mass balance based on the formula and combined it with the snow-ice density transformation to get the future glacier mass balance value of the URB. The results show that the glacier mass balance is the most negative in the SSP5.8-5 pattern, followed by SSP2.4-5, and is the least negative in SSP1.2-6. The variability in glacier mass balance among the three models is present and in good agreement with future changes in area and volume, with an accelerated rate of change in 2035 and little fluctuation after 2075.

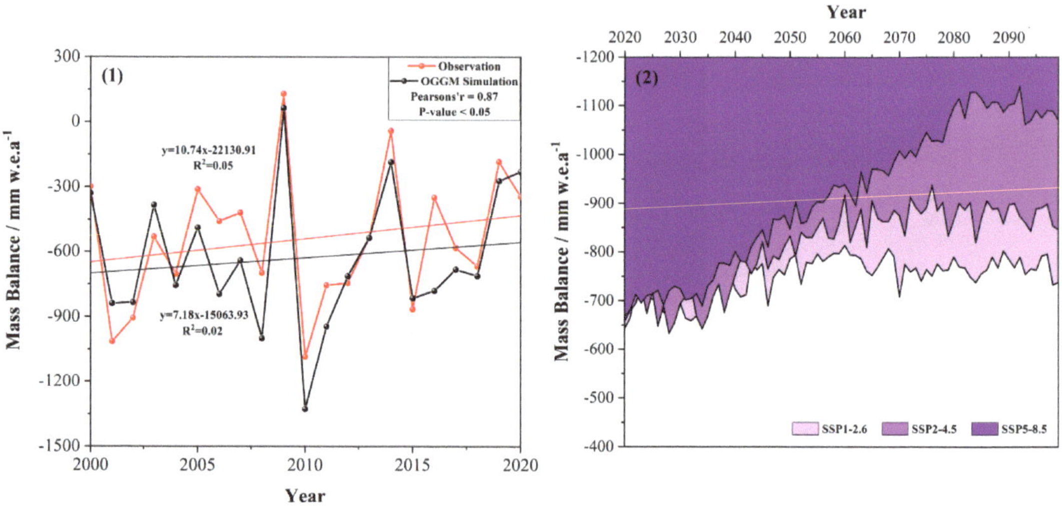

Figure 10. Validation and prediction of glacier mass balance for the OGGM model. (**1**) Comparison of OGGM-based and observed glacier mass balance for 2000–2020. (**2**) Future glacier mass balance values based on glacier area and volume from the OGGM model.

5.3. Glacier Loss Comparison in Typical Regions

For the entire Tianshan mountain range, the glacier mass deficit within the URB is relatively dramatic, with much higher values of negative mass balance than many typical basins, as well as the mass balance of the monitored glaciers. The factors influencing the ablation and mass balance of glaciers are sophisticated (Table 5). First, the degree and speed of glacier ablation are different when the size and type of glacier are different. The larger the size and the more concentrated the distribution of the glacier, the less intense the ablation. Large debris-covered glaciers are less prone to ablation, and such glaciers are most strongly developed in the Tomur region of the Tianshan Mountains. The Urumqi Valley belongs to the typical glacier in a small area and scattered distribution, where fast ablation is inevitable. Second, the combination of aspect and slope affects the development of glaciers mainly by changing the amount of solar radiation received. The mass balance is exacerbated by a negative balance in the N/N-direction, which is usually favored by small slopes for the accumulation of glaciers. The vast majority of glaciers in the Urumqi Valley are orientated N and have little topographic relief, and there are no high mountain systems around them that can block solar radiation. Third, the altitude zone in which the glacier is located is also a key factor affecting the ablation of glaciers, as glaciers are not widely developed at lower altitudes, and the ablation rate is generally faster in the region below 4500 m. The glacier studied in this paper belongs to this situation. Fourth, there is a change of water and heat because of the significant impact of westerly winds from the Atlantic Ocean. In this paper, the effects of temperature and precipitation are discussed in Section 5.4 as a separate section.

Table 5. Comparison of glacier mass balance of monitored glaciers and typical basins in western China.

Region	Typical Glacier	Period	Basin Mass Balance/m w.e.a^{-1}	Period	Single Mass Balance/m w.e.a^{-1}	Source
Urumqi Valley	UHG	2000–2022 2000–2010 2010–2022	−0.65 ± 0.11 −0.67 ± 0.12 −0.63 ± 0.11	2000–2022 2000–2010 2010–2022	−0.69 ± 0.12 −0.71 ± 0.12 −0.67 ± 0.12	This study
Mt.Tomor	Qingbingtan No.72			1964–2008 2008–2014	−0.20 −0.38	[46,47]
	Keqikaer			1999–2003 2004–2006	−0.22 −0.44	[48]
Kuitun River Basin	Haxilegen No.51			2004–2006 2010–2011	−0.38 −0.68	[49]
Yili River Basin	Ts.Tuyuksuyskiy	2003–2018	−0.42 ± 0.16	2003–2018	−0.50	[50]
Kaidu River Basin	——	2001–2016	−0.48	——	——	[51]
Manas River Basin	——	2000–2016	−0.58	——	——	[52]
Northern Tin Shan	——	2000–2020	−0.39 ± 0.04	——	——	[53]

5.4. Contribution of Meteorological Conditions to Glacier Ablation

Climate is the basic factor in assessing glacier change. It is shown that temperature is the leading causes of the glacier change [54,55]. The amount of glacial melt needs to increase by 40~50% of precipitation to compensate for every 1 °C increase in the mean summer temperature [56]. Figure 11 illustrates the temperatures and precipitation obtained from the Daxigou meteorological station for the years 1958–2022, showing that the Urumqi Valley has experienced warming and wetting potential during the last half-century. The average temperature from 1959 to 2022 is −4.75 °C in this region. The average temperature in 1959–2000 and 2000–2022 is 5.02 °C and 4.23 °C, which is an increase of 0.20 °C/10a and 0.07 °C/10a, respectively. The average annual precipitation from 1960 to 2021 is 473.80 mm in this region. The average precipitation in 1959–2000 and 2000–2022 is 444.64 mm and 528.13 mm, which has a increasing rate of 19.38 mm/10a and 29.95 mm/10a, respectively.

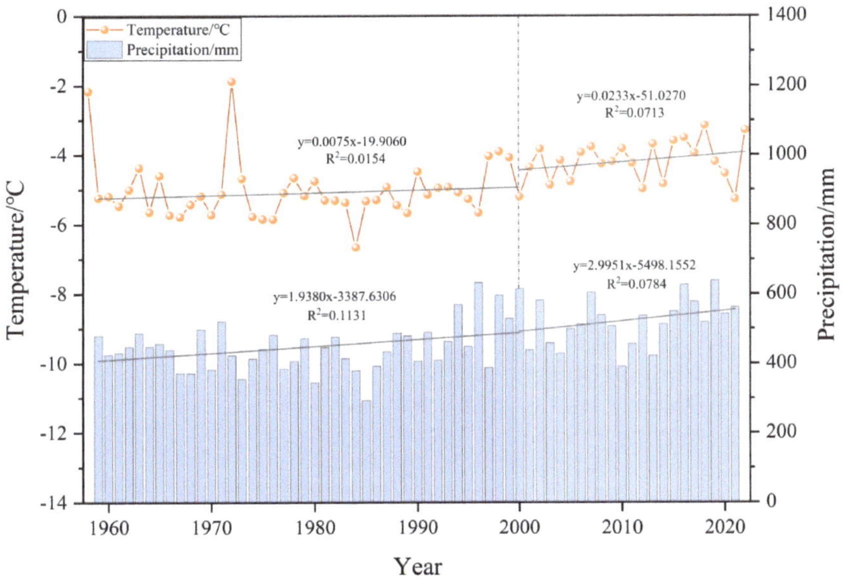

Figure 11. Variations in annual mean temperature and precipitation observed at the Daxigou Meteorological Station from 1959 to 2022.

In order to further explore the dominant factors and approximate time lag of temperature and precipitation on past glacier changes in the URB, we calculated the correlation coefficients between glacier mass balance delayed by 1–6 years and climate factors, respectively (Figure 12). The highest correlation between summer temperatures and the magnitude of glacier loss can be seen, which means that the higher the summer temperatures, the more glacier loss. In addition, because the annual precipitation accumulation in the region is mainly contributed by summer precipitation, the correlation coefficients obtained for annual and summer precipitation with glacier mass balance do not differ much. In terms of the delay period, there is the highest correlation coefficient between glacier mass balance with a three-year lag and summer temperature. The highest correlation coefficient between glacier mass balance and annual precipitation was found in the two-year lag, and the correlation coefficient between glacier mass balance and winter precipitation was negative at that time, which proved that winter precipitation not only did not cause the glacier loss, but contributed to the accumulation of the glacier, but only with a minor contribution. Therefore, the response of glaciers to climate in the URB has a lag of 2–3 years.

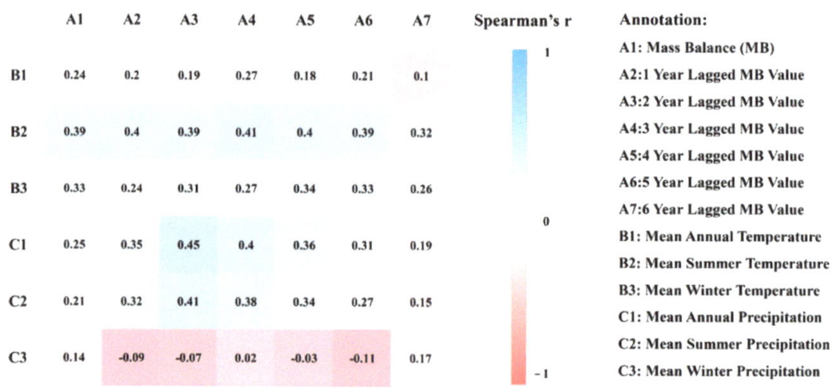

Figure 12. Correlation analysis of different lags of glacial mass balance with air temperature and precipitation.

6. Conclusions

This study aimed to quantify the glacier changes in the Urumqi Valley in the Tianshan Mountains, and the area changes of glaciers were assessed by multi-source remote sensing imagery and analyzed in conjunction with topographic factors. In addition, the mass balance of the glaciers in the basin was estimated using geodetic methods and validated with the aid of traditional glaciological methods, and the following main conclusions were obtained:

(1) In 2022, 108 glaciers were mapped in the entire Urumqi Valley, a decrease of nearly one-third compared to 150 in 1964. The glacier area in the basin decayed rapidly from 48.68 $\pm$ 4.59 km^2 in 1964 to 21.61 $\pm$ 0.26 km^2 in 2022, with a shrinkage rate of 0.47 $\pm$ 0.04 km^2·a^{-1} and an overall relative rate of change in glacier area of 55.6% (0.96%·a^{-1}). Glacier area shrinkage has increased since 1990 and is higher than the average annual change rate (0.96%·a^{-1}) over the past 58 years. The majority of the glaciers are orientated towards N (N\NE\NW) and are situated in the elevation zone between 3800–4000 m. The glacier area oriented N had the most ablation, accounting for 58% of the total area, and the largest percentage of glacier area (over 50%) was between 3900–4000 m. Different combinations of topographic factors are associated with differential glacier area changes

(2) The glacier elevation in the Urumqi Valley declined by 44.64 $\pm$ 0.69 m during 2000–2022, with an average annual decline of 2.31 $\pm$ 0.05 m·a^{-1} in 2000–2010 and 1.80 $\pm$ 0.01 m·a^{-1} in 2010–2022. The volume of glaciers in the Urumqi Valley has decreased

by 0.24 km^3 over the past 22 years. The average annual glacier mass balance of URB is -0.65 ± 0.11 m w.e.a^{-1} during 2000–2022. Of this, the average glacier mass balance was -0.67 ± 0.12 m w.e.a^{-1} before 2010 and was -0.63 ± 0.11 m w.e.a^{-1} after 2010. The average geodetic mass balance of the monitored UHG-1 (-0.69 ± 0.11 m w.e.a^{-1}) somewhat deviates from the observed result (-0.65 m w.e.a^{-1}), but the geodetic method result in this paper can be used to reflect the changes of glaciers in the region;

(3) Based on the OGGM model simulations, SSP5.8-5 model has the fastest glacier area and volume losses and the rate of glacier area and volume loss in 2020–2100 undergoes a fast and then slow process. The fastest rate of glacier area loss occurs between 2030 and 2035; the rate of glacier area loss slows between 2042 and 2050; and the lowest rate of glacier area loss occurs between 2070 and 2080, when glacier area will no longer be decreasing. The highest and lowest inflection points for glacier volume loss are in 2035 and 2070, respectively. The glacier mass balance based on the OGGM model can correspond well with the observed values, so it confirms that the results of this paper's prediction of glacier changes in the Urumqi Valley are credible. Moreover, Different SSP scenarios show different changes in glacier mass balance, with the fastest negative mass balance in the SSP5.8-5 model, followed by SSP2.4-5, and the slowest negative mass balance in SSP1.2-6.

(4) Although the glacier area change and mass balance are spatially heterogeneous, glacier melting in the basin has shown an accelerated shrinkage trend in recent years compared to other regions. Higher temperatures are the dominant factor contributing to accelerated glacier loss, and increased (summer) precipitation can also make glacier mass balance values more negative. Winter precipitation will contribute almost no amount to glacier accumulation. There is a lag of roughly 2–3 years in the response of glaciers to climate in the Urumqi Valley.

Author Contributions: Conceptualization, L.W.; methodology, S.Y., S.L. and Y.X.; software, S.Y. and X.J.; validation, S.Y., X.J. and S.L.; formal analysis, S.Y. and X.J.; investigation, L.W. and S.Y.; resources, S.L.; data curation, L.W., S.Y., K.C., S.L. and X.J.; writing—original draft preparation, L.W., S.Y. and Y.X.; writing—review and editing, S.Y. and K.C.; visualization, K.C.; supervision, L.W., S.Y., K.C., S.L. and X.J.; project administration, L.W.; funding acquisition, L.W. All authors have read and agreed to the published version of the manuscript.

Funding: This research was funded by the Third Comprehensive Scientific Expedition of Xinjiang Uyghur Autonmous Region (2022xjkk0802), the Science and Technology program of Gansu Province (Grant No 23ZDFA017, 22ZD6FA005), the State Key Laboratory of Cryospheric Science (SKLCS-ZZ-2023).

Data Availability Statement: The original contributions presented in the study are included in the article, further inquiries can be directed to the corresponding author.

Acknowledgments: Thanks to NASA, USGS, and Copernicus Open Access Hub for providing the remote sensing images used in this study; thanks to the partners who went to the field together for field measurements.

Conflicts of Interest: The authors declare no conflicts of interest.

References

1. Benn, D.I.; Bolch, T.; Hands, K.; Gulley, J.; Luckman, A.; Nicholson, L.I.; Quincey, D.; Thompson, S.; Toumi, R.; Wiseman, S. Response of Debris-Covered Glaciers in the Mount Everest Region to Recent Warming, and Implications for Outburst Flood Hazards. *Earth-Sci. Rev.* **2012**, *114*, 156–174. [CrossRef]
2. Gardner, A.S.; Moholdt, G.; Cogley, J.G.; Wouters, B.; Arendt, A.A.; Wahr, J.; Berthier, E.; Hock, R.; Pfeffer, W.T.; Kaser, G.; et al. A Reconciled Estimate of Glacier Contributions to Sea Level Rise: 2003 to 2009. *Science* **2013**, *340*, 852–857. [CrossRef] [PubMed]
3. Yao, T.; Wu, G.; Xu, B.; Wang, W.; Gao, J.; An, B. Asian Water Tower Change and its Impacts. *Bull. Chin. Acad. Sci.* **2019**, *34*, 1203–1209. [CrossRef]
4. Hugonnet, R.; Mcnabb, R.; Berthier, E.; Menounos, B.; Nuth, C.; Girod, L.; Farinotti, D.; Huss, M.; Dussaillant, I.; Brun, F.; et al. Accelerated Global Glacier Mass Loss in the Early Twenty-First Century. *Nature* **2021**, *592*, 726–731. [CrossRef] [PubMed]
5. IPCC. *Working Group I Contribution to the IPCC Fifth Assessment Report Climate Change 2013: The Physical Science Basis Summary for Policymakers*; Stocker, T.F., Qin, D., Plattner, G.-K., Tignor, M., Allen, S.K., Boschung, J., Nauels, A., Xia, Y., Bex, V., Midgley, P.M., Eds.; Cambridge University Press: Cambridge, UK; New York, NY, USA, 2013; p. 1535.

6. IPCC. Summary for Policymakers. In *Climate Change 2021: The Physical Science Basis. Contribution of Working Group I to the Sixth Assessment Report of the Intergovernmental Panel on Climate Change*; Masson-Delmotte, V., Zhai, P., Pirani, A., Connors, S.L., Péan, C., Berger, S., Caud, N., Chen, Y., Goldfarb, L., Gomis, M.I., et al., Eds.; Cambridge University Press: Cambridge, UK, 2021.
7. Bosson, J.B.; Huss, M.; Cauvy-Fraunié, S.; Clément, J.C.; Costes, G.; Fischer, M.; Poulenard, J.; Arthaud, F. Future emergence of new ecosystems caused by glacial retreat. *Nature* **2023**, *620*, 562–569. [CrossRef]
8. Chen, Y.; Opała-Owczarek, M.; Chen, F.; Owczarek, P.; Zhang, H.; Wang, S.; Hu, M.; Satylkanov, R.; Ermenbaev, B.; Zulfiyor, B.; et al. Tree-ring perspective on the past and future mass balance of a glacier in Tien Shan (Central Asia): An example from the Tuyuksu glacier, Kyrgyzstan. *Palaeogeogr. Palaeoclimatol. Palaeoecol.* **2023**, *625*, 111696. [CrossRef]
9. Wang, S.; Yao, T.; Pu, J.; Wang, J. Historical reconstruction and future projection of mass balance and ice volume of Qiyi Glacier, Northeast Tibetan Plateau. *J. Hydrol. Reg. Stud.* **2023**, *47*, 101403. [CrossRef]
10. Gobbi, M.; Ambrosini, R.; Casarotto, C.; Diolaiuti, G.; Ficetola, G.F.; Lencioni, V.; Seppi, R.; Smiraglia, C.; Tampucci, D.; Valle, B.; et al. Vanishing permanent glaciers: Climate change is threatening a European Union habitat (code 8340) and its poorly known biodiversity. *Biodivers. Conserv.* **2023**, *30*, 2267–2276. [CrossRef]
11. Yang, S.; Wang, F.; Xie, Y.; Zhao, W.; Bai, C.; Liu, J.; Xu, C. Delineation Evaluation and Variation of Debris-Covered Glaciers Based on the Multi-Source Remote Sensing Images, Take Glaciers in the Eastern Tomur Peak Region for Example. *Remote Sens.* **2023**, *15*, 2575. [CrossRef]
12. Wang, P.; Li, Z.; Xu, C.; Xing, W.; Zhou, P.; Zhang, H. Multi-decadal variations in glacier flow velocity and the influencing factors of Urumqi Glacier No. 1 in Tianshan Mountains, Northwest China. *J. Arid Land* **2017**, *9*, 900–910. [CrossRef]
13. Rankl, M.; Kienholz, C.; Braun, M. Glacier Changes in the Karakoram Region Mapped by Multimission Satellite Imagery. *Cryosphere* **2014**, *8*, 977–989. [CrossRef]
14. Brun, F.; Berthier, E.; Wagnon, P.; Kääb, A.; Treichler, D. A Spatially Resolved Estimate of High Mountain Asia Glacier Mass Balances from 2000 to 2016. *Nat. Geosci.* **2017**, *10*, 668–673. [CrossRef] [PubMed]
15. Kääb, A.; Berthier, E.; Nuth, C.; Gardelle, J.; Arnaud, Y. Contrasting patterns of early 21st century glacier mass change in the Himalaya. *Nature* **2012**, *488*, 495–498. [CrossRef] [PubMed]
16. Rounce, D.R.; Hock, R.; Shean, D.E. Glacier Mass Change in High Mountain Asia Through 2100 Using the Open-Source Python Glacier Evolution Model (PyGEM). *Front. Earth Sci.* **2020**, *7*, 331. [CrossRef]
17. Farinotti, D.; Huss, M.; Fürst, J.; Landmann, J.; Machguth, H.; Maussion, F.; Pandit, A. A consensus estimate for the ice thickness distribution of all glaciers on Earth. *Nat. Geosci.* **2019**, *12*, 168. [CrossRef]
18. Shean, D.E.; Bhushan, S.; Montesano, P.; Rounce, D.R.; Arendt, A.; Osmanoglu, B. A Systematic, Regional Assessment of High Mountain Asia Glacier Mass Balance. *Front. Earth Sci.* **2020**, *7*, 475909. [CrossRef]
19. Vijay, S.; Braun, M.H. Seasonal and interannual variability of Columbia Glacier, Alaska (2011–2016): Ice velocity, mass flux, surface elevation and front position. *Remote Sens.* **2017**, *9*, 635. [CrossRef]
20. Maussion, F.; Butenko, A.; Champollion, N.; Dusch, M.; Eis, J.; Fourteau, K.; Gregor, P.; Jarosch, A.; Landmann, J.; Oesterle, F.; et al. The open global glacier model (OGGM) v1.1. *Geosci. Model Dev.* **2019**, *12*, 909–931. [CrossRef]
21. Marzeion, B.; Jarosch, A.H.; Hofer, M. Past and future sea-level change from the surface mass balance of glaciers. *Cryosphere* **2012**, *6*, 1295–1322. [CrossRef]
22. Zhao, W.; Li, Z.; Mu, J.; Li, H.; Yang, S.; Xu, C. Projection of glacier area and volume in the Sawir Mountains modeled by the OGGM. *J. Glaciol. Geocryol.* **2023**, *45*, 497–508. [CrossRef]
23. Wang, P.; Li, Z.; Zhou, P.; Li, H.; Yu, G.; Xu, C.; Wang, L. Rapid decrease of observed mass balance in the Urumqi Glacier No. 1, Tianshan Mountains, central Asia. *Quat. Int.* **2017**, *349*, 135–141. [CrossRef]
24. Han, T.D.; Ding, Y.J.; Ye, B.S.; Liu, S.Y.; Jiao, K.Q. Mass balance characteristics of Urumqi Glacier No. 1, Tien Shan, China. *Ann. Glaciol.* **2006**, *43*, 323–328. [CrossRef]
25. Wu, L.H.; Li, H.L.; Wang, L. Application of a degree day model for determination of mass balance of Urumqi Glacier No. 1, Eastern Tianshan, China. *J. Earth Sci.* **2011**, *22*, 470–481. [CrossRef]
26. Zhang, G.F.; Li, Z.Q.; Wang, W.D.; Wang, W.B. Comparative study on observed mass balance between east and west branch of Urumqi Glacier No. 1, Eastern Tianshan, China. *Sci. Cold Arid Reg.* **2013**, *5*, 316–323. [CrossRef]
27. Xu, C.; Li, Z.; Wang, P.; Anjum, M.N.; Li, H.; Wang, F. Detailed comparison of glaciological and geodetic mass balances for Urumqi Glacier No. 1, eastern Tien Shan, China, from 1981 to 2015. *Cold Reg. Sci. Technol.* **2018**, *155*, 137–148. [CrossRef]
28. Xu, C.; Li, Z.; Li, H.; Wang, F.; Zhou, P. Long-range terrestrial laser scanning measurements of annual and intra-annual mass balances for Urumqi Glacier No. 1, eastern Tien Shan, China. *Cryosphere* **2019**, *13*, 2361–2383. [CrossRef]
29. Li, H.; Wang, P.; Li, Z.; Jin, S.; He, J. Summertime surface mass balance and energy balance of Urumqi Glacier No. 1, Chinese Tien Shan, modeled by linking COSIMA and in-situ measured meteorological records. *Clim. Dyn.* **2022**, *61*, 765–787. [CrossRef]
30. Wang, P.; Li, H.; Li, Z.; Liu, Y.; Xu, C.; Mu, J.; Zhang, H. Seasonal Surface Change of Urumqi Glacier No. 1, Eastern Tien Shan, China, Revealed by Repeated High-Resolution UAV Photogrammetry. *Remote Sens.* **2021**, *13*, 3398. [CrossRef]
31. Yang, M.; Li, Z.; Anjum, M.N.; Kayastha, R.; Kayastha, R.B.; Rai, M.; Zhang, X.; Xu, C. Projection of streamflow changes under CMIP6 scenarios in the Urumqi River head watershed, Tianshan Mountain, China. *Front. Earth Sci.* **2022**, *10*, 857854. [CrossRef]
32. Liu, S.; Guo, W.; Xu, J. The Second Glacier Inventory Dataset of China (Version 1.0) (2006–2011). A Big Earth Data Platform for Three Poles. 2012. Available online: https://cstr.cn/18406.11.glacier.001.2013.db (accessed on 11 March 2024).

33. Li, Z.Q.; Li, H.L.; Chen, Y.N. Mechanisms simulation of accelerated shrinkage of continental glaciers: A case study of Urumqi Glacier No 1 in Eastern Tianshan central Asia. *J. Earth Sci.* **2011**, *22*, 423–430. [CrossRef]
34. Kaushik, S.; Dharpure, J.K.; Joshi, P.K.; Ramanathan, A.L.; Singh, T. Climate change drives glacier retreat in Bhaga basin located in Himachal Pradesh, India. *Geocarto Int.* **2019**, *35*, 1179–1198. [CrossRef]
35. Bolch, T.; Menounos, B.; Wheate, R. Landsat-based Inventory of Glaciers in Western Canada, 1985–2005. *Remote Sens. Environ.* **2010**, *114*, 127–137. [CrossRef]
36. Nuimura, T.; Sakai, A.; Taniguchi, K.; Nagai, H.; Lamsal, D.; Tsutaki, S.; Kozawa, A.; Hoshina, Y.; Takenaka, S.; Omiya, S.; et al. The GAMDAM Glacier Inventory: A quality controlled inventory of Asian glaciers. *Cryosphere* **2015**, *9*, 849–864. [CrossRef]
37. Tielidze, L.G.; Bolch, T.; Wheate, R.D.; Kutuzov, S.S.; Lavrentiev, I.I.; Zemp, M. Supra-glacial debris cover changes in the Greater Caucasus from 1986 to 2014. *Cryosphere* **2020**, *14*, 585–598. [CrossRef]
38. Zhang, Z.; Liu, S.; Zhang, Y.; Wei, J.; Jiang, Z.; Wu, K. Glacier Variations at Aru Co in Western Tibet from 1971 to 2016 Derived from Remote-Sensing Data. *J. Glaciol.* **2018**, *64*, 397–406. [CrossRef]
39. Nuth, C.; Kääb, A. Co-registration and Bias Corrections of Satellite Elevation Data Sets for Quantifying Glacier Thickness Change. *Cryosphere* **2011**, *5*, 271–290. [CrossRef]
40. Bolch, T.; Pieczonka, T.; Benn, D. Multi-decadal mass loss of glaciers in the Everest area (Nepal Himalaya) derived from stereo imagery. *Cryosphere* **2011**, *5*, 349–358. [CrossRef]
41. Pieczonka, T.; Bolch, T.; Buchroithner, M. Generation and evaluation of multi-temporal digital elevation models from different optical sensors. *Photogramm. Remote Sens.* **2011**, *66*, 927–940. [CrossRef]
42. Huss, M. Density Assumptions for Converting Geodetic Glacier Volume Change to Mass Change. *Cryosphere* **2013**, *7*, 877–887. [CrossRef]
43. Han, T.; Liu, S.; Ding, Y.; Jiao, K. A characteristics mass balance of glacier No. 1 at the headwaters of the Urumqi River, Tianshan Mountains. *Adv. Earth Sci.* **2005**, *20*, 298. [CrossRef]
44. Xie, Z.C.; Liu, C.H. *Introduction to Glaciology*; Shanghai Popular Science Press: Shanghai, China, 2010; pp. 1–490.
45. Paterson, W.S.B. *The Physics of Glaciers*, 3rd ed.; Pergamon: New York, NY, USA, 1994.
46. Che, Y.J.; Zhang, M.J.; Li, Z.Q.; Jin, S.; Wang, W.B.; Wang, S.J. Monitoring calculating of mass balance in Qingbingtan Glacier No 72. *J. Glaciol. Geocryol.* **2019**, *41*, 1–14. (In Chinese)
47. Wang, P.; Li, Z.; Li, H.; Wang, W.; Zhou, P.; Wang, L. Characteristics of a partially debris-covered glacier and its response to atmospheric warming in Mt. Tomor, Tien Shan, China. *Glob. Planet. Chang.* **2017**, *159*, 11–24. [CrossRef]
48. Wang, P.; Li, Z.; Li, H.; Wang, W.; Wu, L.; Zhang, H.; Huai, B.; Wang, L. Recent Evolution in Extent, Thickness, and Velocity of Haxilegen Glacier No. 51, Kuytun River Basin, Eastern Tianshan Mountains. *Arct. Antarct. Alp. Res.* **2016**, *48*, 241–252. [CrossRef]
49. Zhang, H.; Li, Z.; Zhou, P.; Zhu, X.; Wang, L. Mass-balance observations and reconstruction for Haxilegen Glacier No. 51, eastern Tien Shan, from 1999 to 2015. *J. Glaciol.* **2018**, *64*, 689–699. [CrossRef]
50. Zhang, R.; Wei, W.; Shang, H.; Yu, S.; Gou, X.; Qin, L.; Bolatov, K.; Mambetov, B.T. A tree ring-based record of annual mass balance changes for the TS.Tuyuksuyskiy Glacier and its linkages to climate change in the Tianshan Mountains. *Quat. Sci. Rev.* **2019**, *205*, 10–21. [CrossRef]
51. Liu, L.; Tian, H.; Zhang, X.; Chen, H.; Zhang, Z.; Zhao, G.; Kang, Z.; Wang, T.; Gao, Y.; Yu, F. Analysis of Spatiotemporal Heterogeneity of Glacier Mass Balance on the Northern and Southern Slopes of the Central Tianshan Mountains, China. *Water* **2022**, *14*, 1601. [CrossRef]
52. Zhao, G.; Zhang, Z.; Liu, L.; Xu, L.; Wang, P.; Li, L.; Ning, S. Changes of glacier mass balance in Manas river basin based on multi-source remote sensing data. *Acta Geogr. Sin.* **2020**, *75*, 98–112. [CrossRef]
53. Yang, X.; Wang, N.; Liang, Q.; Chen, A. Glacier changes on the north slope of Tianshan Mountains in recent 60 years. *Arid. Land Georg.* **2023**, *46*, 1073–1083. [CrossRef]
54. Bolch, T.; Kulkarni, A.; Kääb, A.; Huggel, C.; Paul, F.; Cogley, J.G.; Frey, H.; Kargel, J.S.; Fujita, K.; Scheel, M. The state and fate of Himalayan glaciers. *Science* **2012**, *336*, 310–314. [CrossRef]
55. Che, Y.; Zhang, M.; Li, Z.; Wang, S.; Du, M.; Wang, P.; Wang, J.; Zhou, P. Quantitative evaluation of glacier change and its response to climate change in the Chinese Tien Shan. *Cold Reg. Sci. Technol.* **2018**, *153*, 144–155. [CrossRef]
56. Kang, E. Calculation of energy balance characteristics and material balance changes in the high Asian cryosphere. *J. Glaciol. Geocryol.* **1996**, *S1*, 12–22.

remote sensing

MDPI

Article

Contrasting Changes of Debris-Free Glacier and Debris-Covered Glacier in Southeastern Tibetan Plateau

Chuanxi Zhao [1,2], Zhen He [2], Shengyu Kang [3], Tianzhao Zhang [2,*], Yongjie Wang [2], Teng Li [4], Yifei He [2] and Wei Yang [2]

1 College of Earth and Environmental Sciences, Lanzhou University, Lanzhou 730000, China; zcx@itpcas.ac.cn
2 State Key Laboratory of Tibetan Plateau Earth System, Environment and Resources (TPESER), Institute of Tibetan Plateau Research, Chinese Academy of Sciences, Beijing 100101, China; hezh@itpcas.ac.cn (Z.H.); wyj@itpcas.ac.cn (Y.W.); heyf@itpcas.ac.cn (Y.H.); yangww@itpcas.ac.cn (W.Y.)
3 State Key Laboratory of Water Resources Engineering and Management, Wuhan University, Wuhan 430072, China; kshy0204@whu.edu.cn
4 School of Geospatial Engineering and Science, Sun Yat-Sen University, Zhuhai 519000, China; liteng28@mail.sysu.edu.cn
* Correspondence: zhangtz@itpcas.ac.cn

Abstract: Debris-free and debris-covered glaciers are both extensively present in the southeastern Tibetan Plateau. High-precision and rigorous comparative observational studies on different types of glaciers help us to accurately understand the overall state of water resource variability and the underlying mechanisms. In this study, we used multi-temporal simultaneous UAV surveys to systematically explore the surface elevation change, surface velocity, and surface mass balance of two representative glaciers. Our findings indicate that the thinning rate in the debris-free Parlung No. 4 glacier UAV survey area was consistently higher than that in the debris-covered 24K glacier in 2020–2021 (-1.16 ± 0.03 cm/d vs. -0.36 ± 0.02 cm/d) and 2021–2022 (-0.69 ± 0.03 cm/d vs. -0.26 ± 0.03 cm/d). Moreover, the surface velocity of the Parlung No. 4 glacier was also consistently higher than that of the 24K glacier across the survey period, suggesting a more dynamic glacial state. The surface mass balance of the Parlung No. 4 glacier (2020–2021: -1.82 ± 0.09 cm/d; 2021–2022: -1.30 ± 0.09 cm/d) likewise outpaced that of the 24K glacier (2020–2021: -0.81 ± 0.07 cm/d; 2021–2022: -0.70 ± 0.07 cm/d) throughout the observation period, which indicates that the debris cover slowed the glacier's melting. Additionally, we extracted the melt contribution of the ice cliff area in the 24K glacier and found that the melt ratio of this 'hotspot' area ranged from 10.4% to 11.6% from 2020 to 2022. This comparative analysis of two representative glaciers provides evidence to support the critical role of debris cover in controlling surface elevation changes, glacier dynamics, and surface mass balance.

Keywords: UAV; debris-free glacier; debris-covered glacier; southeastern Tibetan Plateau

1. Introduction

Since the start of the 21st century, maritime glaciers within the southeastern (SE) Tibetan Plateau have exhibited higher rates of mass loss compared to other regions across High-Mountain Asia [1–5]. The predominant glacier types in this region include both debris-free and debris-covered glaciers. Notably, the estimated proportion of the debris-covered area in the SE Tibetan Plateau is close to 20% of the total glacierized area, which exceeds the global average value (~4.4–7.3%) [6,7]. Therefore, a better understanding of the characteristics of different glaciers and the changes they experience is essential to comprehensively evaluate the situation regarding water resource availability in this region [8,9].

Several studies were conducted in alpine regions, with observational and modeling approaches being adopted for the study of glaciers (with or without debris cover). Some found that the presence of debris influences the responses of glaciers to global warming.

When the debris thickness exceeds a few centimeters, it plays a protective role in mitigating glacier ablation [10–15]. Therefore, the ablation rates of most debris-covered glaciers (mean global debris thickness > 10 cm) [16] are supposed to be weaker than those of debris-free glaciers. Nevertheless, numerous remote sensing studies utilizing satellite imagery revealed comparable rates of thinning for both debris-free and debris-covered glaciers (i.e., "debris coverage anomaly") [1,17–20]. Similarly, this anomaly was verified in the SE Tibetan Plateau, where the debris-covered glaciers are widely developed [21,22]. However, almost all of the above conclusions are based on satellite remote sensing data, which may be lacking in resolution. Moreover, some researchers highlighted the significant impact of 'hotspot areas' (such as ice cliffs and supraglacial ponds) on the mass loss of debris-covered glaciers. Due to the relatively small size of these "hotspots" areas, high-resolution data are required for accurate assessments of the contribution of ablation [23–31].

Since Unpiloted Aerial Vehicles (UAVs) overcome the disadvantages of the poor spatial representativeness of in situ observations and insufficient precision from satellite remote sensing data, the use of UAVs has gradually become one of the main techniques for monitoring glacier changes [24,32–37]. High-resolution digital elevation models (DEMs) and orthomosaics can be readily acquired from UAV imagery through the application of structure-from-motion (UAV-SfM) combined with multi-view stereo photogrammetry [38,39]. In the SE Tibetan Plateau, some researchers also carried out monitoring experiments on debris-covered glaciers and debris-free glaciers with high-resolution photogrammetric measurements [40–46]. However, there are no synchronized monitoring results based on UAV surveys for debris-covered glacier and debris-free glacier changes in the SE Tibetan Plateau. Comparative studies that accurately estimate the surface mass balance changes in two types of glaciers (debris-covered vs. debris-free) by using synchronized and high-precision UAV data are also relatively rare in the glaciology community.

In this study, we simultaneously compared the glacier changes between the debris-free Parlung No. 4 glacier (abbreviated as Parlung No. 4) and the debris-covered 24K glacier (abbreviated as 24K) in the SE Tibetan Plateau from 20 August 2020 to 22 June 2022. The objective of this study was to accurately estimate and compare the variability in mass loss and dynamic states of two typical glaciers to deepen our understanding of the response of maritime glaciers in the SE Tibetan Plateau to global warming.

2. Study Area

Parlung No. 4 and 24K are located in the SE Tibetan Plateau (Figure 1a); the distance between the two glaciers is only ~120–130 km. They are primarily influenced by the moisture from the Bay of Bengal Vortex (during spring) and the Indian Summer Monsoon (during summer) [47–49]. The monthly distribution of precipitation displays a double-peak pattern, with two distinct peaks occurring in the spring and summer [48]. Specifically, Parlung No. 4 (29°13.57′N, 96°55.19′E) is located at the source of the Parlung Zangbo River on the southeastern side of Mount Gangrigabu (Figure 1c). This glacier has a length of ~8 km and an area of ~11 km^2, typical for a debris-free glacier in the SE Tibetan Plateau. The Parlung No. 4 catchment area is ~25 km^2 (Figure 1c), and the glacier area accounts for 44% of the total area (Figure 1c). Moreover, 24K (29°45.59′N, 95°43.11′E) is also located on the northern slope of Mount Gangrigabu (Figure 1b). It is ~24 km away from Bomi City [15] and has a length of ~3 km (area: ~3 km^2). The debris-covered area accounts for 41% of the 24K area, and the spatial distribution of its thickness shows a decreasing trend (Figure A1) [46], which is typical of a debris-covered glacier in the SE Tibetan Plateau. The 24K catchment area is 14 km^2 (Figure 1b), and the glacier area is 21% of the total area (Figure 1e). Despite being in close proximity to each other, there is a striking difference in the climatic characteristics of the two glaciers. During the 2016 ablation period, the average temperature of 24K was 5.4 °C higher than that of Parlung No. 4, and the total precipitation of 24K was approximately nine times higher than that of Parlung No. 4 (1696.8 vs. 189.6 mm) [15].

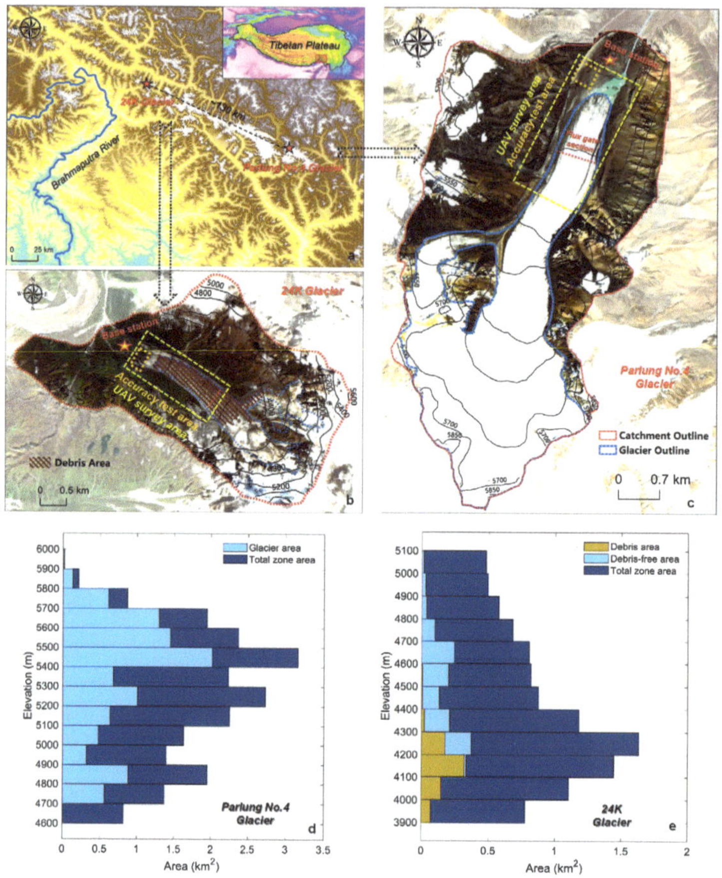

Figure 1. (**a**) Location of the study area. (**b,c**) Maps showing the topography and UAV survey information for the 24K and Parlung No. 4 catchments. (**d,e**) The altitudinal distribution of the glacierized and non-glacierized areas of the Parlung No. 4 and 24K catchments.

3. Data and Methods

3.1. UAV Flights and Data Processing

High-resolution images of the survey areas were obtained by using unpiloted aerial vehicles (UAVs) during six field studies conducted from 20 August 2020 to 22 June 2022. The specific dates and details of these studies are provided in Table 1. To capture the annual glacier surface elevation change and surface displacement, we utilized the eBee Plus UAV equipped with the GNSS Post-Processed Kinematic (PPK) functionality in all of the surveys.

Table 1. UAV flight information of two glaciers.

Flight Time	Flight Type	UAV Name	Glacier Name	Image Number	Resolution (cm)	Area (km^2)
25 August 2020	PPK	eBee Plus	Parlung No. 4	425	9.8	4.0
20 August 2020			24K	160	12.6	3.7
29 July 2021	PPK	eBee Plus	Parlung No. 4	442	9.4	4.6
23 July 2021			24K	426	9.2	4.4
18 June 2022	PPK	eBee Plus	Parlung No. 4	529	9.1	3.9
22 June 2022			24K	249	11.8	5.4

This UAV is a fixed-wing aircraft equipped with a 20MP camera called the SenseFly S.O.D.A. For flight planning purposes, the eMotion3® flight management software (version 3.5.0) was employed. During the survey period, the longitudinal and lateral image overlaps for the eBee Plus were set at 65% and 80%, respectively. By maintaining a consistent survey height above the glacier surface, the flight lines of both UAVs ensured a uniform ground resolution for each survey. To enhance the accuracy of the UAV-based structure-from-motion (SfM) reconstruction, a stationary base station was used. The GNSS data collected from the base station were then appended to the Exchangeable Image File Format metadata of each geotagged image [41]. This data integration process was a part of the PPK correction workflow, which was followed in an attempt to improve the precision of the UAV-SfM reconstruction. The geotagged images obtained from the UAV surveys were further utilized in the creation of orthomosaics and digital elevation models (DEMs) using the SfM-based photogrammetric software Pix4Dmapper (version 4.3.31). The accuracy of all UAV-SfM outputs was indirectly assessed by comparing the horizontal and vertical errors in the UAV-SfM outputs (2020 vs. 2021; 2021 vs. 2022). The horizontal errors were estimated by measuring the displacement of benchmarks (>30 boulders on stable non-glacier areas), while the vertical errors were calculated by counting the surface elevation change value of stable terrain, as outlined in Figure 1b,c.

3.2. Surface Elevation Change and Surface Velocity

After employing the PPK technologies, the acquired DEMs' offsets were found to be minor [41], negating the need for the co-registration of DEMs when calculating glacier surface elevation change. For each period, we derived surface elevation change results via DEM differential analysis in ArcGIS 10.4. The horizontal displacement of each glacier was obtained by using 20 cm resolution DEM hillshades using ImGRAFT (a normalized cross-correlation algorithm) [50] within a search window of 10 × 10 pixels (2 × 2 m).

3.3. Surface Mass Balance of Glacier Ablation Area

The thinning/surface elevation change ($\mathrm{d}h$, in m) of the calculated area is equivalent to the ablation/surface mass balance ($\dot{b}$, in m) plus the emergence velocity (ω, in m):

$$\mathrm{d}h = \dot{b} + \omega \tag{1}$$

$$\omega = \frac{Q}{S} \tag{2}$$

$$Q = \mu \cdot H \cdot D \cdot L \tag{3}$$

where S (in m^2) is the calculation zone area, and Q (in m^3) is the ice flow through a given profile of the glacier. μ is the coefficient for the conversion of surface velocities to depth-averaged velocities. Following Miles et al. [25] and Zhao et al. [46], this coefficient was estimated to be 0.9; H (in m) is the thickness of the ice for the corresponding profile, D (in m) is the component of the normal surface displacement to the cross-section, and L (in m) is the breadth of the profile.

To evaluate the uncertainty of the surface mass balance ($\sigma_{\dot{b}}$) for the calculation area, the following equation was used:

$$\sigma_{\dot{b}} = \sqrt{\sigma_{\mathrm{d}h}{}^2 + \sigma_{\omega}{}^2} \tag{4}$$

The uncertainties of dh (and for D) were estimated by calculating the mean difference in surface elevation and the mean displacement from the non-glacial experimental areas. By averaging the values over all periods, they were determined to be 0.09 m ($\sigma_{\mathrm{d}h}$) and 0.09 m (σ_D), respectively.

The uncertainty of the emergency velocity (σ_ω) was given by the following:

$$\frac{\sigma_\omega}{\omega} = \sqrt{\left(\frac{\sigma_Q}{Q}\right)^2 + \left(\frac{\sigma_S}{S}\right)^2} \quad (5)$$

where the uncertainty of the area of the zone (S) was calculated to be ±20 m from the outlines using the 'buffer method' [25,51]; the uncertainty of the ice flow through a glacier profile (σ_Q) was calculated as follows:

$$\frac{\sigma_q}{q} = \sqrt{\left(\frac{\sigma_D}{D}\right)^2 + \left(\frac{\sigma_\mu}{\mu}\right)^2 + \left(\frac{\sigma_H}{H}\right)^2} \quad (6)$$

where the uncertainty of the ratio μ (column-averaged velocity/surface velocity) is estimated to be 0.1 [25,52]. The uncertainty in H was ~37 m (26%) for Parlung No. 4 [53] and was assumed to be equal to 12 m for the corrected uncertainty in ice thickness for 24K [46].

3.4. Hotspot Area Ablation Contribution

To quantify the proportion of the contribution of ablation in the "hotspots" areas, we used the merged ice cliff outline method [25,46] and the mean hotspot area enhancement factors (1.96) [46]. Only 24K exhibited developed ice cliffs, and there were few developed supraglacial ponds. Regarding the merged ice cliffs outline, the specific operation is to georegister the 2020 and 2022 orthomosaics of the glacier areas by using the reference orthomosaics (July 2021). Further details are described in Zhao et al. (2023) [46]. After completing the registration, ice cliff outlines for August 2020, July 2021, and June 2022 were manually sketched based on orthomosaics. Then, we combined the 2020 ice cliff outline with the 2021 ice cliff outline and the 2021 ice cliff outline with the 2022 ice cliff outline. The hotspot area ablation contributions between 2020 and 2022 were estimated based on the previous enhancement factor.

4. Results

4.1. Surface Elevation Changes of Two Glaciers

Utilizing the high-resolution DEMs derived from the UAV data, we quantified the magnitude and spatial distribution of glacier surface elevation changes on Parlung No. 4. As depicted in Figure 2a,b, this analysis covered two distinct periods: from August 2020 to July 2021 (338 days) and from June 2022 to July 2021 (323 days). During 2020–2021, the mean surface elevation change within the UAV survey area of Parlung No. 4 was −3.93 ± 0.10 m (−1.16 ± 0.03 cm/d). In 2021–2022, the mean surface elevation change in the UAV survey area of Parlung No. 4 was −2.23 ± 0.10 m (−0.69 ± 0.03 cm/d), and the thinning rate in 2021–2022 was lower than that in 2020–2021. Based on the surface elevation change regarding the vertical test area, we obtained the vertical errors of the surface elevation change results for 2020–2021 (0.10 m) and 2021–2022 (0.10 m). These minimal error margins underscore the reliability and precision of the surface elevation change results obtained for the glacier areas.

During the period between August 2020 and July 2021 (337 days), UAV-based observations revealed that the surface elevation change for 24K was −1.20 ± 0.07 m, equating to a daily rate of −0.36 ± 0.02 cm/d, as illustrated in Figure 3a. In the period from July 2021 to June 2022 (334 days), 24K exhibited a mean surface elevation reduction of −0.86 ± 0.10 m, with a mean daily rate of −0.26 ± 0.03 cm/d (Figure 3b). Consistent with the situation of Parlung No. 4, the thinning rate in 2021–2022 was also lower than that in 2020–2021. Compared with Parlung No. 4, the thinning rate of the 24K ablation area was consistently lower than that of the Parlung No. 4 area, which was approximately 3.3 times as large as that of 24K in 2020–2021 and ~2.6 times as large as that of 24K in 2021–2022. Furthermore, the vertical accuracy of the DEMs for 24K remained high, as evidenced by an error value of 0.07 m in 2020–2021 and 0.10 m in 2021–2022.

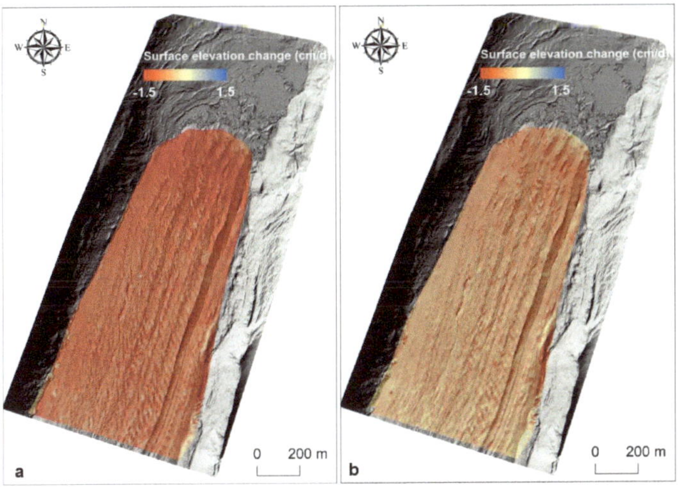

Figure 2. Annual surface elevation change for Parlung No. 4 between DEMs for August 2020–July 2021 (**a**) and surface elevation change for July 2021–June 2022 (**b**).

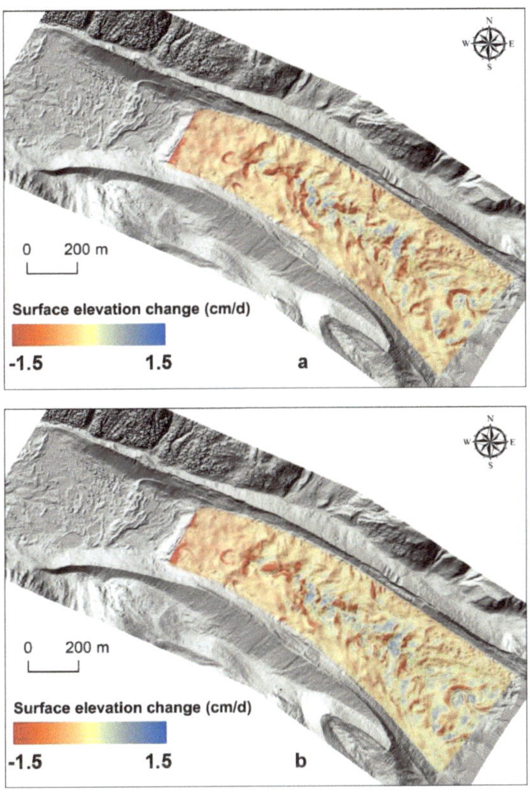

Figure 3. Annual surface elevation change for 24K between DEMs for August 2020–July 2021 (**a**) and surface elevation change for July 2021–June 2022 (**b**).

Figure 4 illustrates the gradient relationship between the surface elevation change rate and altitude for each glacier at 6 m elevation bands (24K) and 10 m elevation bands (Parlung No. 4). It was found that the surface elevation change rate of Parlung No. 4 is constantly higher than that of 24K throughout all observation periods, and the magnitude of the thinning rate of 24K is only slightly above 0, apart from in the terminal ice cliffs area. Additionally, for both glaciers, a decline in thinning magnitude with increasing altitude was noted.

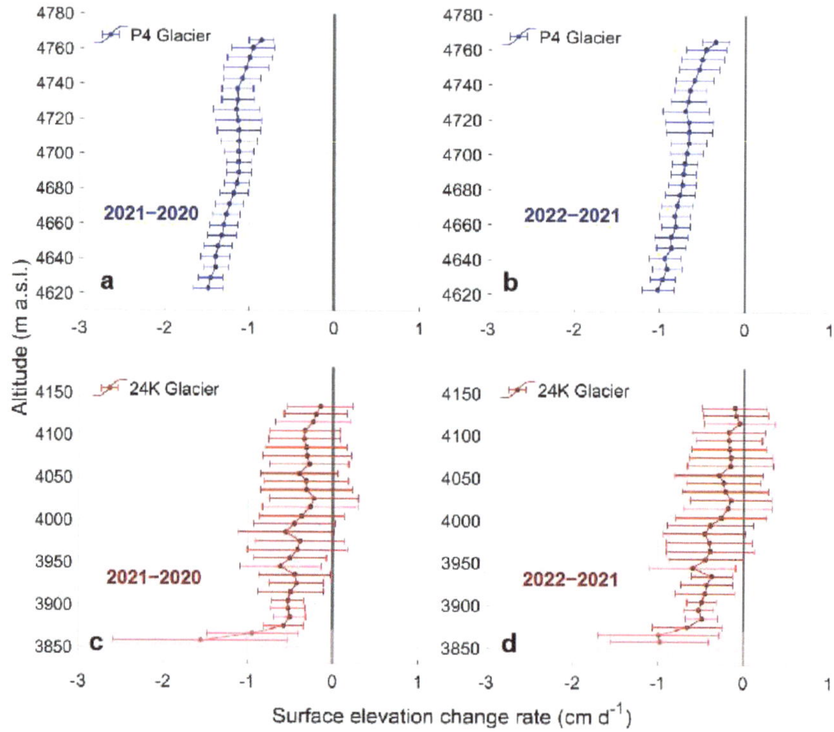

Figure 4. Average surface elevation change rates within 6 m elevation bands for Parlung No. 4 and 10 m elevation bands for 24K (represented by dots) and their corresponding standard deviations (shown as horizontal error bars) for the periods of August 2020 to July 2021 (**a**,**c**) and July 2021 to June 2022 (**b**,**d**). The red-shaded sections in the figure represent the terminal ice cliffs of 24K.

4.2. Surface Velocities of the Two Glaciers

The surface velocity within Parlung No. 4's UAV survey area reached 4.02 ± 0.04 cm/d, with a mean displacement of 13.59 ± 0.13 m, from August 2020 to July 2021. A subsequent assessment showed a marginally reduced surface velocity of 3.91 ± 0.02 cm/d (mean displacement: 12.63 ± 0.08 m) between July 2021 and June 2022, suggesting little variation across the two periods. In addition, the surface velocity of Parlung No. 4 showed a spatial pattern of increasing with altitude in both periods (Figure 5a,b). In this study, the horizontal error (XY), calculated by using the benchmarks in the non-glacial area of the Parlung No. 4 catchment, was 0.07 m in 2020–2021 and 0.05 m in 2021–2022, which proves that the post-processed data from the UAV surveys have high horizontal accuracy. The mean value of the movement velocity of Parlung No. 4 in the vertical test area was found to be 0.13 m in 2021–2020 and 0.08 m in 2021–2022, which evidences that the surface velocity results are reliable.

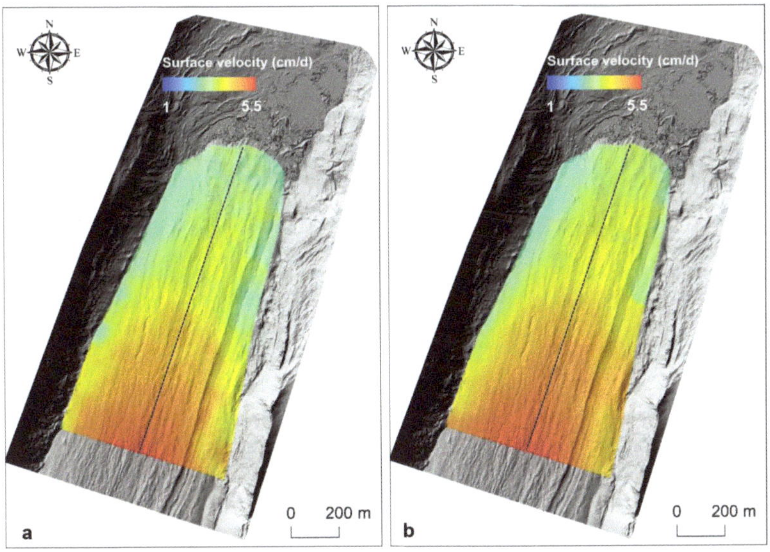

Figure 5. Surface velocity of Parlung No. 4 for August 2020–July 2021 (**a**) and surface elevation change for July 2021–June 2022 (**b**). The black dashed line represents the central flowline.

The average surface velocity between August 2020 and July 2021 for 24K was 2.45 ± 0.02 cm/day (mean displacement: 8.26 ± 0.07 m). The daily surface velocity from July 2021 to June 2022 was 2.22 ± 0.02 cm/d (mean displacement: 7.41 ± 0.06 m), and the daily velocity in 2021–2022 was slightly lower than the velocity in 2020–2021. In addition, Figure 6a,b also demonstrates that the spatial pattern of 24K is characterized by a surface velocity that increases with the altitude. In this study, the horizontal error (XY) of the post-processed data sourced from the benchmarks in the non-glacial area was 0.04 m for 2020–2021 and 0.05 m for 2021–2022, which reflected a relatively preferable horizontal accuracy. The mean surface velocity for 2020–2021 in the vertical test area was 0.07 m; the value for 2021–2022 was 0.06 m.

We extracted and compared the surface velocities of two glaciers' central flowlines in 2020–2021 and 2021–2022, respectively (Figure 7), and found that the surface velocity patterns of both glaciers are consistent, with both showing higher surface velocities at higher altitudes. In addition, the Parlung No. 4 surface velocity is significantly greater than that of 24K (~two times). The inter-annual difference in the velocity of Parlung No. 4 is not significant, whereas the velocity of 24K in 2021–2022 is smaller than the surface velocity in 2020–2021.

4.3. *Surface Mass Balance Patterns of the Two Glaciers*

Based on the thinning results and our computation of the overall ice flux in the UAV survey area for the two glaciers, we calculated the average surface mass balance for the two glaciers (Equations (1)–(3)). We defined two flux gate sections in the upper part of the UAV survey area for the two glaciers (Figure 1b,c). Ice flow replenishment was then calculated for the ablation areas (below the flux gate section/surface mass balance calculation area) for the debris-free glacier (Parlung No. 4) and debris-covered glacier (24K) (Figure 8). In 2020–2021, the average rates of ice flow replenishment were 0.66 ± 0.08 cm/d in the Parlung No. 4 ablation area and 0.45 ± 0.06 cm/d in the 24K ablation area. The average ice flow replenishment was 0.61 ± 0.09 cm/d for the Parlung No. 4 ablation area and 0.44 ± 0.06 cm/d for the 24K ablation area between 2021 and 2022. Taking into consideration the ice fluxes, we further calculated the overall ablation amount for these two representative glaciers (Figure 8). In 2020–2021, the average surface mass balance change rate was −1.82 ± 0.09 cm/d for the Parlung No. 4 area and −0.81 ± 0.07 cm/d for the 24K area. The average surface mass balance change rate was

−1.30 ± 0.09 cm/d in the Parlung No. 4 ablation area and −0.70 ± 0.07 cm/d in the 24K ablation area in 2021–2022. Overall, the amplitude of the ice flow replenishment for the two glaciers is roughly similar. However, there is a significant difference between both glaciers in the magnitude of ablation, with the ablation rate of the Parlung No. 4 area being approximately 1.9–2.2 times that of the 24K area.

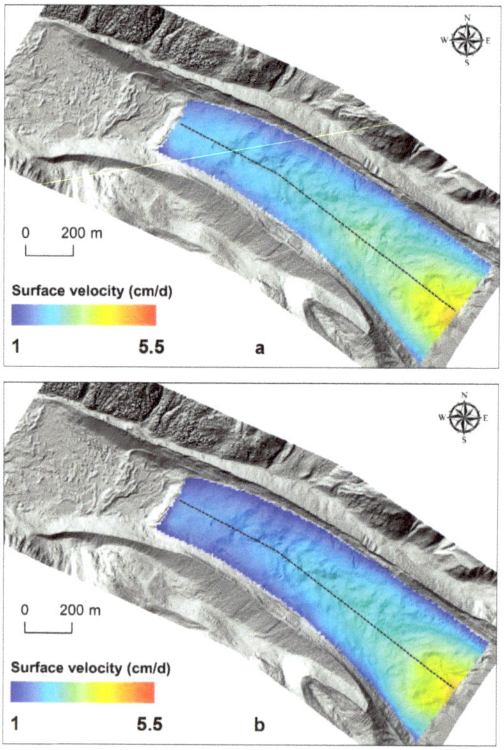

Figure 6. Surface velocity of 24K for August 2020–July 2021 (**a**) and surface elevation change for July 2021–June 2022 (**b**). The black dashed line represents the central flowline.

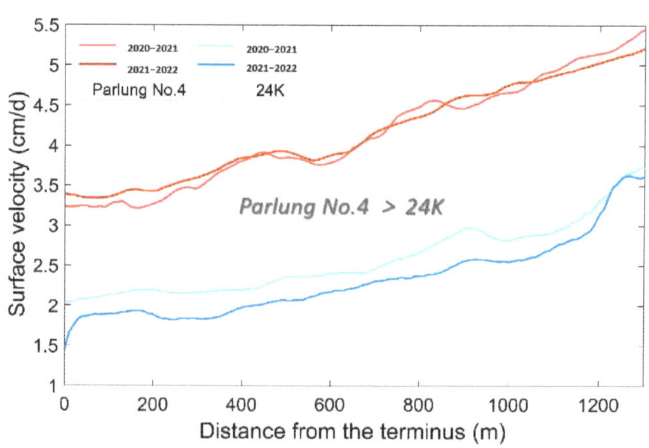

Figure 7. The surface velocities of central flowlines for Parlung No. 4 (red lines) and 24K (blue lines).

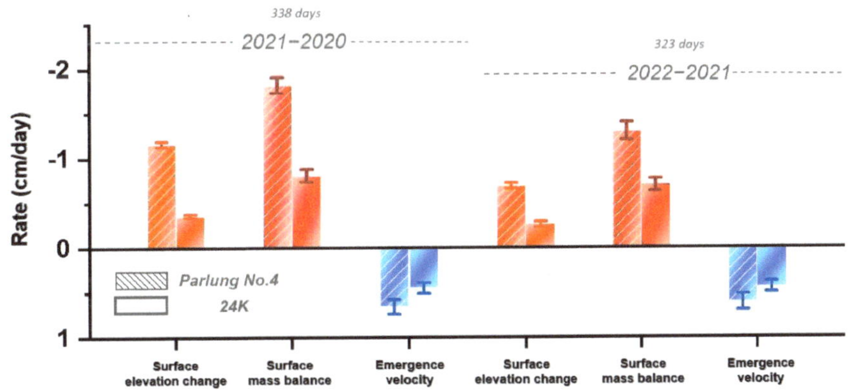

Figure 8. Comparison of surface ablation (red bar), surface elevation changes (green bar), and emergence velocity (blue bar) of two different glaciers from 2020 to 2022.

5. Discussion

5.1. Contrasting Melt Pattern of the Two Glaciers

Figures 8 and 9 show that the melt rate of Parlung No. 4 is consistently higher than that of 24K, and the melt rate of both glaciers was higher in 2020–2021 than in 2021–2022. To gather meteorological data, we installed automatic weather stations in both glaciers [15]. Utilizing the data collected from these stations, we calculated the mean daily air temperature and positive cumulative temperature for both glaciers. Upon comparing the average air temperatures and positive cumulative temperatures of the two glaciers, it became apparent that significantly lower values were consistently recorded for Parlung No. 4 than for 24K (the Parlung No. 4 terminus is ~800 m higher than that of 24K). More specifically, from 2020 to 2021, the mean air temperature and positive cumulative temperature of Parlung No. 4 were −1.02 °C and 802.01 °C·d, while those of 24K were 1.82 °C and 1195.93 °C·d, respectively. In 2021–2022, the mean air temperature and positive cumulative temperature of Parlung No. 4 were −2.59 °C and 630.85 °C·d, while those of 24K were 1.30 °C and 1061.33 °C·d, respectively. Therefore, the melt rate of 24K should be higher than that of Parlung No. 4, but this is in opposition to the actual findings recorded for the two glaciers. This high-precision comparative observation confirms the role of debris cover in inhibiting ablation [10–15]. It is essential to fully consider the factor of debris cover for future large-scale reconstruction and prediction studies of glacier change and catchment runoff in the SE Tibetan Plateau.

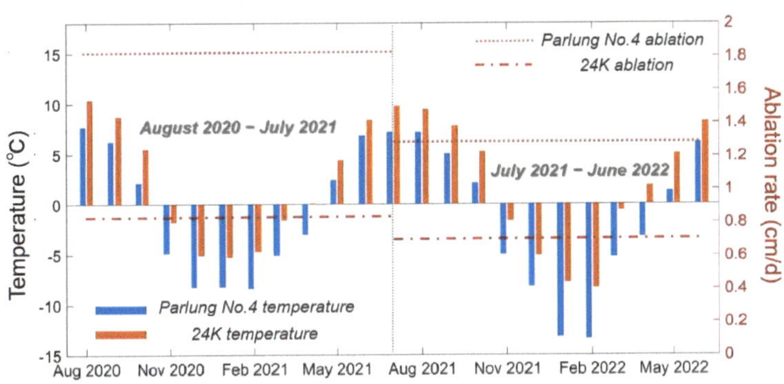

Figure 9. Comparison of the temperatures and ablation rates of the two glaciers in 2020−2021 and 2021–2022.

Additionally, the mean air temperature and positive cumulative temperature of both glaciers were higher during the 2020–2021 period in comparison to the 2021–2022 period. For this reason, the melt rates recorded for both glaciers in 2020–2021 are higher than the melt rates recorded in 2021–2022. However, we found that the melt rate of Parlung No. 4 in 2020–2021 is 1.40 times higher than that of the same glacier in 2021–2022, while the melt rate of 24K in 2020–2021 is 1.16 times higher than that of the same glacier in 2021–2022 (i.e., the response of Parlung No. 4 to climate change is more sensitive than that of 24K). Our findings are in agreement with some studies that suggest that debris-covered glaciers are insensitive to climate change (e.g., Anderson and Anderson, 2016; Vincent et al., 2016) [14,20]. Based on the results of this comparative study, we suggest that the 'debris-cover anomaly' phenomenon [19,20] does not imply that debris is ineffective in buffering ablation; rather, it may be related to areas referred to as 'hotspots', where debris-covered glaciers commonly develop [27,28,54,55]. The ablation 'hotspot' area within the 24K UAV survey area accounted for approximately 5.9% of the UAV survey area during the period from 2020 to 2021 (Figure 10a), and its melt ratio within the UAV survey area was found to reach 11.6% (Figure 10a). In the subsequent period, from 2021 to 2022, the 'hotspot' area ratio within the 24K UAV survey area was 5.3%, with a corresponding melt ratio of up to 10.4% (Figure 10b). These findings indicate that the 'hotspot' area contributes greatly to the overall melt of 24K, accounting for at least 10% of the total ablation. This partial compensation helps mitigate the restraining effect of debris cover on the overall ablation process. The proportion of 'hotspot' areas in the 24K UAV survey area is relatively small, while it is slightly higher in the neighboring glacier (23K Glacier: 6.8–7.2%) [46]. However, the ablation contribution of the 23K glacier (a debris-covered glacier adjacent to 24K) 'hotspot' area reached 31.5%, which is due to the debris thickness of the 23K glacier being higher than that of 24K (mean thicknesses: 47.1 cm vs. 24.2 cm), resulting in a low melt rate in the 23K debris cover area [46]. Compared with the debris-covered glaciers in other regions, the melt ratio of the 24K 'hotspot' area is still small; for example, the Changri Nup glacier in the middle Himalayas has a hotspot area melt contribution of 23–24% [24]. The 'hotspot' area ablation contributions of the Lirung and Langtang Glaciers in Nepal were also calculated, and they are 36.43% and 19.84%, respectively, more than that of 24K [56]. The thinner debris thickness of 24K and its smaller 'hotspot' area are the driving factors causing the low percentage of ablation contribution in the 24K 'hotspot' area.

5.2. Glacier Changes of the Two Glaciers over Two Decades

We extracted the glacier surface elevation changes for Parlung No. 4 and 24K for the periods of 2000–2004, 2005–2009, 2010–2014, and 2015–2019 from the work of Hugonnet et al. (2021) [5]. As the observation periods of 2020–2021 and 2021–2022 do not align with complete calendar years, we normalized the observed thinning/surface velocity amplitudes to annual rates by utilizing the conversion ratio (number of days in the observation period/365). For Parlung No. 4 and 24K, the ablation area exhibited an increasing trend in thinning rate from 2000 to 2019, ranging from −1.48 m/a to −3.08 m/a and −0.59 m/a to −1.21 m/a, respectively, as illustrated in Figures 11 and 12. Both glaciers demonstrate similar response characteristics (i.e., increasing thinning rate) in response to climate change. These trends provide insight into the glaciers' responses to climate change [57]. However, the thinning rate of Parlung No. 4 is always higher than that of 24K (Figures 11 and 12), and the growth in the thinning of Parlung No. 4 is also greater. These findings demonstrate the key role of debris cover in characterizing glaciers' responses to climate change.

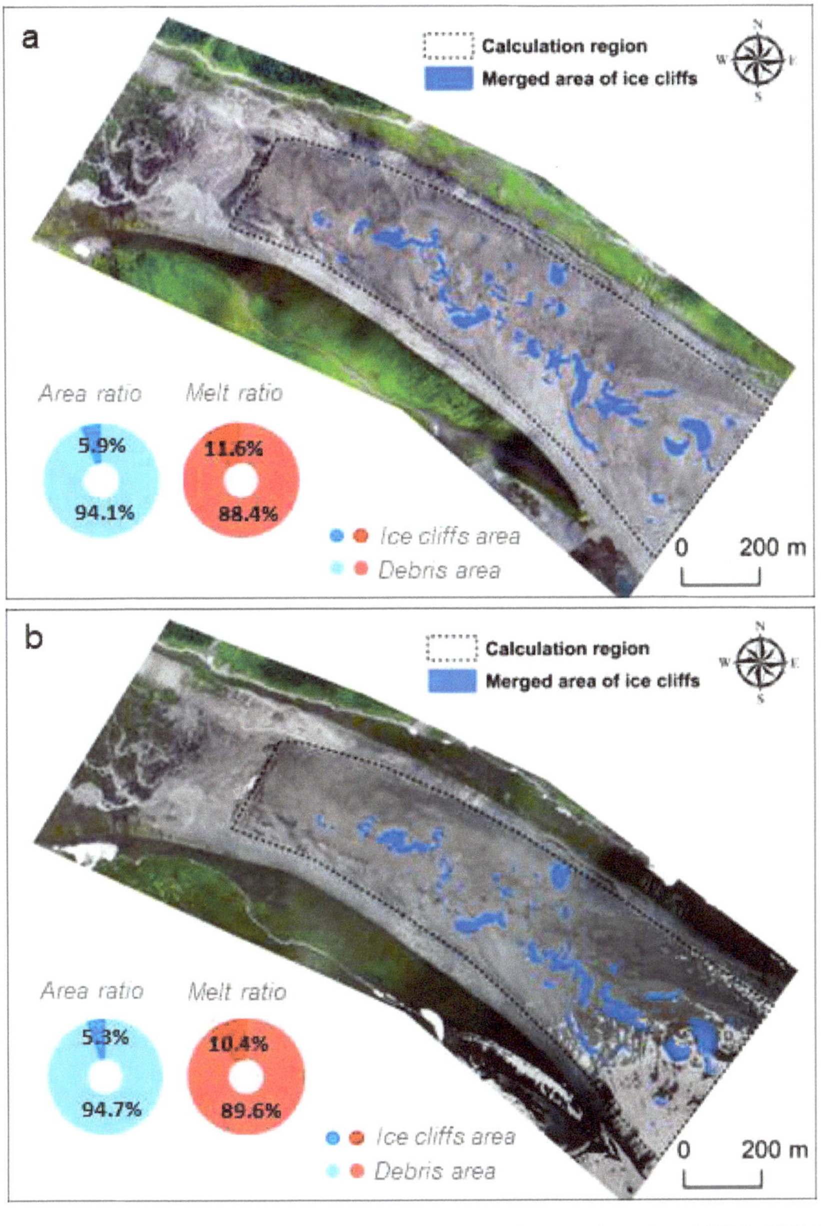

Figure 10. Ice cliff area spatial distribution, area ratio, and melt ratio information of 24K for 2020–2021 (**a**) and 2021–2022 (**b**).

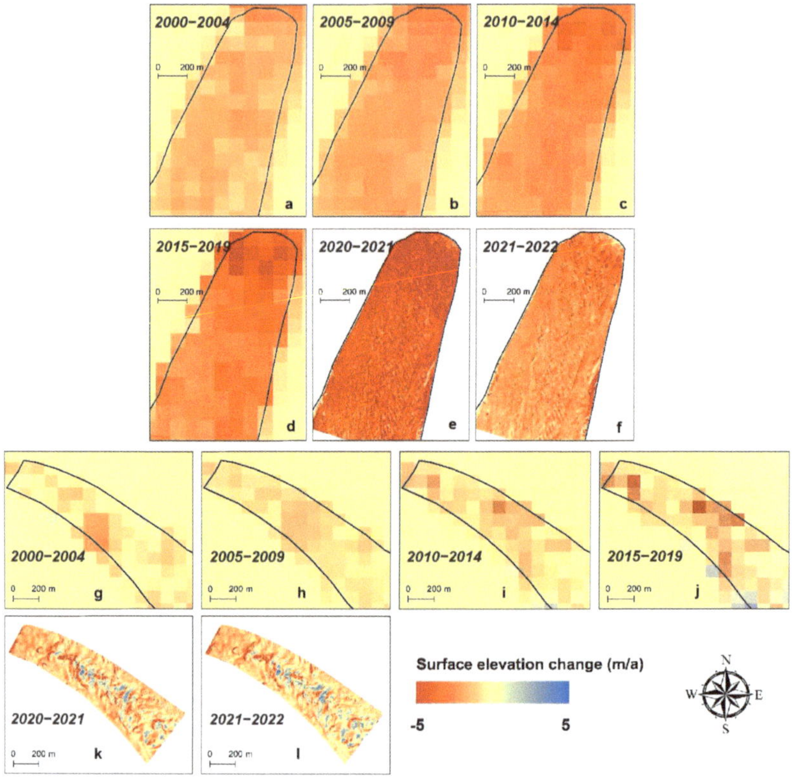

Figure 11. Annual rates of surface elevation change for 2000.01–2004.12 (**a**,**g**), 2005.01–2009.12 (**b**,**h**), 2010.01–2014.12 (**c**,**i**), 2015.01–2019.12 (**d**,**j**), 2020.08–2021.07 (**e**,**k**), and 2021.07–2022.06 (**f**,**l**) for Parlung No. 4 (**a**–**f**) and 24K (**g**–**l**).

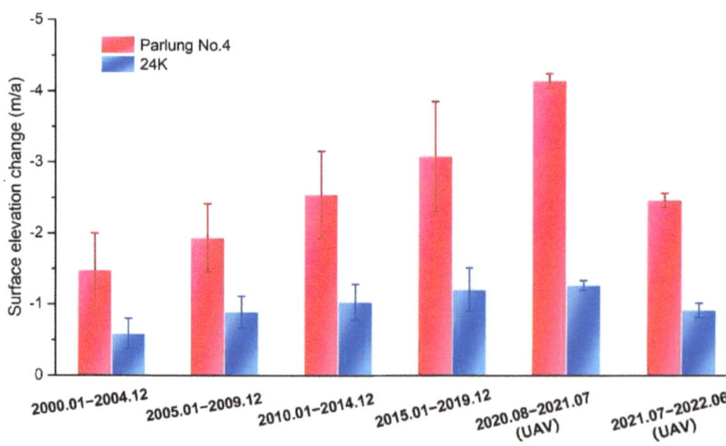

Figure 12. Surface elevation changes for the UAV survey area of Parlung No. 4 and 24K since the start of the 21st century.

We compared the surface velocities of the two glaciers during all periods (1999–2003, 2013–2015, 2020–2021, and 2021–2022) using the dataset provided by Dehecq et al. (2015) [58] and UAV data (Figures 11 and 12). Regarding the surface velocity of both glaciers, there has been a decreasing trend since the early 2000s, indicating a weakening dynamic state. Specifically, the annual surface velocity of Parlung No. 4 was 22.06 m/a in 1999–2003, which significantly reduced to 8.32 m/a in 2013–2015. Similarly, the annual surface velocity of 24K decreased from 16.75 m/a in 1999–2003 to 7.71 m/a in 2013–2015. The most plausible reason for the weakening trend of the dynamic state derives from the decreasing ice thickness, leading to a reduction in glacier basement shear stress [52,59,60]. As two typical glaciers in the SE Tibetan Plateau, their dynamic changes also imply that the glaciers in the whole region are in a weakening state. Weakened glacier dynamics contribute to a reduction in ice flow replenishment, accompanied by a constant or increasing ablation intensity, which accelerates the thinning rate of the glaciers, thus damaging the glaciers' health and sustainability [61].

It is worth noting that the surface velocity of Parlung No. 4 was slightly higher than that of 24K during all periods (Figures 13 and 14). The surface velocity of the debris-covered glacier in this study consistently exhibited weaker dynamics compared to the debris-free glacier. We compared the average ice thickness in the aerial survey areas of the two glaciers and found that the two glaciers were close to each other (Parlung No. 4: 128.1 m; 24K: 106.7 m). In addition, we compared the mean slopes of the aerial surveys and found that the slope of 24K is nearly twice as steep as that of Parlung No. 4 (Parlung No. 4: 6°; 24K: 11°). Overall, based on the above analyses, we conclude that the dynamic state of 24K should not be weaker than that of Parlung No. 4. However, the high-precision results suggest that the dynamic state of 24K is weaker. So, this result can be attributed to the presence of debris cover, which inhibits melting and suppresses basal sliding in glaciers [62]. In addition, the glacial lake of Parlung No. 4 only emerged in recent years, which may have had an influence on the dynamic state of the glacier [63,64]. Studies found that some debris-covered glaciers are in a much weaker dynamic state [46,55,65]; these findings may contribute to the observed 'debris-cover anomaly' phenomenon [16,66]. Compared with the nearly stagnant glacier (ablation area), 24K is still in a relatively strong dynamic state. Compared with the neighboring 23K glacier, though they are in the same climatic setting, the higher slope of 24K is responsible for its stronger dynamic state [46]. The great amount of precipitation and thin debris cover in the 24K area are factors that facilitate basal lubrication [15,46] and, thus, make the movement of this glacier faster. Previous studies found that the developed supraglacial ponds are strongly negatively correlated with driving stresses [60]. In the 24K area, there are almost no developed supraglacial ponds, which supports the conclusion that this glacier possesses a relatively strong power state.

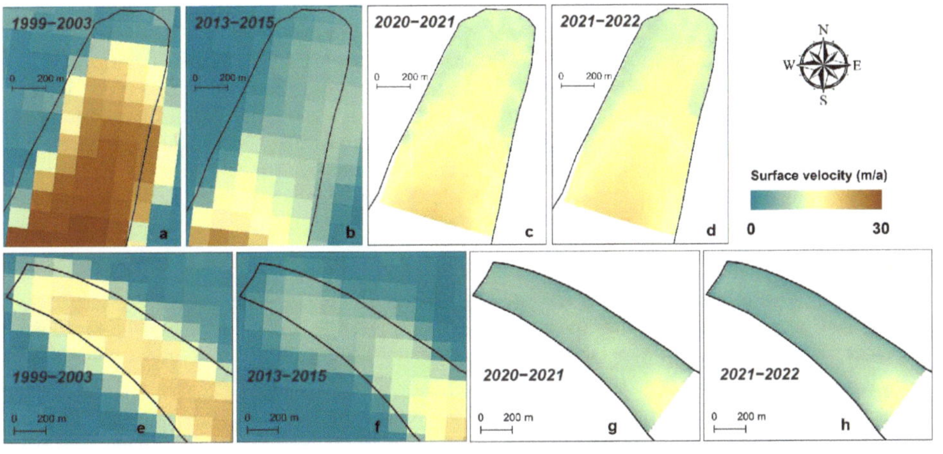

Figure 13. (**a**,**b**) Annual surface velocities for 1999–2003 (**a**,**e**), 2013–2015 (**b**,**f**), 2020–2021 (**c**,**g**), and 2021–2022 (**d**,**h**) for Parlung No. 4 (**a**–**d**) and 24K (**e**–**h**).

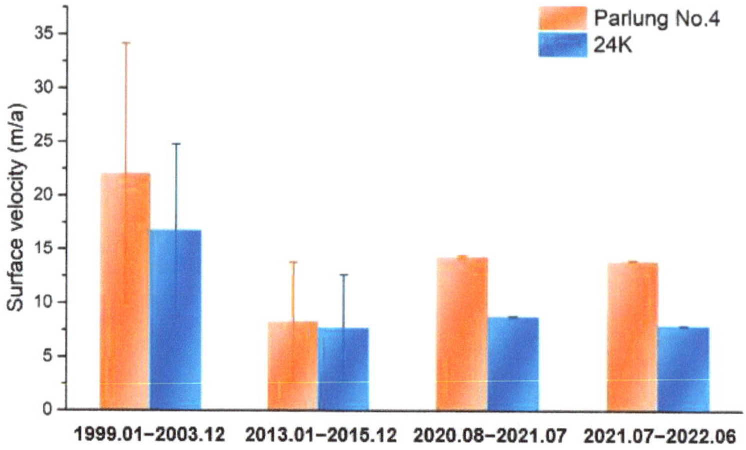

Figure 14. Surface velocities for the UAV survey area of Parlung No. 4 and 24K since the start of the 21st century.

6. Conclusions

We explored the mass loss and dynamic states of two representative glaciers in the SE Tibetan Plateau based on repeated UAV data. Our high-precision results suggest that the two types of glaciers have contrasting responses to climate change. Our conclusions are summarized as follows:

Despite the terminus altitude of Parlung No. 4 being much higher than that of 24K (an 800 m difference), the thinning rate of Parlung No. 4 was consistently higher than that of 24K during all periods (2.7–3.2 times). In 2020–2021, the surface elevation change rates of Parlung No. 4 and 24K were -1.16 ± 0.03 cm/d and -0.36 ± 0.02 cm/d, respectively. In 2021–2022, the surface elevation change rates of Parlung No. 4 and 24K were -0.69 ± 0.03 cm/d and -0.26 ± 0.03 cm/d, respectively.

The dynamic state of the debris-free Parlung No. 4 glacier is slightly stronger than that of the debris-covered 24K glacier. In 2020–2021, the mean surface velocities of Parlung No. 4 and 24K were 4.02 ± 0.04 cm/d and 2.45 ± 0.02 cm/day, respectively. In 2021–2022, the mean surface velocities of Parlung No. 4 and 24K were 3.91 ± 0.02 cm/d and 2.22 ± 0.02 cm/d, respectively.

We calculated the surface mass balance of the two different types of glaciers in the SE Tibetan Plateau based on high-precision data, which helped confirm the key role of debris cover. In all periods, the temperature and positive cumulative temperature for Parlung No. 4 were lower than those for 24K, but the ablation rate of Parlung No. 4 was approximately 1.9–2.2 times that of 24K. In addition, both glaciers had lower temperatures and positive cumulative temperatures in 2021–2022 than in 2020–2021. The difference in ablation between these two periods for 24K was small, whereas the difference in ablation for Parlung No. 4 in these periods was large (i.e., the response of Parlung No.4 to climate change is more sensitive than that of 24K). The dominant controlling factor for the above differences is debris cover.

Author Contributions: W.Y., C.Z., Z.H., T.Z., Y.H. and Y.W. conducted the fieldwork. C.Z. and S.K. designed the study and completed the data analysis. C.Z., S.K., T.Z. and T.L. contributed to writing and revising the paper. W.Y. supervised the study. All authors have read and agreed to the published version of the manuscript.

Funding: This study was supported by the National Natural Science Foundation of China (grant nos. 41961134035, 42271138, 42271312, 42130306, and 42206249) and the Science and Technology Plan Projects of Tibet Autonomous Region (grant nos. XZ202301ZY0028G, and XZ202301ZY0022G).

Data Availability Statement: The data presented in this study are available upon request from the corresponding author. The data are not publicly available due to privacy.

Acknowledgments: We thank Hugonnet et al. (2021) [5] and Dehecq et al. (2015) [58] for their datasets of glacier surface elevation change and velocity. We would like to sincerely acknowledge and express our deep appreciation to the editor, and to four anonymous referee for their thorough review of this work.

Conflicts of Interest: The authors declare no conflicts of interest.

Appendix A

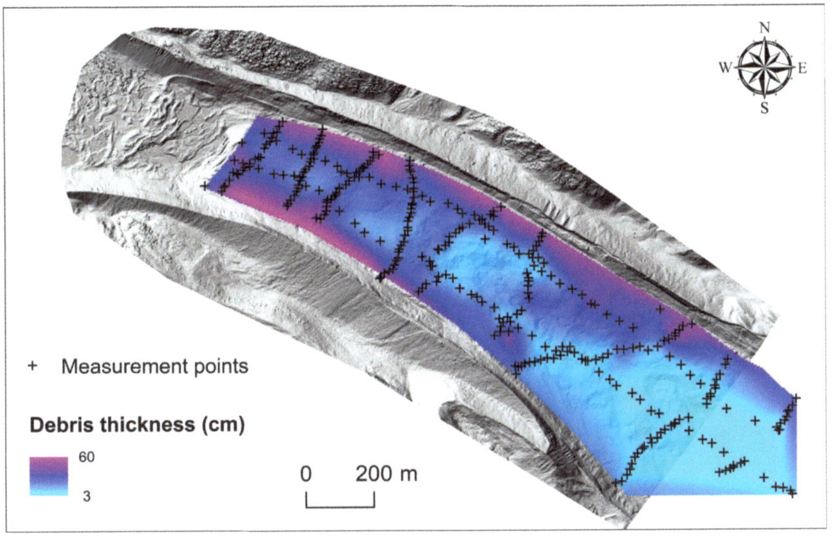

Figure A1. Debris measurement points and debris thickness spatial distribution pertaining to 24K.

References

1. Kääb, A.; Berthier, E.; Nuth, C.; Gardelle, J.; Arnaud, Y. Contrasting patterns of early twenty-first-century glacier mass change in the Himalayas. *Nature* **2012**, *488*, 495–498. [CrossRef]
2. Yao, T.D.; Thompson, L.; Yang, W.; Yu, W.S.; Gao, Y.; Guo, X.J.; Yang, X.X.; Duan, K.Q.; Zhao, H.B.; Xu, B.Q.; et al. Different glacier status with atmospheric circulations in Tibetan Plateau and surroundings. *Nat. Clim. Chang.* **2012**, *2*, 663–667. [CrossRef]
3. Brun, F.; Berthier, E.; Wagnon, P.; Kääb, A.; Treichler, D. A spatially resolved estimate of High Mountain Asia glacier mass balances from 2000 to 2016. *Nat. Geosci.* **2017**, *10*, 668–673. [CrossRef]
4. Shean, D.E.; Bhushan, S.; Montesano, P.; Rounce, D.R.; Arendt, A.; Osmanoglu, B. A systematic, regional assessment of High Mountain Asia Glacier mass balance. *Front. Earth Sci.* **2020**, *7*, 363. [CrossRef]
5. Hugonnet, R.; McNabb, R.; Berthier, E.; Menounos, B.; Nuth, C.; Girod, L.; Farinotti, D.; Huss, M.; Dussaillant, I.; Brun, F.; et al. Accelerated global glacier mass loss in the early twenty-first century. *Nature* **2021**, *592*, 726–731. [CrossRef] [PubMed]
6. Scherler, D.; Wulf, H.; Gorelick, N. Global Assessment of Supraglacial Debris-Cover Extents. *Geophys. Res. Lett.* **2018**, *45*, 11798–11805. [CrossRef]
7. Herreid, S.; Pellicciotti, F. The state of rock debris covering Earth's glaciers. *Nat. Geosci.* **2020**, *13*, 621–627, Correction in *Nat. Geosci.* **2020**, *13*, 711. [CrossRef]
8. Zhang, Y.; Hirabayashi, Y.; Fujita, K.; Liu, S.Y.; Liu, Q. Heterogeneity in supraglacial debris thickness and its role in glacier mass changes of the Mount Gongga. *Sci. China Earth Sci.* **2016**, *59*, 170–184. [CrossRef]
9. Zhang, Y.; Gu, J.; Liu, S.Y.; Wang, X.; Jiang, Z.L.; Wei, J.F.; Zheng, Y.J. Spatial pattern of the debris-cover effect and its role in the Hindu Kush-Pamir-Karakoram-Himalaya glaciers. *J. Hydrol.* **2022**, *615*, 128613. [CrossRef]
10. Hagg, W.; Mayer, C.; Lambrecht, A.; Helm, A. Sub-debris melt rates on southern inylchek glacier, central tian shan. *Geogr. Ann. Ser. A-Phys. Geogr.* **2008**, *90A*, 55–63. [CrossRef]
11. Nakawo, M.; Yabuki, H.; Sakai, A. Characteristics of Khumbu Glacier, Nepal Himalaya: Recent change in the debris-covered area. *Ann. Glaciol.* **1999**, *28*, 118–122. [CrossRef]

12. Nicholson, L.; Benn, D.I. Calculating ice melt beneath a debris layer using meteorological data. *J. Glaciol.* **2006**, *52*, 463–470. [CrossRef]
13. Reid, T.D.; Brock, B.W. An energy-balance model for debris-covered glaciers including heat conduction through the debris layer. *J. Glaciol.* **2010**, *56*, 903–916. [CrossRef]
14. Anderson, L.S.; Anderson, R.S. Modeling debris-covered glaciers: Response to steady debris deposition. *Cryosphere* **2016**, *10*, 1105–1124. [CrossRef]
15. Yang, W.; Yao, T.D.; Zhu, M.L.; Wang, Y.J. Comparison of the meteorology and surface energy fluxes of debris-free and debris-covered glaciers in the southeastern Tibetan Plateau. *J. Glaciol.* **2017**, *63*, 1090–1104. [CrossRef]
16. Rounce, D.R.; Hock, R.; McNabb, R.W.; Millan, R.; Sommer, C.; Braun, M.H.; Malz, P.; Maussion, F.; Mouginot, J.; Seehaus, T.C.; et al. Distributed global debris thickness estimates reveal debris significantly impacts glacier mass balance. *Geophys. Res. Lett.* **2021**, *48*, e2020GL091311. [CrossRef] [PubMed]
17. Vincent, C.; Ramanathan, A.; Wagnon, P.; Dobhal, D.P.; Linda, A.; Berthier, E.; Sharma, P.; Arnaud, Y.; Azam, M.F.; Jose, P.G.; et al. Balanced conditions or slight mass gain of glaciers in the Lahaul and Spiti region (northern India, Himalaya) during the nineties preceded recent mass loss. *Cryosphere* **2013**, *7*, 569–582. [CrossRef]
18. Brun, F.; Wagnon, P.; Berthier, E.; Jomelli, V.; Maharjan, S.B.; Shrestha, F.; Kraaijenbrink, P.D.A. Heterogeneous influence of glacier morphology on the mass balance variability in high mountain Asia. *J. Geophys. Res. Earth* **2019**, *124*, 1331–1345. [CrossRef]
19. Pellicciotti, F.; Stephan, C.; Miles, E.; Herreid, S.; Immerzeel, W.W.; Bolch, T. Mass-balance changes of the debris-covered glaciers in the Langtang Himal, Nepal, from 1974 to 1999. *J. Glaciol.* **2015**, *61*, 373–386. [CrossRef]
20. Vincent, C.; Wagnon, P.; Shea, J.M.; Immerzeel, W.W.; Kraaijenbrink, P.; Shrestha, D.; Soruco, A.; Arnaud, Y.; Brun, F.; Berthier, E.; et al. Reduced melt on debris-covered glaciers: Investigations from Changri Nup Glacier, Nepal. *Cryosphere* **2016**, *10*, 1845–1858. [CrossRef]
21. Neckel, N.; Loibl, D.; Rankl, M. Recent slowdown and thinning of debris-covered glaciers in south-eastern Tibet. *Earth Planet. Sci. Lett.* **2017**, *464*, 95–102. [CrossRef]
22. Ke, L.H.; Song, C.Q.; Yong, B.; Lei, Y.B.; Ding, X.L. Which heterogeneous glacier melting patterns can be robustly observed from space? A multi-scale assessment in southeastern Tibetan Plateau. *Remote Sens. Environ.* **2020**, *242*, 111777. [CrossRef]
23. Thompson, S.; Benn, D.I.; Mertes, J.; Luckman, A. Stagnation and mass loss on a Himalayan debris-covered glacier: Processes, patterns and rates. *J. Glaciol.* **2016**, *62*, 467–485. [CrossRef]
24. Brun, F.; Wagnon, P.; Berthier, E.; Shea, J.M.; Immerzeel, W.W.; Kraaijenbrink, P.D.A.; Vincent, C.; Reverchon, C.; Shrestha, D.; Arnaud, Y. Ice cliff contribution to the tongue-wide ablation of Changri Nup Glacier, Nepal, central Himalaya. *Cryosphere* **2018**, *12*, 3439–3457. [CrossRef]
25. Miles, E.S.; Willis, I.; Buri, P.; Steiner, J.F.; Arnold, N.S.; Pellicciotti, F. Surface pond energy absorption across four Himalayan glaciers accounts for 1/8 of total catchment ice loss. *Geophys. Res. Lett.* **2018**, *45*, 10464–10473. [CrossRef] [PubMed]
26. Miles, E.S.; Steiner, J.F.; Buri, P.; Immerzeel, W.W.; Pellicciotti, F. Controls on the relative melt rates of debris-covered glacier surfaces. *Environ. Res. Lett.* **2022**, *17*, 064004. [CrossRef]
27. Buri, P.; Miles, E.S.; Steiner, J.F.; Immerzeel, W.W.; Wagnon, P.; Pellicciotti, F. A physically based 3-D model of ice cliff evolution over debris-covered glaciers. *J. Geophys. Res. Earth* **2016**, *121*, 2471–2493. [CrossRef]
28. Buri, P.; Miles, E.S.; Steiner, J.F.; Ragettli, S.; Pellicciotti, F. Supraglacial ice cliffs can substantially increase the mass loss of debris-covered glaciers. *Geophys. Res. Lett.* **2021**, *48*, e2020GL092150. [CrossRef]
29. Kneib, M.; Miles, E.S.; Buri, P.; Fugger, S.; McCarthy, M.; Shaw, T.E.; Zhao, C.X.; Truffer, M.; Westoby, M.J.; Yang, W.; et al. Sub-seasonal variability of supraglacial ice cliff melt rates and associated processes from time-lapse photogrammetry. *Cryosphere* **2022**, *16*, 4701–4725. [CrossRef]
30. Sato, Y.; Fujita, K.; Inoue, H.; Sunako, S.; Sakai, A.; Tsushima, A.; Podolskiy, E.A.; Kayastha, R.; Kayastha, R.B. Ice cliff dynamics of debris-covered trakarding glacier in the Rolwaling region, nepal Himalaya. *Front. Earth Sci.* **2021**, *9*, 623623. [CrossRef]
31. Mishra, N.B.; Miles, E.S.; Chaudhuri, G.; Mainali, K.P.; Mal, S.; Singh, P.B.; Tiruwa, B. Quantifying heterogeneous monsoonal melt on a debris-covered glacier in Nepal Himalaya using repeat uncrewed aerial system (UAS) photogrammetry. *J. Glaciol.* **2022**, *68*, 288–304. [CrossRef]
32. Immerzeel, W.W.; Kraaijenbrink, P.D.A.; Shea, J.M.; Shrestha, A.B.; Pellicciotti, F.; Bierkens, M.F.P.; de Jong, S.M. High-resolution monitoring of Himalayan glacier dynamics using unmanned aerial vehicles. *Remote Sens. Environ.* **2014**, *150*, 93–103. [CrossRef]
33. Brun, F.; Buri, P.; Miles, E.S.; Wagnon, P.; Steiner, J.; Berthier, E.; Ragettli, S.; Kraaijenbrink, P.; Immerzeel, W.W.; Pellicciotti, F. Quantifying volume loss from ice cliffs on debris-covered glaciers using high-resolution terrestrial and aerial photogrammetry. *J. Glaciol.* **2016**, *62*, 684–695. [CrossRef]
34. Mölg, N.; Bolch, T.; Walter, A.; Vieli, A. Unravelling the evolution of Zmuttgletscher and its debris cover since the end of the Little Ice Age. *Cryosphere* **2019**, *13*, 1889–1909. [CrossRef]
35. Westoby, M.J.; Rounce, D.R.; Shaw, T.E.; Fyffe, C.L.; Moore, P.L.; Stewart, R.L.; Brock, B.W. Geomorphological evolution of a debris-covered glacier surface. *Earth Surf. Process. Landf.* **2020**, *45*, 3431–3448. [CrossRef]
36. Anderson, L.S.; Armstrong, W.H.; Anderson, R.S.; Buri, P. Debris cover and the thinning of Kennicott Glacier, Alaska: In situ measurements, automated ice cliff delineation and distributed melt estimates. *Cryosphere* **2021**, *15*, 265–282. [CrossRef]

37. Kneib, M.; Miles, E.S.; Buri, P.; Molnar, P.; McCarthy, M.; Fugger, S.; Pellicciotti, F. Interannual dynamics of ice cliff populations on debris-covered glaciers from remote sensing observations and stochastic modeling. *J. Geophys. Res. Earth* **2021**, *126*, e2021JF006179. [CrossRef] [PubMed]
38. Westoby, M.J.; Brasington, J.; Glasser, N.F.; Hambrey, M.J.; Reynolds, J.M. 'Structure-from-Motion' photogrammetry: A low-cost, effective tool for geoscience applications. *Geomorphology* **2012**, *179*, 300–314. [CrossRef]
39. Benoit, L.; Gourdon, A.; Vallat, R.; Irarrazaval, I.; Gravey, M.; Lehmann, B.; Prasicek, G.; Gräff, D.; Herman, F.; Mariethoz, G. A high-resolution image time series of the Gorner Glacier-Swiss Alps-derived from repeated unmanned aerial vehicle surveys. *Earth Syst. Sci. Data* **2019**, *11*, 579–588. [CrossRef]
40. Che, Y.J.; Wang, S.J.; Yi, S.H.; Wei, Y.Q.; Cai, Y.C. Summer mass balance and surface velocity derived by unmanned aerial vehicle on debris-covered region of Baishui River Glacier No. 1, Yulong Snow Mountain. *Remote Sens.* **2020**, *12*, 3280. [CrossRef]
41. Yang, W.; Zhao, C.X.; Westoby, M.; Yao, T.D.; Wang, Y.J.; Pellicciotti, F.; Zhou, J.M.; He, Z.; Miles, E. Seasonal dynamics of a temperate Tibetan Glacier revealed by high-resolution UAV Photogrammetry and In Situ Measurements. *Remote Sens.* **2020**, *12*, 2389. [CrossRef]
42. Fu, Y.; Liu, Q.; Liu, G.X.; Zhang, B.; Zhang, R.; Cai, J.L.; Wang, X.W.; Xiang, W. Seasonal ice dynamics in the lower ablation zone of Dagongba Glacier, southeastern Tibetan Plateau, from multitemporal UAV images. *J. Glaciol.* **2022**, *68*, 636–650. [CrossRef]
43. Wu, K.P.; Liu, S.Y.; Zhu, Y.; Xie, F.M.; Gao, Y.P.; Qi, M.M.; Miao, W.F.; Duan, S.M.; Han, F.Z.; Grünwald, R. Monitoring the surface elevation changes of a monsoon temperate glacier with repeated UAV Surveys, Mainri Mountains, China. *Remote Sens.* **2022**, *14*, 2229. [CrossRef]
44. Xu, S.Y.; Fu, P.; Quincey, D.; Feng, M.L.; Marsh, S.; Liu, Q. UAV-based geomorphological evolution of the Terminus Area of the Hailuogou Glacier, Southeastern Tibetan Plateau between 2017 and 2020. *Geomorphology* **2022**, *411*, 108293. [CrossRef]
45. He, Z.; Yang, W.; Wang, Y.J.; Zhao, C.X.; Ren, S.T.; Li, C.H. Dynamic changes of a thick debris-covered glacier in the southeastern Tibetan Plateau. *Remote Sens.* **2023**, *15*, 357. [CrossRef]
46. Zhao, C.; Yang, W.; Miles, E.; Westoby, M.; Kneib, M.; Wang, Y.; He, Z.; Pellicciotti, F. Thinning and surface mass balance patterns of two neighbouring debris-covered glaciers in the southeastern Tibetan Plateau. *Cryosphere* **2023**, *17*, 3895–3913. [CrossRef]
47. Ye, D.; Gao, Y. *Meteorology of the Tibetan Plateau*; Science Press: Beijing, China, 1979; pp. 49–61.
48. Yang, W.; Yao, T.D.; Guo, X.F.; Zhu, M.L.; Li, S.H.; Kattel, D.B. Mass balance of a maritime glacier on the southeast Tibetan Plateau and its climatic sensitivity. *J. Geophys. Res. Atmos.* **2013**, *118*, 9579–9594. [CrossRef]
49. Yang, W.; Guo, X.F.; Yao, T.D.; Zhu, M.L.; Wang, Y.J. Recent accelerating mass loss of southeast Tibetan glaciers and the relationship with changes in macroscale atmospheric circulations. *Clim. Dyn.* **2016**, *47*, 805–815. [CrossRef]
50. Messerli, A.; Grinsted, A. Image georectification and feature tracking toolbox: ImGRAFT. *Geosci. Instrum. Methods Data Syst.* **2015**, *4*, 23–34. [CrossRef]
51. Bolch, T.; Menounos, B.; Wheate, R. Landsat-based inventory of glaciers in western Canada, 1985–2005. *Remote Sens. Environ.* **2010**, *114*, 127–137. [CrossRef]
52. Cuffey, K.M.; Paterson, W.S.B. *The Physics of Glaciers*, 4th ed.; Academic Press: Amsterdam, The Netherlands, 2010; 704p.
53. Farinotti, D.; Huss, M.; Fürst, J.J.; Landmann, J.; Machguth, H.; Maussion, F.; Pandit, A. A consensus estimate for the ice thickness distribution of all glaciers on Earth. *Nat. Geosci.* **2019**, *12*, 168–173. [CrossRef]
54. Steiner, J.F.; Pellicciotti, F.; Bur, P.; Miles, E.S.; Immerzeel, W.W.; Reid, T.D. Modelling ice-cliff backwasting on a debris-covered glacier in the Nepalese Himalaya. *J. Glaciol.* **2015**, *61*, 889–907. [CrossRef]
55. Kneib, M.; Fyffe, C.L.; Miles, E.S.; Lindemann, S.; Shaw, T.E.; Buri, P.; McCarthy, M.; Ouvry, B.; Vieli, A.; Sato, Y.; et al. Controls on ice cliff distribution and characteristics on debris-covered glaciers. *Geophys. Res. Lett.* **2023**, *50*, e2022GL102444. [CrossRef]
56. Buri, P.; Pellicciotti, F. Aspect controls the survival of ice cliffs on debris-covered glaciers. *Proc. Natl. Acad. Sci. USA* **2018**, *115*, 4369–4374. [CrossRef] [PubMed]
57. Jouberton, A.; Shaw, T.E.; Miles, E.; McCarthy, M.; Fugger, S.; Ren, S.; Dehecq, A.; Yang, W.; Pellicciotti, F. Warming-induced monsoon precipitation phase change intensifies glacier mass loss in the southeastern Tibetan Plateau. *Proc. Natl. Acad. Sci. USA* **2022**, *119*, e2109796119. [CrossRef] [PubMed]
58. Dehecq, A.; Gourmelen, N.; Trouve, E. Deriving large-scale glacier velocities from a complete satellite archive: Application to the Pamir-Karakoram-Himalaya. *Remote Sens. Environ.* **2015**, *162*, 55–66. [CrossRef]
59. Dehecq, A.; Gourmelen, N.; Gardner, A.S.; Brun, F.; Goldberg, D.; Nienow, P.W.; Berthier, E.; Vincent, C.; Wagnon, P.; Trouvé, E. Twenty-first century glacier slowdown driven by mass loss in High Mountain Asia. *Nat. Geosci.* **2019**, *12*, 22. [CrossRef]
60. Wu, K.P.; Liu, S.Y.; Zhu, Y.; Liu, Q.; Jiang, Z.L. Dynamics of glacier surface velocity and ice thickness for maritime glaciers in the southeastern Tibetan Plateau. *J. Hydrol.* **2020**, *590*, 125527. [CrossRef]
61. Miles, E.; McCarthy, M.; Dehecq, A.; Kneib, M.; Fugger, S.; Pellicciotti, F. Health and sustainability of glaciers in High Moun tain Asia. *Nat. Commun.* **2021**, *12*, 2868. [CrossRef] [PubMed]
62. Delaney, I.; Anderson, L.S. Debris cover limits subglacial erosion and promotes till accumulation. *Geophys. Res. Lett.* **2022**, *49*, e2022GL099049. [CrossRef] [PubMed]
63. Liu, Q.; Mayer, C.; Wang, X.; Nie, Y.; Wu, K.P.; Wei, J.F.; Liu, S.Y. Interannual flow dynamics driven by frontal retreat of a lake-terminating glacier in the Chinese Central Himalaya. *Earth Planet. Sci. Lett.* **2020**, *546*, 116450. [CrossRef]
64. Sato, Y.; Fujita, K.; Inoue, H.; Sakai, A.; Karma. Land- to lake-terminating transition triggers dynamic thinning of a Bhutanese glacier. *Cryosphere* **2022**, *16*, 2643–2654. [CrossRef]

65. Anderson, L.S.; Armstrong, W.H.; Anderson, R.S.; Scherler, D.; Petersen, E. The causes of debris-covered glacier thinning: Evidence for the importance of ice dynamics from Kennicott Glacier, Alaska. *Front. Earth Sci.* **2021**, *9*, 680995. [CrossRef]
66. Berthier, E.; Vincent, C. Relative contribution of surface mass-balance and ice-flux changes to the accelerated thinning of Mer de Glace, French Alps, over 1979–2008. *J. Glaciol.* **2012**, *58*, 501–512. [CrossRef]

remote sensing

MDPI

Article

Monitoring of Supraglacial Lake Distribution and Full-Year Changes Using Multisource Time-Series Satellite Imagery

Dongyu Zhu [1,2], Chunxia Zhou [1], Yikai Zhu [1], Tao Wang [1] and Ce Zhang [2,3,4,*]

1 Chinese Antarctic Center of Surveying and Mapping, Wuhan University, Wuhan 430079, China; zdy1229@whu.edu.cn (D.Z.); zhoucx@whu.edu.cn (C.Z.); zhuyikai@whu.edu.cn (Y.Z.); taowang@whu.edu.cn (T.W.)
2 Lancaster Environment Centre, Lancaster University, Lancaster LA1 4YB, UK
3 UK Centre for Ecology & Hydrology, Library Avenue, Lancaster LA1 4AP, UK
4 School of Geographical Sciences, University of Bristol, Bristol BS8 1QU, UK
* Correspondence: c.zhang9@lancaster.ac.uk

Abstract: Change of supraglacial lakes (SGLs) is an important hydrological activity on the Greenland ice sheet (GrIS), and storage and drainage of SGLs occur throughout the year. However, current studies tend to split SGL changes into melt/non-melt seasons, ignoring the effect of buried lakes in the exploration of drainage, and the existing threshold-based approach to SGL extraction in a synthetic aperture radar (SAR) is influenced by the choice of the study area mask. In this study, a new method (Otsu–Canny–Otsu (OCO)), which accesses the features of SGLs on optical and SAR images objectively, is proposed for full-year SGL extraction with Google Earth Engine (GEE). The SGLs on the Petermann Glacier were monitored well by OCO throughout 2021, including buried lakes and more detailed rapid drainage events. Some SGLs' extent varied minimally in a year (area varying by 10–25%) while some had very rapid drainage (a rapid drainage event from July 26 to 30). The SGL extraction results were influenced by factors such as the mode of polarization, the surface environment, and the depth of the lake. The OCO method can provide a more comprehensive analysis for SGL changes throughout the year.

Keywords: supraglacial lakes; full-year monitoring; OCO method; buried lake; Google Earth Engine; Petermann Glacier

1. Introduction

The Greenland ice sheet (GrIS) is the second largest ice sheet in the world after the Antarctic ice sheet. In recent years, with the influence of global warming, there has been a drastic change in glaciers worldwide [1–4], and the mass loss of the GrIS has also become increasingly prominent in this century [5]. One of the main reasons for GrIS mass loss is surface melting. Studies have shown that surface runoff accounts for nearly 40% of the GrIS's mass loss between 2000 and 2008 [6], and 84% of the increase in mass loss from 2009 to 2012 came from increased surface melting. To study the process of ice surface melting, scholars have discussed it in terms of different objects, including surface runoff, moulins, snow melt, and supraglacial lakes (SGLs) [7–11].

SGLs, which are a storage site for liquid water, are one of the central representations of surface melting [12]. The development of SGLs affects the stability of ice sheets by forming fractures [13] and inducing basal slip [14]. In recent years, SGLs have been increasing in area and depth and are gradually expanding inland of the ice sheet [10,11]. The process of storing and draining of SGLs often occurs during the melt season, and this process can be slow or rapid ($\leq$4 days) [15]. In the non-melt season, a part of the liquid water only freezes on the surface and forms a buried lake. Dunmire et al. detected that a buried lake also stores and drains during the non-melt season, with a potential effect on the ice sheet mass balance [16]. To comprehensively monitor changes in SGLs, multisource remote sensing satellites are used to capture the spatiotemporal variations. Typical data used for detecting SGLs information include optical data, synthetic aperture radar (SAR) data, snow radar, global navigation satellite system (GNSS) [17], advanced topographic laser altimeter system (ATLAS), etc.

Although it is well established that storage and drainage occurs in SGLs throughout the year, most studies tended to a single analysis of the melt season or the non-melt season when testing the annual season of SGL development [18–20]. In recent years, several studies on SGLs have focused on buried lakes and have proposed approaches based on optical and SAR imagery to extract buried lakes in the non-melt season. Optical images are used for lake extraction in the melt season and locating the SGLs' extent in SAR images. For SGL extraction in SAR, there are mainly two categories: deep learning [16,21] and threshold segmentation [18,19,22]. Compared to deep learning, the histogram-based threshold segmentation approach has less subjectivity and workload. However, the current threshold setting method relies on the extent of the water area in the SAR image, and a smaller extent of the lake can cause errors in extraction [22]. To explore the complete development of SGLs, we have designed an optical and SAR-based algorithm for SGL extraction named the Otsu–Canny–Otsu (OCO) method, which can be effectively combined with optical and SAR data to analyze changes in SGL growth over a year. The new model uses the edge information in SAR imagery to extract the segmentation threshold and do post-classification judgements to assess a lake. The OCO method is applicable to any size lakes in SAR imagery and also applies to water-free SAR masks.

The OCO method was designed by improving the Canny edge Otsu algorithm, which was proposed by Kolli et al. in 2022 [23]. The Canny edge Otsu algorithm combines the Otsu algorithm and the Canny edge detector, and it aims to enhance the features at the water edges and better extract the water area in SAR images. The Otsu algorithm and the Canny edge detector are commonly used for image segmentation and edge detection. The Otsu algorithm selects the threshold naturally by looking for the maximum variance between classes in the data histogram [24,25], whereas the Canny edge detector detects edge regions by Gaussian filtering and the gradient magnitude between neighbor pixels [26,27]. However, for the Canny edge detection algorithm, using a single threshold for edge detection over a large area is incomplete, whereas SAR images tend to have localized features [26,27]. In addition, SAR images of SGLs are influenced by the topography and surface environment, which cause each individual lake to have different boundary characteristics in SAR. To overcome the regional environmental differences for each SGL extraction, the OCO method introduces the Otsu–Canny edge detection algorithm [28,29] to the Canny edge Otsu algorithm. The Otsu–Canny edge detection algorithm uses the Otsu threshold to complete the segmentation of automatic thresholds, thereby overcoming the subjective dual-threshold settings in Canny edge detection [30]. As the OCO method was designed based on SGLs' SAR images, the extraction results must be post-processed accordingly.

In this study, we used the OCO method and multisource remote sensing data to monitor the SGLs on the Petermann Glacier (PG) of Greenland in 2021. For satellite images, Sentinel-1 and Sentinel-2 satellite imagery were used to draw the extent of the lakes; a snow radar was used to verify the extraction results. All image data processing for this experiment, including the pre-processing of satellite images and the extraction of SGLs, were based on the Google Earth Engine (GEE) platform. The new lake extraction model could extract the SGLs well, but the result would also be affected by satellite imaging. Based on multisource satellite data and the OCO method, the main objectives of this study are as follows:

(1) Expand the image source of extracted SGL information by aggregating the datasets of Sentinel-1 and Sentinel-2;
(2) Create a new model that has less subjectivity and higher accuracy to extract lake information from optical and SAR images; and

(3) Analyze the dynamic processes in SGLs with different characteristics (such as buried lakes) and explore the time-series storage and drainage events.

2. Material

2.1. Study Site

The PG in northern GrIS was chosen as the study area for the experiment (Figure 1). PG is one of the largest outlet glaciers in the north of Greenland, with active surface meltwater [31]. Additionally, this study area contains sufficient optical and SAR satellite images and snow radar passes to favor the validation of the OCO method. Given that one aim of this experiment was to investigate the full-year storage and drainage process, a total of 12 SGLs were included in this study as typical cases (Regions A, B, and C). Fully drained lakes and buried lakes were included in these typical study areas for comparison. As shown in Figure 1b, the lake situation was similar for Regions A and B, which were visible in the optical and SAR images. For Region C, however, the lake could only be observed in the SAR image. Region C was chosen to highlight the importance of SAR imagery in SGL monitoring.

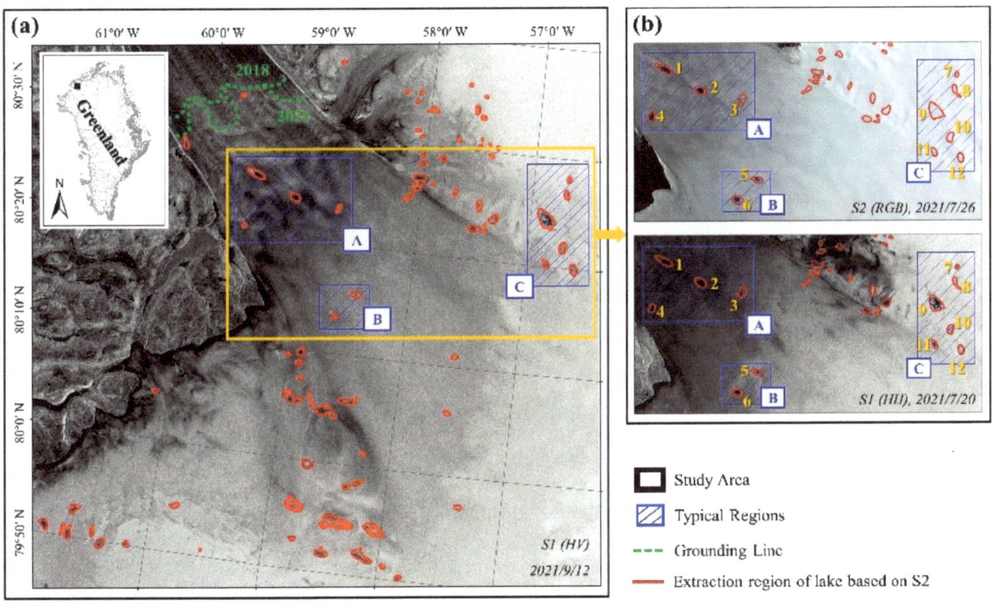

Figure 1. Diagram of the study area (**a**). Comparison of optical and SAR images of the typical study area in the PG (**b**). The typical regions included 12 SGL study objects with IDs in orange.

2.2. Satellite Data

The Sentinel series are the Earth observation missions developed by the European Space Agency. In this study, Sentinel-1 and Sentinel-2 were used to extract information of the SGLs. Sentinel-1 and Sentinel-2 images in 2021 were screened for water identification in SAR images, melt/non-melt season imaging comparisons of the SGLs, extraction of monthly changes for the SGLs, and studies of typical drainage phenomena. All Sentinel images were processed in GEE.

The Sentinel-1 mission performs C-band (at a center frequency of 5.405 GHz) SAR imaging, which has some penetration of the shallow snow covering the water [19,32]. Considering image coverage and data quality, the ground range detected (GRD) format with the interferometric wide (IW) swath mode and 10 m pixel spacing from Sentinel-1 SAR images were used in this study. The Sentinel-2 mission comprises a constellation of two polar-orbiting satellites (Sentinel-2A and Sentinel-2B). It is a high-resolution multispectral imaging satellite system carrying a multispectral instrument (MSI) for land surface monitoring. The MSI covers 13 spectral bands and 290 km swath width. For this experiment, we selected Sentinel-2 images with less than 10% cloudiness and applied bands (B2 blue and B4 red) with 10 m spatial resolution for the extraction of SGLs. Sentinel-2 images offer accurate information on supraglacial lake (SGL) extent in favorable weather. Sentinel-1 images, unaffected by weather and lighting, excel in cloud penetration, suitable for diverse environmental observations. Additionally, SAR can penetrate the snow cover, revealing partially buried lakes. Combining Sentinel-1 and Sentinel-2 data not only expands the dataset but also improves SGL localization and year-round monitoring precision.

Snow radar data were used to verify the accuracy and reliability of the extracted lake results that are produced by IceBridge Snow Radar L1B Geolocated Radar Echo Strength Profiles, Version 2 (IRSNO1B) by NASA's Operation IceBridge (OIB), and radar echograms were taken from the Center for Remote Sensing of Ice Sheets ultrawide-band snow radar over the GrIS. The snow radar operates over the frequency range from ~2 to 6.5 GHz, and it can detect information beneath snow layers to tens of meters in depth [33]. The greater penetration of the snow radar can be used to check the authenticity of SAR-extracted lakes.

3. Methods

3.1. Lake Extraction from Optical Images

Sentinel-2 images used for this experiment were all from GEE's "COPERNICUS/S2" databases, which contains the Level-1C orthorectified top-of-atmosphere (TOA) reflectance product of the Sentinel-2 satellite, which is widely used for SGL extraction from optical images. In the experiment, the TOA images of Sentinel-2 were pre-processed for the subsequent lake extraction, including date screening, cloud volume filtering (<10%), cloud removal (using the "QA60" band), and clipping. After the pre-treatment, the area with a normalized difference water index adapted for ice ($NDWI_{ice}$) > 0.25 was extracted as the initial extraction area of the lake [34,35].

$$NDWI_{ice} = \frac{Blue - Red}{Blue + Red} \tag{1}$$

The *blue* and *red* bands in Equation (1) correspond to the reflectance in the blue and red bands for Sentinel-2, respectively. We eliminated SGLs smaller than 5 pixels (30 m resolution) and linear features narrower than 2 pixels, which have relatively large uncertainties, to improve the extraction accuracy for SGLs further [15,36]. Considering the possible influence of clouds in the images and the irrelevant location migration of the SGLs on the ice sheet surface, we superimposed all the lake areas extracted by $NDWI_{ice} > 0.25$ for the melt season (from May to September) during the study period and selected the area with more than five recurrence times as the area with the largest water mask.

3.2. Otsu–Canny–Otsu Method for SGL Extraction

Extraction using the OCO model is divided into the following five steps: Sentinel-1 data acquisition, Sentinel-1 data pre-processing, mask fabrication for SAR extraction area based on the optical image, lake extraction in SAR based on the OCO method, and postprocessing for the lake area (Figure 2).

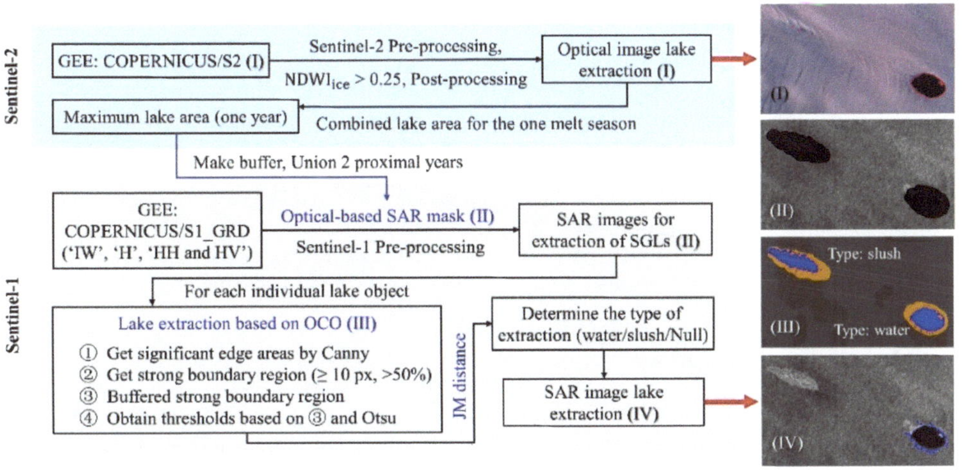

Figure 2. SGL extraction from SAR images with the OCO method based on optical masks. The important steps are in blue font. The examples shown on the right are the extraction process of Lakes 4 and 5 on 3 September 2019.

(1) Sentinel-1 data acquisition: Sentinel-1 images were selected from GEE's "COPERNICUS/S1_GRD" product. The parameter "instrumentMode" is "IW", which contains HH and HV polarization. Image quality is "H" with a resolution of 10 × 10 m. The GRD product in GEE already contains most of the pre-processing steps (thermal noise removals, data calibration, multilooking, and range-Doppler terrain correction) of Sentinel-1 [37].
(2) Sentinel-1 data pre-processing: The topographic correction of the study area was conducted using the Greenland Mapping Project, a Greenland digital elevation model (DEM) product that comes with the GEE platform (https://developers.google.com/earth-engine/datasets/catalog/OSU_GIMP_DEM) (accessed on 6 November 2023), which enables the SAR images to have better feature recognition in areas with large topographic relief. To minimize the influence of the original noise of SAR images on feature extraction, the Lee sigma filter was applied to the images [37,38].
(3) Mask fabrication of the SAR extraction area based on optical image: The lake exhibits low backscatter intensity in the SAR image, appearing nearly black [39]. However, many non-water regions in SAR also exhibit low backscatter intensity due to topographic features. To pinpoint the area where the SGL appears in SAR images, we utilized the combined maximum lake area from two consecutive melt seasons (May to September each year) in the optical image as the SAR extraction area [18,19].

This experiment adopted the concept of maximum lake area, which is the maximum area covered by the lake extracted in the optical image during one melt season. The possible lake extent between two melt seasons can be obtained by superimposing the maximum lake area for the two proximal years. Before making the masking area for SAR extraction, applying a buffer for maximum lake area was necessary due to slush surrounding some lakes having similar backscatter values to the lakes, thereby complicating the differentiation of slush and water [18]. The size of the radius of the buffer zone is determined by the size of the individual lakes. The final SAR masking area is obtained by superimposing the two-buffering maximum lake area of the proximal years. The maximum lake areas for 2020, 2021, and 2022 were used in this experiment.

(4) Lake extraction in SAR based on the OCO method: The SGLs were extracted from Sentinel-1 with HH and HV polarization. HH polarization was used for lake extraction of SAR during the melt season, and HV polarization was used to extract the lake

that was covered by snow during the non-melt season [40]. The experiment first clipped the pre-processed image of the SAR using the masked area of the SAR in (3) and calculated the input value of the Canny algorithm using OTSU for a single lake area, resulting in remarkable edge areas in the SAR image. The edge areas must be bordered in GEE. Second, fragmented remarkable edges of less than 10 pixels were removed, and the top 50% of the edge strength was selected as the strong boundary region. Third, a two-pixel buffer was made outward of the strong boundary region, which contained the two most varied classes in SAR images of a single lake mask. Finally, the buffered strong masked regions were used as the basis for the extraction of Otsu thresholds to obtain classification thresholds for the two classifications in SAR images. This threshold was used to separate the remarkable regions in the SAR image [18].

(5) Post-processing for lake area: After acquiring areas that differ considerably from the surrounding environment, whether the extracted area contains water (low backscattering coefficient) or slush (high backscattering coefficient) must be determined. Three regions of a single lake were selected for comparison in the experiment. The analyzed regions were the high backscatter region of the extracted region (classes g), the low backscatter region of the extracted region (classes l), and the backscattering intensity of the 30 pixels expanded around the extracted region (classes a). The experiment used the Jeffries–Matusita distance (JM distances) to discuss the differences between the three regions and to distinguish between areas of open water and slush, as well as partially misdivided areas without water (Equations (2) and (3)). The JM distance is an indicator of difference degree between classes. It ranges from 0 to 2, and the larger it is, the more considerable the difference between classes [41]. In our experimental statistics, we considered that a discernible difference existed between the classes when JM > 1 for the two types of features [42].

$$\text{Water}: \ \text{JM}(g,l) > 1 \cap \text{JM}(a,g) < 1 \cap \text{JM}(a,l) > 1 \tag{2}$$

$$\text{Slush}: \ \text{JM}(g,l) > 1 \cap \text{JM}(a,g) > 1 \cap \text{JM}(a,l) < 1 \tag{3}$$

JM(,) indicates the JM distance between two classes; g, l, and a indicate the three classes for distinction. They are the high backscatter region of the extracted region, the low backscatter region of the extracted region, and the backscattering intensity of the 30 pixels expanded around the extracted region, respectively. The primary objective of this approach is to assess if there is a real feature in the area obtained based on the OCO method, and if this feature is a lake. After determining the presence of lakes, areas smaller than 20 pixels were removed to reduce the effect of SAR noise, and the small hole in the water was clumped. The final processed data were the area of the SGLs extracted based on SAR images.

4. Results

4.1. Lake Detection

It is challenging to evaluate the performance of the OCO method due to the absence of ground truth. Here, we compared the OCO extraction results with the manual extraction results with the HH-polarized SAR image of 12 September 2021 in the study area (Figure 3). As shown in the four examples in Figure 3, the lake extracted from the SAR image by the OCO method matched the results of the manual interpretation. The lake acquired by the OCO method contained more scattered pixels—some of them were noise from the SAR imagery, and some of them were details of water that were not refined by the visual interpretation. There were also omissions in the OCO-extracted results, e.g., the obvious SGL in the middle of Figure 3d was not extracted with the OCO method.

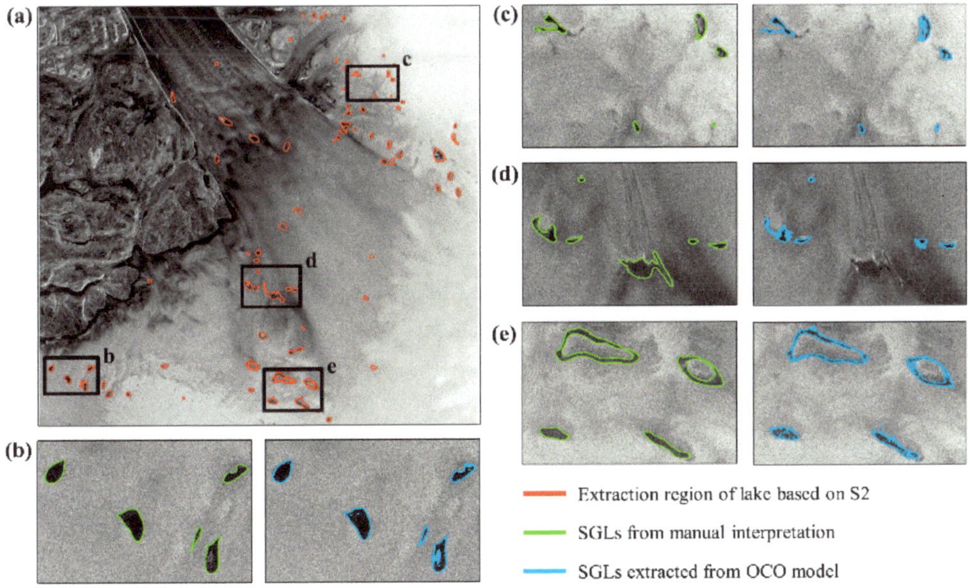

Figure 3. SGL extraction comparison between the OCO method (blue area) and manual interpretation (green area). The base image is a Sentinel-1 image of the study area on 12 September 2021 in HH polarization. (**a**) displays the optical extraction results of SGLs in the study area, while (**b**–**e**) illustrates four typical cases of SGLs extraction of manual and OCO model in SAR images.

Our study included 85 lake mask areas, and only two lake objects were not decoded by the OCO method from the 72 lakes that existed in the visual interpretation of this SAR image. In order to verify the accuracy of the OCO model, a confusion matrix based on the OCO extraction results and the manual extraction results was computed, and the calculated pixel range was the maximum lake area. By calculating the true positive (TP), false negative (FN), false positive (FP), and true negative (TN) rates in the confusion matrix, the performance of the algorithm could be measured as follows: overall accuracy was 0.885, precision was 0.929, recall was 0.749, the F1 score was 0.829, and the Kappa coefficient was 0.745.

To compare the differences in lake extraction between optical and SAR images in the melt season, we extracted 12 typical SGLs from optical and SAR images of the study area on 26 July 2021, respectively, with the SAR image using HH polarization. Lake 6 results were supplemented with the 24 July SAR image as the 26 July SAR image could not cover all 12 lake objects. Table 1 shows the SGL extraction results of the optical image using $NDWI_{ice}$, the SGL extraction results of the SAR image based on the OCO method, and the SGL area of visual interpretation based on SAR images. For the 12 SGLs on 26 July 2021 (supplemented on 24 July), six SGLs were extracted from the optical image, appearing in Regions A and B; 12 SGLs were obtained in SAR-based manual extraction; and 11 SGLs were obtained in the SAR-based OCO method, for which Lake 3 was not extracted. Comparing the SGL extraction results of optical and SAR images showed that optical and SAR images could extract the lakes in Regions A and B, but only the SAR image could extract the lake in Region C, which was completely covered by snow. The water extraction area of the optical image was generally larger than that of the SAR image, with a difference basically below 0.3 km^2. Region B showed a greater difference between optical and SAR water extraction areas than Region A did. Except for Lakes 7 and 12 that had a larger difference, the difference between the lake area acquired from the automatic OCO method and the manually mapped one was around 20%.

Table 1. Comparison of SGL area extraction based on Sentinel-1 and Sentinel-2. The three columns in JM distance represent the pairwise differences between classes a, g, and l (Equations (2) and (3)).

Region	Lake ID	Sentinel 2 Area	Sentinel 1				
			Manual Area	OCO Method Area	JM-Distance		
					g&l	a&g	a&l
A	1	1.153	1.052	1.235	1.487	0.07	1.352
	2	0.969	0.959	0.953	1.511	0.013	1.502
	3	0.222	0.222	NA	1.096	0.092	0.676
	4	0.813	0.845	0.776	1.86	0.07	1.939
B	5	0.494	0.215	0.250	1.677	0.002	1.712
	6	0.901	0.660	0.581	1.822	0.013	1.891
C	7	NA	0.241	0.215	1.855	0.463	1.995
	8	NA	0.204	0.108	1.659	0.569	1.959
	9	NA	2.446	2.163	1.988	0.122	1.998
	10	NA	0.228	0.177	1.88	0.187	1.981
	11	NA	0.261	0.244	1.886	0.13	1.974
	12	NA	0.137	0.197	1.321	0.228	1.644

Lake 3 was not successfully extracted when determining the existence of a lake in SAR images based on the JM distance. The JM distance between the backscattering intensity around the lake and inside the lake for Lake 3 was small, i.e., only 0.676, which did not satisfy the judgment that JM distance > 1. The other 11 lakes' JM distances were all in the range of 1.3 to 1.9, with good separation.

4.2. Full-Year SGLs Changes in 2021

The experiment monitored and extracted monthly SGL data from January to December in 2021. The results included optical and SAR-extracted data for 12 typical lakes, which were divided into three regions (Regions A, B, and C) for separate discussions (Figure 4). For SAR images, the data for May, June, July, August, and September were in HH polarization; and October, November, December, the following year's January, February, March, and April were in HV polarization.

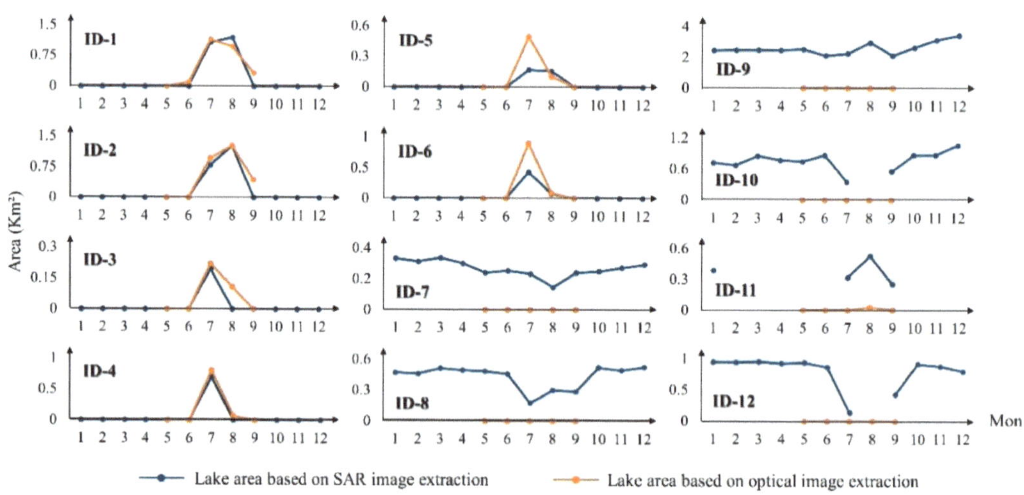

Figure 4. Monthly area changes of 12 SGLs in 2021 from SAR and optical images. The SAR-extracted area breakpoints (Lake ID-9, ID-10, ID-11, and ID-12) originated from areas of the lake that were visually discernible but could not be extracted by the OCO method.

The SGL changes were similar in Regions A and B. The lakes in these areas developed and drained during the melt season, with the peak occurring in July and August. The development of the SGL was well monitored by optical and SAR imagery during the melt season, with the same trends in the two types of imagery. The SAR and optical extraction results were slightly different, with the area extracted from SAR being smaller than that extracted from the optical images.

Large differences were observed between Region C and Regions A and B. The optical images only detected the area of three SGLs (Lake 10, 11, and 12) in August. However, the extraction of the SAR images showed that all six lakes in Region C were already present in January 2021, and only a slight change in area occurred from January to June for those lakes. For Region C, SAR imagery was slightly effective in detecting images during the melt season, especially in August when water was difficult to detect completely. The SGLs were still detectable in SAR imagery in September, October, November, and December after the melt season. The change in lake area after the melt season from the pre-melt season was small, with less than 25%. In particular, the area of Lakes 8 and 12 changed by less than 10% before and after the melt season. The experiments also compared the results from May to September in Region C for HH and HV polarization, and the area difference between the two polarizations was in the range of 15–20%, with no remarkable improvement in the extracted lakes.

4.3. Lake Drainage Monitoring

To discuss the time series changes of the lake at a finer scale, Lakes ID-4 and ID-5 were selected for the experiment to explore the changes in storage and discharge from 1 July to 31 August 2021. The experiment combined SAR and optical data. Fewer optical images were available, with only July 26 available in July and seven date images available in August. SAR imagery was denser than the optical imagery over the area of the studied SGLs, with 10 and 11 images for July and August, respectively.

Figure 5 excerpts the Sentinel-1 HH polarization images of Lakes ID-4 and ID-5 at several typical time points during the storage and drainage period. Lake ID-4 experienced a dramatic drainage event from 26 to 30 July, accompanied by the production of slush around the lake. The change in area of Lake ID-5 was more moderate, with drainage occurring from 30 July to 7 August. Lake ID-5 was scattered in extent at the beginning of the melting, with the water gradually pooling into a whole lake during the storage process.

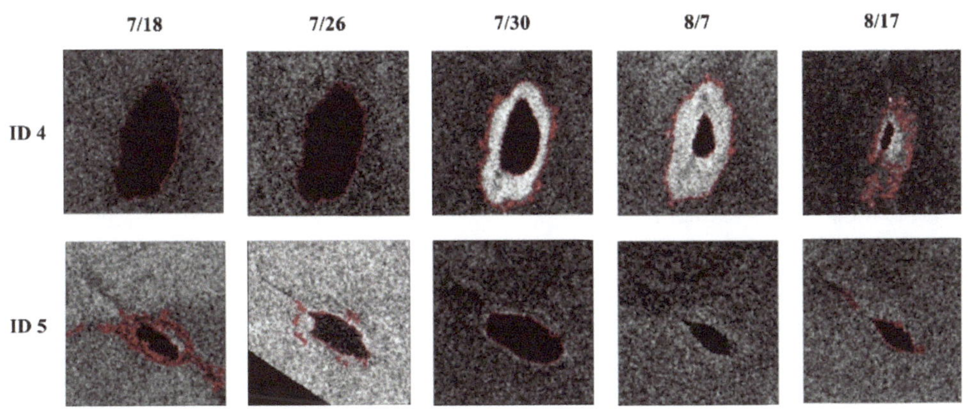

Figure 5. Sentinel-1 HH polarization images of typical lake (Lakes 4 and 5) drainage.

The changes of water area occurring in July and August 2021 for two typical SGLs, ID-4 and ID-5, are shown in Figure 6. Lake ID-5 began to accumulate water on July 18, reaching the maximum area of water detected by SAR on 18 and 30 July, and the water completely disappeared on 25 August. Between 30 July and 5 August, the lake underwent a relatively strong drainage event, with the water area dropping from 0.343 km^2 to 0.108 km^2. The area of Lake ID-5 remained around 0.1 km^2 after 5 August until water could not be extracted in the SAR in 23 August. At the beginning of the melt, on 18 July, Lake ID-5 was monitored in SAR to a larger area of 0.345 km^2. During this period, the SGL was in the early stages of catchment, and some shallow meltwater was found, all of which was included in the SAR imagery as water areas.

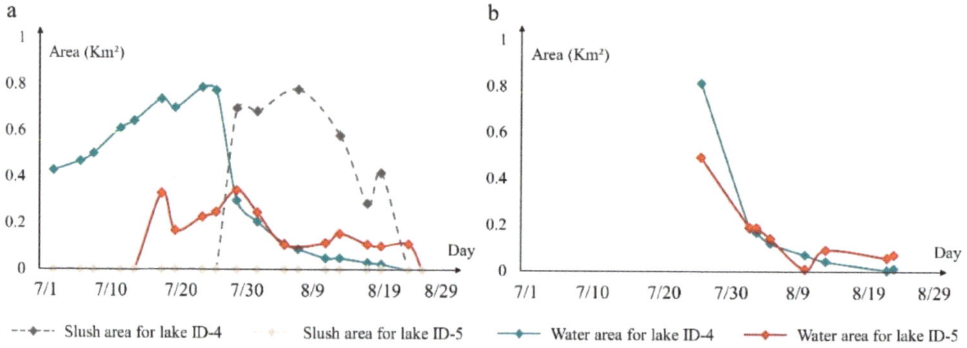

Figure 6. Water and slush area variation of the typical lake (Lakes 4 and 5) drainage based on SAR (**a**) and optical (**b**) extraction.

Drainage was more pronounced at Lake ID-4 than at ID-5. The results of the area change showed that Lake ID-4 was drained from 26 to 30 July, with the area dropping from 0.776 km^2 to 0.300 km^2, a 61.3% reduction in area over the four days. Extensive slush formed from the drainage of the SGL could be clearly detected in the SAR image on 30 July (Figure 6a). After 30 July, the area of the SGL continued to decline and completely disappeared on August 23. During the drainage period, the area of slush also changed. From 30 July to 7 August, the slush area tended to increase, reaching a maximum area of around 0.7 km^2. After 7 August, the slush area decreased considerably until it disappeared.

5. Discussion

5.1. Advantages of Multisource Remote Sensing

Optical and SAR imagery could be effectively used to extract information about SGLs. However, given the characteristics of optical and SAR imaging, flaws and differences were found in the extraction results between the two types. The area extracted from $NDWI_{ice}$-based optical images of the lake was simple and straightforward, but the quality of some images was poor because of the weather and snow cover, resulting in fewer data sources being available for SGL extraction. SAR satellites have better penetration and are effective in avoiding cloud effects on images and detecting thin snow-covered lakes. C-band SAR images are generally not hindered by atmospheric effects and are capable of imaging through tropical clouds and rain showers. Its penetration capability with snow on the ground is limited, usually around 1 to 2 m [43], and the specific depth of snow penetration is related to the physical properties of the snow and the radar's incidence angle [44].

Optical images can only identify exposed lakes on the ice sheet and cannot detect the lake covered by ice floes and thin snow. This kind of buried lake is widespread in the GrIS at all times of the year in melt and non-melt seasons. SAR imagery can detect buried lakes covered by shallow snow partially, but lakes buried under a thicker ice cover are difficult to detect from SAR imagery. For monitoring validation of buried lakes, snow radar data collected by NASA's OIB were chosen. Snow radar data have greater penetration and can be used to estimate the cross section of glaciers. Figure 7a shows the echogram results obtained by a snow radar on 22 March 2017, when lakes had no optical visibility.

Comparing Sentinel-1 results from 23 March of the same year showed that Sentinel-1 could detect well the water located under the ice surface. The monitored water was covered with snow at a thickness of about 3 m (Figure 7b,c). The comparison showed that the SAR-based approach to lake extraction expanded the data for buried lakes, thus providing additional data support for a comprehensive analysis of the seasonal variability of SGLs.

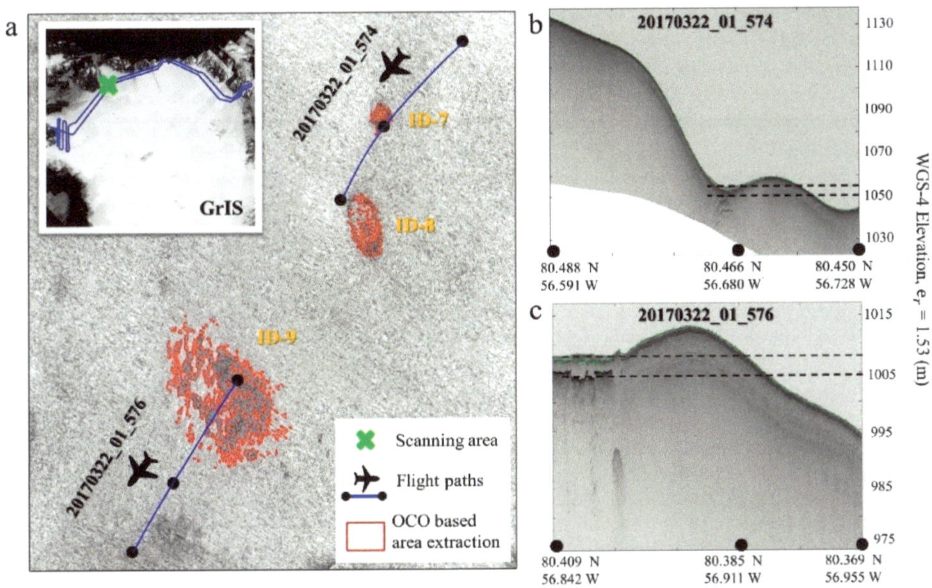

Figure 7. Snow radar data validation of SAR-extracted SGL results. A snow radar passed through the buried lake extraction location in this study area, and the base image was a Sentinel-1 HH polarization image on 22 March 2017 (**a**). The snow radar echogram results of the two flight paths show the presence of water in these regions (**b**,**c**).

This study used Sentinel-1 and Sentinel-2 images to monitor SGL transformations during the melt and non-melt seasons. The experimental method combined the characteristics of SAR imagery with optical imagery, addressing the limitations of optical images in winter and the greater environmental noise of SAR images in part. Additionally, it is important to note that due to the limited penetration capability of C-band SAR imagery, there may be more buried lakes in areas where C-band SAR signals cannot reach [45].

5.2. Factors Affecting Water Extraction in SAR

HH and HV polarization are two common types of polarization used in SAR imagery, and the choice of polarization can affect the effectiveness of lake extraction [46,47]. HV polarization often shows a better contrast between water and snow in areas of buried water. The reason for this situation is that HH polarization is more susceptible to the radar signal with the top snow layers, resulting in less information acquired in HH polarization than in HV polarization for buried lakes [32]. For open water, the backscattering in HH polarization is more stable than in HV backscattering [48]. Hence, HH polarization can distinguish open-water and high-water content snow in more cases than HV polarization can.

SAR imagery has the advantage of being available in all weather and being able to penetrate thin snow; however, it is susceptible to external factors, such as the angle of incidence, the melting state of the ice surface, wind, etc., when detecting SGLs on the ice sheet [18,49,50]. These external influences can lead to variations in the backscatter intensity of the SAR image, resulting in the inability to extract lakes from a SAR image. Many anomalies in the extraction of SGLs from SAR imagery occurred in this experiment. For example, during lake extraction of month-by-month SAR imagery for 2021, the SAR image in August of Region C was poorly imaged, with the entire zone exhibiting lower backscatter intensity (Figure 4). During drainage monitoring of Lakes ID-4 and ID-5 in late July and early August, no information was obtained for Lake ID-5 in the 7 August SAR image (Figure 5). Furthermore, in lake detection of Section 4.1, Lake ID-3 was not identified due to the poor separation (JM(a, l) = 0.676) of the lake and surrounding area backscatter intensities (Table 1). In summary, the most important reason for not being able to extract water from SAR images is that water has a backscatter intensity similar to that of the surrounding environment, making it difficult to distinguish water from the environment in SAR images.

Some differences were also found between the optical results and the SAR results in lake extraction at nearly the same time. We compared the SGLs in Regions A and B, which could be extracted from the optical image (imaging time 18:10 UTM) and SAR (imaging time 11:36 UTM) in the study area on 26 July 2021 (Figure 8), and the results showed that the difference was mainly in Lake ID-5, with a 2.30-fold difference in extraction between the two types of images. We calculated the water depth distribution of Lake ID-5 using the water depth formula for optical imagery [12,51] and found that the main difference in water extraction area between optical and SAR imagery came from shallow water areas, mainly below <0.75 m. A similar situation occurred for Lake ID-3, which was not identified in the SAR image. The SAR lake extraction area of some of the lakes was also found to be slightly smaller than the optical lake extraction area in Section 4.3 (Figure 4). On the basis of these results, we inferred that the depth of the open lake would influence the imaging results of the water area in the SAR [18]. The temporal difference between Sentinel-1 and Sentinel-2 images in this test could also be the cause of this error.

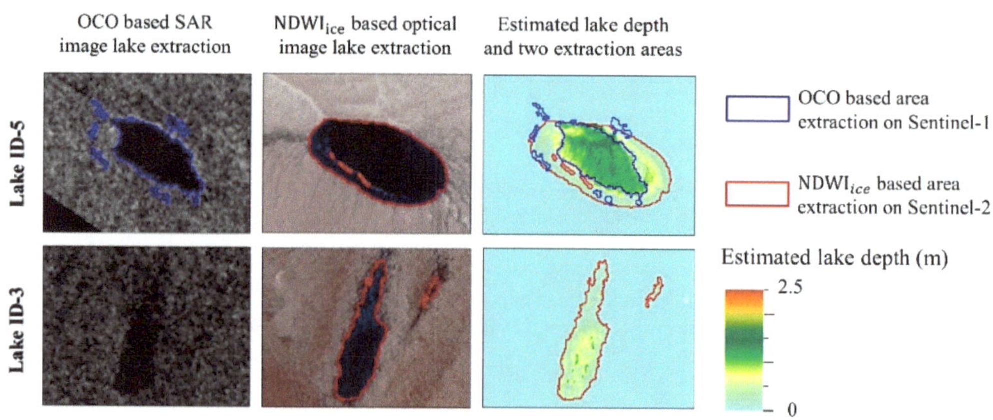

Figure 8. Comparison of area differences in lake extraction with depth distribution. Sentinel-1 (HH polarization) and Sentinel-2 (true color display) images come from 26 July 2021.

SGL extraction in SAR is influenced by external natural factors and the lake's own properties, which may lead to some challenges in accurate SAR-based lake area extraction for time series.

5.3. Drainage Monitoring

Optical image-based SGLs drainage has been widely used in SGL change detection [52]. In Section 4.3, we monitored the drainage process of two lakes, i.e., Lakes ID-4 and ID-5, from July to August 2021. Lakes ID-4 and ID-5 had two types of drainage processes, fast and slow, respectively (Figures 5 and 6). Although rapid drainage of SGLs could also be detected using the optical images in the experiment (Figure 6b), the limited amount of available optical images did not allow for a better reproduction of SGL development. Rapid drainage of SGLs in SAR is often accompanied by a sudden increase in the intensity of backscatter around the lake (Figure 5). The reason for this phenomenon is that rapid drainage leaves behind some rough heterogeneous surface consisting of patches of slush and ice, producing high backscatter values. Over time, these slush and ice blocks smooth out, and the backscatter values match those of the surrounding environment. Thus, rapid drainage can also be monitored by high backscatter from lake anomalies, but this is not always featured in the imagery [18].

During optical and SAR image monitoring of the SGLs in Region C in 2021, the SGLs were barely visible in the optical images of Region C, with only a small area showing during the peak melt season (Figure 4). However, the detection of SGLs in SAR imagery revealed that the lake was always present in Region C and that some of the lakes did not change considerably in area before and after the melt season (Figure 4). The change of area of the lake in Region C was less than 25% before and after the melt season and even less than 10% in Lakes 8 and 12 due to the presence of buried lakes. Some lakes do not drain nor freeze completely as they enter winter; they are stored as liquid water under the frozen snow and ice layers, forming a buried lake [53]. Therefore, it is ambiguous to judge SGL drain by the area change in the optical image. It may underestimate the actual area of water stored on the ice sheet and overestimate the drainage of SGLs at the end of the melt season.

5.4. Limitations and Prospects

SGL extraction results in SAR images based on the OCO method in the experiment are always dichotomous; however, the boundary between the backscattering intensities of water and non-water in a region is challenging to characterize with a single threshold. At the same time, the availability of SAR imagery is not completely established and is affected by factors, which can also affect the results of the time-dependent detection of SGLs. Therefore, we will analyze how to maximize SAR images and reduce the noise associated with threshold extraction methods in following research. The main purpose of this experiment was to verify the feasibility of the OCO method and the importance of SAR imagery in SGL monitoring. In future studies, the study area will be expanded, and SGL changes across the cryosphere will be discussed by fusing the time series of multisource remote sensing data.

6. Conclusions

SGL changes are an important monitoring object of ice sheets that reflects the volume change of ice sheet surface melting. With the development of remote sensing technology, the monitoring of SGL changes based on remote sensing observation has been gradually improved. Optical and SAR remote sensing has been widely used in the detection of SGLs. However, given the vulnerability of optical imagery to the weather, the amount of optical data that can be practically applied is limited. Meanwhile, information on buried lakes is not available in winter from optical imagery. Although SAR has good penetration and a large volume of data, it presents a backscattered intensity signal with a dense "homospectral foreign object" situation over a large area. Effective extraction of SGL information from optical and SAR imagery allows improved monitoring of SGL development throughout the year.

This study proposed a new model named the OCO method. It combines optical and SAR satellites and the Otsu algorithm and Canny edge detector to investigate the development of SGLs throughout the year. Optical images can locate SGLs more accurately, and SAR images have a degree of penetration of clouds and snow. Based on optical and SAR data, the changes in the SGLs can be better monitored throughout the year. In terms of extraction methods, $NDWI_{ice}$ is used to extract the lake area from the optical images, and the OCO method is used to extract the SGLs' area from the SAR images masked by the maximum lake area of optical images. The OCO method performs feature extraction for each individual SGL based on the Otsu–Canny edge detection algorithm and the Canny edge Otsu algorithm, and it is built to be more objective and targeted.

The experiment validated the accuracy of the OCO method for SGLs extraction, together with monitoring the full-year variations of typical lakes based on Sentinel-1 and Sentinel-2 images in 2021. Benchmarked against the manual SGLs extraction result, the OCO method performed well overall accuracy of 0.885 and F1 score of 0.829. In the OCO method for SGL extraction, SAR images can be affected by external natural factors, such as incidence, surface melting, wind, and other factors, resulting in the inability to extract SGLs properly. The difference between the SAR results of extracting SGLs based on the OCO method and the visual interpretation was around 20%. During the same period, the depth of the lake may influence the extraction range of SGLs in different types of imagery. Experimental results indicated that shallow lakes were challenging to observe in SAR images (<0.75 m in this experiment). SAR imagery had higher data density in the study area than optical imagery did and could monitor SGL development during the non-melt season, which was the result that could not be acquired with optical imagery. For the 12 typical SGLs in the study area in 2021, some of the lakes were stored and drained during the melt season (Region A and B), but some SGLs did not change considerably before and after the melt season (Region C). The SGLs in Region C, although no longer detectable on optical imagery during the non-melt season, were consistently detectable in SAR imagery with 10–25% changes in area. Validation of the snow radar data revealed the presence of buried lakes under ice in this area. As SAR imagery had more intensive data in the study area than optical imagery did, this experiment also detected a rapid drainage event of the SGLs over a four-day period.

Combining optical and SAR imagery for monitoring changes allows for better discussion of the annual and seasonal variation of SGLs, allowing for analysis of hydrological events with more intensive time series. The OCO method proposed in this study can combine the advantages of optical and SAR imagery and consider the buried lakes during winter storage of water. This approach avoids the overestimation of the discharge of the SGLs after the melt season that would result from monitoring the lake using optical images only.

Author Contributions: Conceptualization, D.Z. and C.Z. (Chunxia Zhou); Methodology, D.Z., C.Z. (Chunxia Zhou) and C.Z. (Ce Zhang); Software, Y.Z.; Validation, Y.Z.; Investigation, Y.Z. and T.W.; Resources, C.Z. (Ce Zhang); Data curation, T.W.; Writing—original draft, D.Z.; Writing—review & editing, C.Z. (Chunxia Zhou), Y.Z. and C.Z. (Ce Zhang); Supervision, C.Z. (Ce Zhang); Funding acquisition, C.Z. (Chunxia Zhou) and C.Z. (Ce Zhang). All authors have read and agreed to the published version of the manuscript.

Funding: This work was supported by the Natural Environment Research Council (grant number NE/T004002/1) and was partially funded by the National Key Research and Development Program of China (2021YFC2803303) and the National Natural Science Foundation of China (NSFC) (42171133, 41776200).

Data Availability Statement: Data available on request due to restrictions privacy. The data presented in this study are available on request from the first author.

Conflicts of Interest: The authors declare no conflict of interest.

References

1. Van Wychen, W.; Hallé, D.A.M.; Copland, L.; Gray, L. Anomalous Surface Elevation, Velocity, and Area Changes of Split Lake Glacier, Western Prince of Wales Icefield, Canadian High Arctic. *Arct. Sci.* **2022**, *8*, 1288–1304. [CrossRef]
2. Van Wychen, W.; Bayer, C.; Copland, L.; Brummell, E.; Dow, C. Radarsat Constellation Mission Derived Winter Glacier Velocities for the St. Elias Icefield, Yukon/Alaska: 2022 and 2023. *Can. J. Remote Sens.* **2023**, *49*, 2264395. [CrossRef]
3. Hogg, A.E.; Shepherd, A.; Gourmelen, N.; Engdahl, M. Grounding Line Migration from 1992 to 2011 on Petermann Glacier, North-West Greenland. *J. Glaciol.* **2016**, *62*, 1104–1114. [CrossRef]
4. Zhu, Y.; Zhou, C.; Zhu, D.; Wang, T.; Zhang, T. Interannual Variation of Landfast Ice Using Ascending and Descending Sentinel-1 Images from 2019 to 2021: A Case Study of Cambridge Bay. *Remote Sens.* **2023**, *15*, 1296. [CrossRef]

5. Pattyn, F.; Ritz, C.; Hanna, E.; Asay-Davis, X.; DeConto, R.; Durand, G.; Favier, L.; Fettweis, X.; Goelzer, H.; Golledge, N.R.; et al. The Greenland and Antarctic Ice Sheets under 1.5 °C Global Warming. *Nat. Clim. Chang.* **2018**, *8*, 1053–1061. [CrossRef]
6. Mouginot, J.; Rignot, E.; Bjørk, A.A.; van den Broeke, M.; Millan, R.; Morlighem, M.; Noël, B.; Scheuchl, B.; Wood, M. Forty-Six Years of Greenland Ice Sheet Mass Balance from 1972 to 2018. *Proc. Natl. Acad. Sci. USA* **2019**, *116*, 9239–9244. [CrossRef]
7. Slater, T.; Shepherd, A.; McMillan, M.; Leeson, A.; Gilbert, L.; Muir, A.; Munneke, P.K.; Noël, B.; Fettweis, X.; van den Broeke, M.; et al. Increased Variability in Greenland Ice Sheet Runoff from Satellite Observations. *Nat. Commun.* **2021**, *12*, 1–9. [CrossRef]
8. Hoffman, M.J.; Perego, M.; Andrews, L.C.; Price, S.F.; Neumann, T.A.; Johnson, J.V.; Catania, G.; Lüthi, M.P. Widespread Moulin Formation During Supraglacial Lake Drainages in Greenland. *Geophys. Res. Lett.* **2018**, *45*, 778–788. [CrossRef]
9. Abdalati, W.; Steffen, K. Passive Microwave-derived Snow Melt Regions on the Greenland Ice Sheet. *Geophys. Res. Lett.* **1995**, *22*, 787–790. [CrossRef]
10. Leeson, A.A.; Shepherd, A.; Briggs, K.; Howat, I.; Fettweis, X.; Morlighem, M.; Rignot, E. Supraglacial Lakes on the Greenland Ice Sheet Advance Inland under Warming Climate. *Nat. Clim. Chang.* **2015**, *5*, 51–55. [CrossRef]
11. Zhu, D.; Zhou, C.; Zhu, Y.; Peng, B. Evolution of Supraglacial Lakes on Sermeq Avannarleq Glacier, Greenland Using Google Earth Engine. *J. Hydrol. Reg. Stud.* **2022**, *44*, 101246. [CrossRef]
12. Banwell, A.F.; Willis, I.C.; Macdonald, G.J.; Goodsell, B.; MacAyeal, D.R. Direct Measurements of Ice-Shelf Flexure Caused by Surface Meltwater Ponding and Drainage. *Nat. Commun.* **2019**, *10*, 730. [CrossRef] [PubMed]
13. Banwell, A.F.; MacAyeal, D.R.; Sergienko, O.V. Breakup of the Larsen B Ice Shelf Triggered by Chain Reaction Drainage of Supraglacial Lakes. *Geophys. Res. Lett.* **2013**, *40*, 5872–5876. [CrossRef]
14. Lemos, A.; Shepherd, A.; McMillan, M.; Hogg, A.E. Seasonal Variations in the Flow of Land-Terminating Glaciers in Central-West Greenland Using Sentinel-1 Imagery. *Remote Sens.* **2018**, *10*, 1878. [CrossRef]
15. Williamson, A.G.; Banwell, A.F.; Willis, I.C.; Arnold, N.S. Dual-Satellite (Sentinel-2 and Landsat 8) Remote Sensing of Supraglacial Lakes in Greenland. *Cryosphere* **2018**, *12*, 3045–3065. [CrossRef]
16. Dunmire, D.; Banwell, A.F.; Wever, N.; Lenaerts, J.T.M.; Datta, R.T. Contrasting Regional Variability of Buried Meltwater Extent over 2 Years across the Greenland Ice Sheet. *Cryosphere* **2021**, *15*, 2983–3005. [CrossRef]
17. Ghiasi, Y.; Duguay, C.R.; Murfitt, J.; Asgarimehr, M.; Wu, Y. Potential of GNSS-R for the Monitoring of Lake Ice Phenology. *IEEE J. Sel. Top. Appl. Earth Obs. Remote Sens.* **2023**, *17*, 660–673. [CrossRef]
18. Miles, K.E.; Willis, I.C.; Benedek, C.L.; Williamson, A.G.; Tedesco, M. Toward Monitoring Surface and Subsurface Lakes on the Greenland Ice Sheet Using Sentinel-1 SAR and Landsat-8 OLI Imagery. *Front. Earth Sci.* **2017**, *5*, 58. [CrossRef]
19. Benedek, C.; Willis, I. Winter Drainage of Surface Lakes on the Greenland Ice Sheet from Sentinel-1 SAR Imagery. *Cryosph. Discuss.* **2020**, *15*, 1587–1606. [CrossRef]
20. Poinar, K.; Andrews, L. Challenges in Predicting Greenland Supraglacial Lake Drainages at the Regional Scale. *Cryosphere* **2021**, *15*, 1455–1483. [CrossRef]
21. Jiang, L.; Ma, Y.; Chen, F.; Liu, J.; Yao, W.; Shang, E. Automatic High-Accuracy Sea Ice Mapping in the Arctic Using Modis Data. *Remote Sens.* **2021**, *13*, 550. [CrossRef]
22. Li, L.; Chen, Z.; Zheng, L.; Cheng, X. Extraction of Greenland Ice Sheet Buried Lakes Using Multi-Source Remote Sensing Data: With Application to the Central West Basin of Greenland. *Chin. J. Geophys.* **2022**, *65*, 3818–3828.
23. Kolli, M.K.; Opp, C.; Karthe, D.; Pradhan, B. Automatic Extraction of Large-Scale Aquaculture Encroachment Areas Using Canny Edge Otsu Algorithm in Google Earth Engine–the Case Study of Kolleru Lake, South India. *Geocarto Int.* **2022**, *37*, 11173–11189. [CrossRef]
24. Sahoo, P.K.; Soltani, S.; Wong, A.K.C. A Survey of Thresholding Techniques. *Comput. Vis. Graph. Image Process.* **1988**, *41*, 233–260. [CrossRef]
25. Sari, Y.; Prakoso, P.B.; Baskara, A.R. Road Crack Detection Using Support Vector Machine (SVM) and OTSU Algorithm. In Proceedings of the 2019 6th International Conference on Electric Vehicular Technology (ICEVT), Bali, Indonesia, 18–21 November 2019; pp. 349–354. [CrossRef]
26. Tang, T.; Xiang, D.; Liu, H.; Su, Y. A New Local Feature Extraction in SAR Image. In Proceedings of the IEEE 2013 Asia-Pacific Conference on Synthetic Aperture Radar (APSAR), Tsukuba, Japan, 23–27 September 2013; pp. 377–379.
27. Elwan, M.; Amein, A.S.; Mousa, A.; Ahmed, A.M.; Bouallegue, B.; Eltanany, A.S. SAR Image Matching Based on Local Feature Detection and Description Using Convolutional Neural Network. *Secur. Commun. Netw.* **2022**, *2022*, 5669069. [CrossRef]
28. Gayathri Monicka, S.; Manimegalai, D.; Karthikeyan, M. Detection of Microcracks in Silicon Solar Cells Using Otsu-Canny Edge Detection Algorithm. *Renew. Energy Focus* **2022**, *43*, 183–190. [CrossRef]
29. Singh, G.; Reynolds, C.; Byrne, M.; Rosman, B. A Remote Sensing Method to Monitor Water, Aquatic Vegetation, and Invasive Water Hyacinth at National Extents. *Remote Sens.* **2020**, *12*, 4021. [CrossRef]
30. Yang, P.; Song, W.; Zhao, X.; Zheng, R.; Qingge, L. An Improved Otsu Threshold Segmentation Algorithm. *Int. J. Comput. Sci. Eng.* **2020**, *22*, 146–153. [CrossRef]
31. MacDonald, G.J.; Banwell, A.F.; Macayeal, D.R. Seasonal Evolution of Supraglacial Lakes on a Floating Ice Tongue, Petermann Glacier, Greenland. *Ann. Glaciol.* **2018**, *59*, 56–65. [CrossRef]
32. Schröder, L.; Neckel, N.; Zindler, R.; Humbert, A. Perennial Supraglacial Lakes in Northeast Greenland Observed by Polarimetric Sar. *Remote Sens.* **2020**, *12*, 2798. [CrossRef]

33. Koenig, L.S.; Ivanoff, A.; Alexander, P.M.; MacGregor, J.A.; Fettweis, X.; Panzer, B.; Paden, J.D.; Forster, R.R.; Das, I.; McConnell, J.R.; et al. Annual Greenland Accumulation Rates (2009–2012) from Airborne Snow Radar. *Cryosphere* **2016**, *10*, 1739–1752. [CrossRef]
34. Arthur, J.F.; Stokes, C.R.; Jamieson, S.S.R.; Rachel Carr, J.; Leeson, A.A. Distribution and Seasonal Evolution of Supraglacial Lakes on Shackleton Ice Shelf, East Antarctica. *Cryosphere* **2020**, *14*, 4103–4120. [CrossRef]
35. Yang, K.; Smith, L. Supraglacial Streams on the Greenland Ice Sheet Delineated from Combined Spectral-Shape Information in High-Resolution Satellite Imagery. *IEEE Geosci. Remote Sens. Lett.* **2013**, *10*, 801–805. [CrossRef]
36. Arthur, J.F.; Stokes, C.; Jamieson, S.S.R.; Carr, J.R.; Leeson, A.A. Recent Understanding of Antarctic Supraglacial Lakes Using Satellite Remote Sensing. *Prog. Phys. Geogr.* **2020**, *44*, 837–869. [CrossRef]
37. Mullissa, A.; Vollrath, A.; Odongo-Braun, C.; Slagter, B.; Balling, J.; Gou, Y.; Gorelick, N.; Reiche, J. Sentinel-1 Sar Backscatter Analysis Ready Data Preparation in Google Earth Engine. *Remote Sens.* **2021**, *13*, 1954. [CrossRef]
38. Markert, K.N.; Markert, A.M.; Mayer, T.; Nauman, C.; Haag, A.; Poortinga, A.; Bhandari, B.; Thwal, N.S.; Kunlamai, T.; Chishtie, F.; et al. Comparing Sentinel-1 Surface Water Mapping Algorithms and Radiometric Terrain Correction Processing in Southeast Asia Utilizing Google Earth Engine. *Remote Sens.* **2020**, *12*, 2469. [CrossRef]
39. Li, Y.; Yang, K.; Gao, S.; Smith, L.C.; Fettweis, X.; Li, M. Surface Meltwater Runoff Routing through a Coupled Supraglacial-Proglacial Drainage System, Inglefield Land, Northwest Greenland. *Int. J. Appl. Earth Obs. Geoinf.* **2022**, *106*, 102647. [CrossRef]
40. Watanabe, M.; Koyama, C.N.; Hayashi, M.; Nagatani, I.; Tadono, T.; Shimada, M. Refined Algorithm for Forest Early Warning System with ALOS-2/PALSAR-2 ScanSAR Data in Tropical Forest Regions. *Remote Sens. Environ.* **2021**, *265*, 112643. [CrossRef]
41. Dabboor, M.; Howell, S.; Shokr, M.; Yackel, J. The Jeffries–Matusita Distance for the Case of Complex Wishart Distribution as a Separability Criterion for Fully Polarimetric SAR Data. *Int. J. Remote Sens.* **2014**, *35*, 6859–6873.
42. Lohse, J.; Doulgeris, A.P.; Dierking, W. Incident Angle Dependence of Sentinel-1 Texture Features for Sea Ice Classification. *Remote Sens.* **2021**, *13*, 552. [CrossRef]
43. Zheng, L.; Li, L.; Chen, Z.; He, Y.; Mo, L.; Chen, D.; Hu, Q.; Wang, L.; Liang, Q.; Cheng, X. Multi-Sensor Imaging of Winter Buried Lakes in the Greenland Ice Sheet. *Remote Sens. Environ.* **2023**, *295*, 113688. [CrossRef]
44. Murfitt, J.; Duguay, C.R.; Picard, G.; Gunn, G.E. Investigating the Effect of Lake Ice Properties on Multifrequency Backscatter Using the Snow Microwave Radiative Transfer Model. *IEEE Trans. Geosci. Remote Sens.* **2022**, *60*, 1–23. [CrossRef]
45. Koenig, L.S.; Lampkin, D.J.; Montgomery, L.N.; Hamilton, S.L.; Turrin, J.B.; Joseph, C.A.; Moutsafa, S.E.; Panzer, B.; Casey, K.A.; Paden, J.D.; et al. Wintertime Storage of Water in Buried Supraglacial Lakes across the Greenland Ice Sheet. *Cryosphere* **2015**, *9*, 1333–1342. [CrossRef]
46. Murashkin, D.; Spreen, G.; Huntemann, M.; Dierking, W. Method for Detection of Leads from Sentinel-1 SAR Images. *Ann. Glaciol.* **2018**, *59*, 124–136. [CrossRef]
47. Chen, X.; Li, G.; Chen, Z.; Ju, Q.; Cheng, X. Incidence Angle Normalization of Dual-Polarized Sentinel-1 Backscatter Data on Greenland Ice Sheet. *Remote Sens.* **2022**, *14*, 5534. [CrossRef]
48. Kundu, S.; Chakraborty, M. Delineation of Glacial Zones of Gangotri and Other Glaciers of Central Himalaya Using RISAT-1 C-Band Dual-Pol SAR. *Int. J. Remote Sens.* **2015**, *36*, 1529–1550. [CrossRef]
49. Bioresita, F.; Puissant, A.; Stumpf, A.; Malet, J.P. Fusion of Sentinel-1 and Sentinel-2 Image Time Series for Permanent and Temporary Surface Water Mapping. *Int. J. Remote Sens.* **2019**, *40*, 9026–9049. [CrossRef]
50. Gulácsi, A.; Kovács, F. Sentinel-1-Imagery-Based High-Resolutionwater Cover Detection on Wetlands, Aided by Google Earth Engine. *Remote Sens.* **2020**, *12*, 1614. [CrossRef]
51. Moussavi, M.S.; Abdalati, W.; Pope, A.; Scambos, T.; Tedesco, M.; MacFerrin, M.; Grigsby, S. Derivation and Validation of Supraglacial Lake Volumes on the Greenland Ice Sheet from High-Resolution Satellite Imagery. *Remote Sens. Environ.* **2016**, *183*, 294–303. [CrossRef]
52. Legleiter, C.J.; Tedesco, M.; Smith, L.C.; Behar, A.E.; Overstreet, B.T. Mapping the Bathymetry of Supraglacial Lakes and Streams on the Greenland Ice Sheet Using Field Measurements and High-Resolution Satellite Images. *Cryosphere* **2014**, *8*, 215–228. [CrossRef]
53. Law, R.; Arnold, N.; Benedek, C.; Tedesco, M.; Banwell, A.; Willis, I. Over-Winter Persistence of Supraglacial Lakes on the Greenland Ice Sheet: Results and Insights from a New Model. *J. Glaciol.* **2020**, *66*, 362–372. [CrossRef]

remote sensing

MDPI

Article

Photogrammetric Monitoring of Rock Glacier Motion Using High-Resolution Cross-Platform Datasets: Formation Age Estimation and Modern Thinning Rates

Tyler M. Meng [1,*], Roberto Aguilar [1], Michael S. Christoffersen [2], Eric I. Petersen [2], Christopher F. Larsen [2], Joseph S. Levy [3] and John W. Holt [1,4]

1 Lunar and Planetary Laboratory, University of Arizona, Tucson, AZ 85721, USA
2 Geophysical Institute, University of Alaska Fairbanks, Fairbanks, AK 99775, USA
3 Department of Earth and Environmental Geosciences, Colgate University, Hamilton, NY 13346, USA
4 Department of Geosciences, University of Arizona, Tucson, AZ 85721, USA
* Correspondence: tmeng@arizona.edu

Abstract: The availability of remote sensing imagery at high spatiotemporal resolutions presents the opportunity to monitor the surface motion of rock glaciers, a key constraint for characterizing the dynamics of their evolution. In this paper, we investigate four North American rock glaciers by automatically measuring their horizontal surface displacement using photogrammetric data acquired with crewed and uncrewed aircraft along with orbital spacecraft over monitoring periods of up to eight years. We estimate vertical surface changes on these rock glaciers with photogrammetrically generated digital elevation models (DEM) and digitized topographic maps. Uncertainty analysis shows that the imagery with the highest resolution and most precise positioning have the best performance when used with the automated change detection algorithm. This investigation produces gridded velocity fields over the entire surface area of each study site, from which we estimate the age of rock glacier formation using along-flow velocity integration. Though the age estimates vary, the ice within the modern extent of these landforms began flowing between 3000 and 7000 years before present, postdating the last glacial maximum. Surface elevation change maps indicate present-day thinning at the lower latitude/higher elevation sites in Wyoming, while the higher latitude/lower elevation sites in Alaska exhibit relatively stable surface elevations.

Keywords: photogrammetry; rock glacier; kinematics; UAS; airborne; satellite; flow; ablation; Alaska; Wyoming

1. Introduction

1.1. Surface Motion of Rock Glaciers

Mountainous terrains with moderate precipitation and mean annual air temperatures less than or equal to 0 °C often develop ice and lithic-rich landforms known as rock glaciers. The ice units originate under a continuum of surface processes, including the burial of glacial ice by rockfall and ablation lag [1,2], preservation of overlapping debris and snow avalanche deposits [3], and the infiltration/refreezing of liquid water in rocky talus [4,5]. These ice/debris mixtures creep downslope under gravitational driving stresses to form discernible lobate morphologies, often superimposed by ridges, furrows, and flow bands [6]. In general, rock glaciers flow with an average surface displacement on the order of tens of centimeters to meters per year [1,7].

Previous measurements of active rock glacier flow have included both in situ and remote sensing methods. Before the availability of remote sensing data with temporal and spatial resolutions sufficient to track surface movement at the scales of rock glacier creep, surface-based displacement measurements were collected at benchmark positions at various locations along rock glacier surfaces [1,4,8–13]. Early displacement measurements were made by repeatedly calculating the position of surface features in a local coordinate system referenced to stable bedrock points, marking the end of a "movement line". High accuracy and high precision global navigation satellite system (GNSS) positioning has recently allowed for displacement measurements through the monitoring of surface features' absolute positions over known time intervals. These point measurements are only possible at surface locations that are safely accessible and stable for repeated measurement, which limits this form of surface motion data both temporally and spatially.

Recent developments in remote sensing technology have created the opportunity to observe rock glacier flow with higher spatiotemporal resolution. Photogrammetric techniques using data collected via uncrewed aerial systems (UAS), piloted overflights, or optical satellite imagery provide a solution to the spatial limitations of surface-based point measurements of rock glacier surface features, as a photogrammetric orthomosaic encompasses a larger percentage of a rock glacier's surface than a set of ground-based point measurements. In addition to optical imagery, interferometric synthetic aperture radar (InSAR) has been used to measure rock glacier surface displacement with high precision [14–16]; however, such surveys are limited by the orbital geometry of the instrument and the flow direction of each target, as the displacement is measured along the line of sight of the InSAR system. Overall, remote monitoring provides a method to produce regularly repeated measurements of features in rugged and isolated terrain, which benefits time series analysis through the availability of a longer record length and higher sampling frequency. The ability to consistently track rock glacier surface motion has led to significant advances in the understanding of rock glacier kinematics and its relationship with glacier and permafrost dynamics [7,17–24]. New information regarding the extents and magnitudes of rock glacier flow fields addresses Tasks 1, 2, and 3 of the International Permafrost Association Action Group for Rock Glacier Inventories and Kinematics (RGIK) by contributing to a database of rock glacier attributes, including locations, surface areas, and flow velocities [6,25].

With the objective of monitoring rock glacier activity, inferring their flow history, and characterizing their kinematics and dynamics, we present new surface motion measurements on four North American rock glaciers, two in Wyoming and two in Alaska. We use an existing image correlation algorithm to detect feature displacement over time intervals of up to eight years in an analysis using optical imagery collected with UAS, airborne, and satellite platforms. Using the gridded velocity fields of the entire surface area of each rock glacier, we estimate the age of each landform by integrating head-to-toe velocity profiles, then we discuss their relationship with documented glacial advances in the region of each study site. We use the elevation data produced by the photogrammetric processing to estimate surface elevation change and modern rock glacier thinning rates. Our study aims to examine the local heterogeneities in rock glacier evolution by comparing the surface motion and elevation change of two rock glaciers in each geographic region. We achieve these objectives using a combination of UAS, airborne, and satellite imagery with photogrammetric processing and surface change analysis. High-resolution data capturing the three-dimensional change of rock glacier surfaces can provide a foundation for future monitoring campaigns and further investigation of rock glacier dynamics. Our study adds new surface change datasets for four rock glaciers to the existing inventory of rock glacier activity. In addition, it presents novel techniques for evaluating the uncertainty of surface change results and interpreting these results in the context of Quaternary geology.

1.2. Study Areas

1.2.1. Absaroka Mountains, Wyoming

Due to their central location in the contiguous United States and their relative accessibility by road, the small population of rock glaciers in the Absaroka Mountains of northwest Wyoming have been the subject of the longest-lived and most comprehensive studies of any rock glaciers or debris-covered glaciers in North America. Galena Creek Rock Glacier (Figure 1a, henceforth referred to as "Galena Creek") has been the particular focus of heated debate over the origins of ice in rock glaciers. The discussion has centered on whether to classify these features as ice-cored (glaciogenic) or ice-cemented (periglacial) rock glaciers [1,26–28]. Surface boulder displacement monitoring began at Galena Creek in the 1960s [1]. Based on geomorphic and geophysical data, along with observed ice exposures, it was concluded that the upper section of Galena Creek is an active ice-cored rock glacier consisting of glacier ice buried beneath an unconsolidated layer of debris. The debate has continued, as skeptics have attempted to refute the evidence of the glacial ice core [26], although this debate was largely settled in the 1990s when a drilling campaign at Galena Creek retrieved an ice core nearly 10 m long indicating the present-day existence of a debris-covered glacier [29]. This drill core contained a unit of bubbly ice with thin layers of gravel and sand, and the isotopic composition of this ice indicated that it originated as a glacier rather than as frozen interstitial meltwater [30]. Further exploration using ground-penetrating radar (GPR) revealed dipping debris bands in the cirque of Galena Creek, suggesting an ongoing ice accumulation process facilitated by the deposition of debris on the surface of the glacier [3]. On this upper portion of the rock glacier where there is a glacial ice unit, the debris is approximately 0.8–1.5 m thick [31]. With these observations in mind, we interpret the upper two-thirds of Galena Creek to be a debris-covered glacier. In comparison, the lower third of Galena Creek has slower surface velocities, a debris mantle thicker than 2 m, and a lower ice concentration, resembling an older ice-cemented rock glacier modified by periglacial processes and possible interactions with the advancing debris-covered glacier, similar to processes that have been documented at other rock glaciers. [12,31–34].

Sulphur Creek Rock Glacier (Figure 1b, henceforth referred to as "Sulphur Creek") lies approximately 3 km southeast of Galena Creek. Despite this proximity to Galena Creek, Sulphur Creek has received relatively less geological research attention due to its larger surface area, higher topographic relief, and more difficult access. Historical photos acquired in 1893 during a surveying expedition led by Thomas A. Jaggar, Jr., showed a clean ice glacier in the cirque of the Sulphur Creek basin, with a thin supraglacial debris layer developing a few hundred meters downslope of the location of the terminus of the present-day debris-free snow and ice [35]. Recent GPR measurements indicate a transition from an alpine debris-covered glacieret to a relatively ice-poor rock glacier as elevation decreases along the length of Sulphur Creek [31]. The same measurements showed that the debris on the glacieret ranges from 0.1–1 m thick, while the debris on the lower glacier is greater than 2 m thick. Both Galena Creek and Sulphur Creek follow the Östrem curve [36], showing evidence of sub-debris ice melt where the debris is thin; at the surface of both sites, streams can be heard flowing at the debris–ice interface. These two rock glaciers provide unique examples of the effect of debris supply and valley geometry on ice units transitioning between glaciers, debris-covered glaciers, and rock glaciers [37].

1.2.2. Wrangell Mountains, Alaska

Sourdough Peak is a mountain in Wrangell–St. Elias National Park that hosts two large lobate rock glaciers. The rock glacier flowing down the peak's southern flank is named Sourdough Rock Glacier (Figure 1d, henceforth referred to as "Sourdough" for simplicity), and the rock glacier flowing down its northwest slope is McCarthy Creek Rock Glacier (Figure 1e, henceforth referred to as "McCarthy Creek"). Sourdough has been surveyed with GPR; these surveys detect a landform thickness of up to 50 m, and the dielectric mixing model indicates volumetric ice concentrations greater than 50 percent based on the radar wave speed within the rock glacier [31]. The ice-free debris thickness measurements at Sourdough are generally greater than 2 m, although runoff can be observed through the sound of localized sub-debris streams despite the thickness of the overburden. The McCarthy Creek site has not been studied with in situ geophysical methods. A surface motion survey at Fireweed Rock Glacier, which is a nearby rock glacier in the Wrangell Mountains, measured velocities exceeding 3.5 $\frac{\mathrm{m}}{\mathrm{yr}}$ [13]. The oversteepened terminus of this rock glacier experiences periodic slope failure events when heavy precipitation swells its proglacial stream, a process that may impact the dynamics of the rock glacier and the characteristics of its velocity field in comparison with Sourdough and McCarthy Creek.

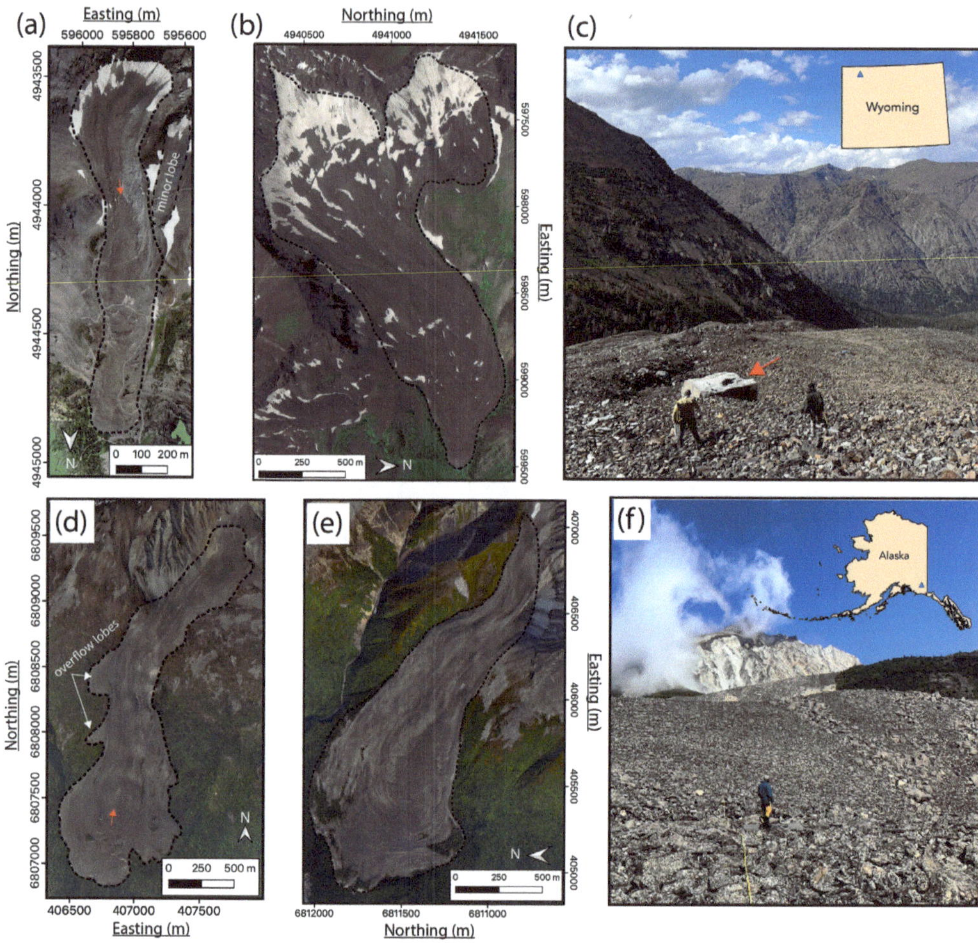

Figure 1. Projected orthomosaics and surface photos showing the rock glaciers targeted in this study. (**a**) Galena Creek Rock Glacier, Wyoming (Galena Creek), UAS image acquired in August 2022 and projected to WGS 84/ UTM Zone 12N. The red arrow shows the location and viewing direction of the photo in panel (**c**). (**b**) Sulphur Creek Rock Glacier, Wyoming (Sulphur Creek), satellite imaged acquired in August 2022 and projected to WGS 84/ UTM Zone 12N. (**c**) Field photo at Galena Creek, showing debris clast size distribution and topographic relief. The red arrow identifies the boulder used for the example in Figure 2. (**d**) Sourdough Rock Glacier, Alaska (Sourdough), airborne image acquired in May 2014 and projected to WGS 84/UTM Zone 7N. The red arrow shows the location and viewing direction of the photo in panel (**f**). (**e**) McCarthy Creek Rock Glacier, Alaska (McCarthy Creek), airborne image acquired in in August 2014 and projected to WGS 84/UTM Zone 7N. (**f**) Field photo at Sourdough. The images have been rotated to make the direction of flow point roughly towards the bottom of the page. All rock glacier outlines presented are the extended delineations [6], including the input talus slopes and front and lateral margins. All subsequent maps use the same projections as those shown here.

2. Materials and Methods

2.1. Photogrammetric Data Acquisition and Processing

For this investigation, our objective was to measure feature displacement with remote imagery over multiple time intervals and compile the longest possible time series of surface displacement for each study site. To detect surface velocities less than 1 $\frac{m}{yr}$ at seasonal intervals, it is necessary to use decimeter-resolution imagery to resolve the details of the surface features as well as to detect displacements on the order of decimeters. The imagery for our two Wyoming sites was collected via UAS, crewed aircraft, and satellite platforms between 2020 and 2022. The Alaska sites were targeted by a crewed airborne photogrammetry campaign between 2014 and 2022. Supplementary Figure S1 details the methodological workflow for measuring the surface displacement and elevation change with these combined data sources. Below, we discuss the advantages and disadvantages of these methods at each site and compare our remote sensing results with surface-based boulder displacement measurements at Galena Creek.

2.1.1. Wyoming

In August 2020 and August 2022, we acquired photogrammetry data covering Galena Creek using a DJI Phantom 4 RTK UAS. We used the DJI GS RTK flight planning software in 2D Photogrammetry mode with terrain awareness. In 2022, eight ground control points (GCP) were deployed and surveyed. For both years, the GCP locations were measured using real-time kinematic positioning (RTK) with Emlid Reach RS2 GNSS receivers. The coordinates of the base station were postprocessed with precise point positioning using the Canadian Spatial Reference System Precise Point Positioning tool. We used Emlid Studio software version 1.3 to apply the postprocessing kinematics to the UAS images.

The photogrammetric processing workflow was carried out using Agisoft Metashape Professional software version 1.7.5 build 13229. After the photos were aligned and the dense clouds were created, digital elevation models (DEM) and orthomosaics were generated for further analysis. The detailed parameters used in the workflow for each flight are provided in the processing reports included with the supplementary materials, and the workflow diagram is shown in Supplementary Figure S1. These orthomosaics have a spatial resolution of 7.9 $\frac{cm}{pixel}$ for 2020 and 5.4 $\frac{cm}{pixel}$ for 2022 (Table 1), and the DEMs have a pixel width double that of their corresponding orthomosaics. In 2022, four points were used as control at Galena Creek, with a root mean square (RMS) error of 0.012 m, and the remaining four points were check point (CP), with an RMS error of 0.068 m. To supplement the two UAS datasets, we purchased satellite imagery from the SkyMap50 system through Soar.Earth, a commercial organization that distributes orbital imagery data. We obtained one complete 41.1 $\frac{cm}{pixel}$ SkyMap50 scene of Galena Creek without clouds, acquired on 10 July 2021.

We compared independent motion measurements at Galena Creek using surface-based and remote sensing methods. Large debris clasts on the Galena Creek surface were marked with paint in the 1960s to measure rock glacier surface motion, and the set of marked clasts was expanded and updated with new paint and bolts in the 1990s [1,29]. The paint markings and identifying symbols on the clasts remain legible. We collected positioning data for 22 identifiable marked boulders using the Emlid Reach RTK system in August 2022 and compared these locations to measurements collected in 1997, 1998, 1999, and 2015 [12,29,38]. Because the measurements from the 1990s and 2015 were acquired with a total station, we converted the local coordinate system used for these earlier datasets to the WGS84 / UTM Zone 12N projected coordinate system using a USGS benchmark and stationary points on stable bedrock for direct comparison with the 2022 RTK measurements.

Table 1. List of acquisition details for the imagery used for the change detection analysis at each field site (GC = Galena Creek; SC = Sulphur Creek; SRG = Sourdough; MC = McCarthy Creek). Each date is provided in YYYYMMDD format.

Region	Site	Date	# of Images	Avg. Camera Error (cm)	Resolution ($\frac{cm}{pixel}$)	# of GCP/CP	GCP/CP RMSE (cm)
Wyoming	GC	20200823 [U]	1076	0.3	7.9	0/0	n/a
	SC	20200825 [A]	269	n/a	10.8	10/3	0.44/28.3
	GC	20210710 [S]	1	n/a	40.1	0/0	n/a
	SC	20210710 [S]	1	n/a	41.0	0/0	n/a
	SC	20220807 [S]	1	n/a	41.3	0/0	n/a
	GC	20220808 [U]	941	0.8	5.4	4/4	0.97/7.59
Alaska	SRG, MC	20140525 [A]	433	16.2	20.0	0/0	n/a
	SRG, MC	20140823 [A]	345	17.2	19.8	0/0	n/a
	SRG, MC	20150523 [A]	546	15.2	20.6	0/0	n/a
	SRG, MC	20150829 [A]	561	18.8	19.3	0/0	n/a
	SRG, MC	20160601 [A]	614	14.5	25.3	0/0	n/a
	SRG	20160817 [A]	494	21.9	24.2	0/0	n/a
	SRG, MC	20190905 [A]	628	65.3	12.1	0/0	n/a
	SRG	20200517 [A]	215	9.7	12.4	0/0	n/a
	SRG, MC	20201018 [A]	520	80.5	12.4	0/0	n/a
	SRG, MC	20210622 [A]	340	11.7	14.8	0/0	n/a
	SRG	20220708 [A]	357	26.2	18.5	0/0	n/a

[U] UAS image; [A] Piloted airborne image; [S] SkyMap50 satellite image.

At Sulphur Creek, airborne imagery was collected on 25 August 2020 by Kestrel Aerial Services using a Canon EOS 5D Mark III DSLR with a Canon EF 50 mm 1.2 lens. The positioning information was recorded by a Garmin Aera 796 synchronized with the camera clock mounted in the panel of the aircraft. For this acquisition, we used ten GCPs to optimize the positioning of the imagery. The individual images and their positions were delivered as georeferenced TIFF files, and these data were processed in Agisoft Metashape to generate an orthomosaic and DEM. The ten points used as control had an RMS error of 0.0004 m, and three points used as check points had an RMS error of 0.283 m. In addition to the 2020 airborne image, we obtained one partial image and one complete image of a cloudless Sulphur Creek available in the SkyMap50 collection. The image from 7 August 2022, contains the full rock glacier, while the lowest 500 m section of Sulphur Creek is cut off at the eastern edge of the 10 July 2021 scene.

2.1.2. Alaska

The photogrammetry data for Sourdough and McCarthy Creek were acquired during eleven piloted overflights between May 2014 and July 2022 (Table 1). The flights were planned with the objective that more than nine overlapping images would cover the target surfaces. The 2014–2016 images were collected with a Nikon D800 DSLR and the 2019–2022 images were collected with a Nikon D850 DSLR, both with a Zeiss Distagon 25 mm lens. Each raw image was collected in NEF format and postprocessed to maximize contrast before conversion to JPG format. Aircraft positions were measured with a Trimble R7 GNSS receiver recording at 5 Hz. Following [39], an intervalometer was used to trigger event markers in the GNSS data associated with each camera flash. These coordinates were transformed from the GNSS antenna to the camera image plane using a triple coordinate rotation of the measured lever arm for the aircraft's antenna/camera configuration.

By interpolating the camera positions from the 5 Hz GNSS data with the event markers, each image was tagged with a position to approximately 10 cm accuracy [39]. From these tags, a camera position file associating each JPG image name and position was generated and used as a reference for the photogrammetric processing steps. We followed the same Agisoft Metashape workflow described in Section 2.1.1. The individual processing reports for each flight are provided in the supplementary information. Due to logistical limitations in the field, no GCPs were used in the photogrammetric processing for the Alaska sites, and there were no surface boulder measurements or GCPs to provide direct in situ validation for these sites. Sourdough was imaged with all eleven flights, while McCarthy Creek was only imaged on eight of the eleven flights due to variable weather conditions in the narrower McCarthy Creek Valley. The acquisition dates and original resolutions of all rock glacier imagery analyzed in this study are provided in Table 1.

2.2. Change Detection Analysis

To automatically measure the horizontal rock glacier surface displacement between image pairs, we used the free correlation image analysis software CIAS [7,40]. This algorithm, available as a compiled IDL program, requires grayscale images with identical extents and resolutions in a Cartesian coordinate system. We preprocessed the images using the Geospatial Data Abstraction Library (GDAL) within the QGIS user interface. Both of these software packages are open source. To preprocess the images, we used the GDAL *Warp* tool to project the raster data to their appropriate Cartesian coordinate systems from the WGS84 geographic coordinates (UTM zone 7N for the Alaska sites and zone 12N for the Wyoming sites). *Warp* was used to clip the images to an appropriate extent and to resample the images to the resolution of the coarsest image in the pair with the cubic spline resampling method. Finally, the grayscale images used for the change detection input were extracted from the RGB data using the GDAL *Translate* tool.

After preprocessing the images, each image pair in the set was analyzed to derive the surface displacement vectors over the time interval between acquisitions. All of the image pairs at each field site were analyzed using a common set of grid points specified by a file containing the Cartesian coordinates of each grid point. The Galena Creek datasets were analyzed on a 5 m $\times$ 5 m grid, while the more extensive Sulphur Creek and Alaskan sites were analyzed on a 10 m $\times$ 10 m grid. All of the image pairs were analyzed using normalized cross-correlation and the normal pyramid matching speed. Our experiments with CIAS determined that a reference block of 45 pixels $\times$ 45 pixels and a search window of 100 pixels $\times$ 100 pixels at each grid point was optimal for correlating surface features and detecting realistic displacements for all of the image resolutions and time periods analyzed (Figure 2). Due to the forest cover surrounding the Alaska sites, which obscured the stable terrain, co-registration was not performed for individual image pairs. Instead, we used the minimum uncertainty in surface displacement as the averaged CIAS-derived displacement value for stable, off-glacier terrain within each scene. We verified these displacement values through manual inspection of stable surface features wherever possible. This added the benefit of decreasing the processing time and avoiding the application of inconsistent uncertainties due to an extra transformation step unique to the processing workflow of each image pair. To convert the displacement results into the surface velocity in $\frac{\text{m}}{\text{yr}}$, we found the precise number of years between images by dividing the number of days between each image pair by 365.25 days per year, then divided the displacement in meters by the time interval in years.

In image pairs for which one or both of the images is a SkyMap50 image targeting Galena Creek or Sulphur Creek, there is a static offset in the images due to limitations in image precision using the positioning of a space-borne camera. To solve this issue of imprecise image co-registration, we first ran the change detection algorithm on the images with their initial positioning information. We selected a subset of displacement vectors over a portion of the images interpreted to be stable bedrock. The mean of this subset of displacement vectors was subtracted from the projected coordinates of the corners of the later image in the image pair to shift the geolocated image using the "-a_ullr" flag in the GDAL *Translate* function to correct for the initial static shift. The largest magnitude of the shifts that were used was 1.56 m for the Galena Creek July 2021 SkyMap50 image; the parameters for each shift are shown in Table 2.

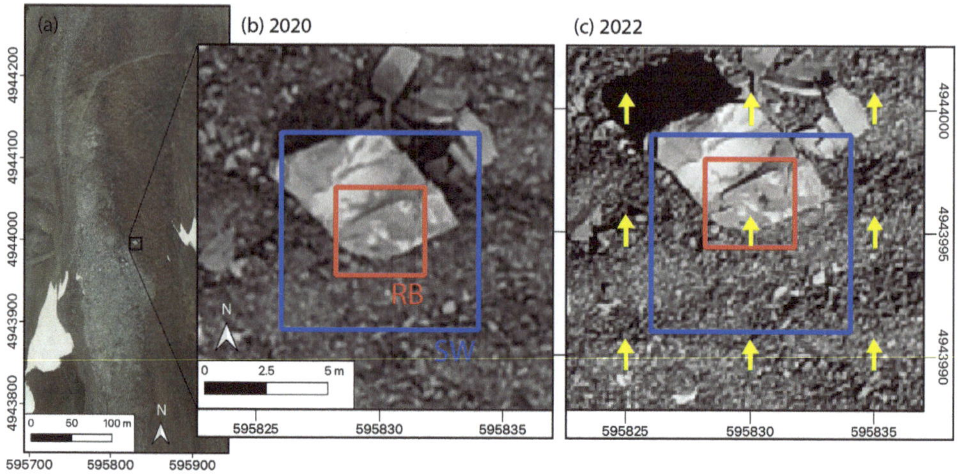

Figure 2. Schematic of the change detection procedure applied to a boulder identified at a central location in the August 2020/August 2022 image pair at Galena Creek (**a**), where the red box labeled RB represents the 45 × 45 pixel reference block and the blue box labeled SW represents the 100 × 100 pixel search window. The movement of RB between 2020 (**b**) and 2022 (**c**) shows the displacement measured as the location of the peak normalized correlation coefficient within the search window. The yellow arrows represent the two-year displacement vectors measured by CIAS; this boulder moved approximately 1.2 m at an azimuth of 358°.

The change detection process was then performed again, this time using the shifted SkyMap50 image with its corresponding unshifted image partner. Its results were verified by manually examining off-glacier stationary features in both the unshifted and shifted images for each pair containing a SkyMap50 scene. The new change detection results were compared with the initial results of the unshifted images after subtracting the mean displacement of the stationary subset for further verification and uncertainty analysis. The added step of translating the second image of the pair, along with the relatively lower resolution of the SkyMap50 imagery (42 $\frac{\text{cm}}{\text{pixel}}$) compared with the drone imagery (<10 $\frac{\text{cm}}{\text{pixel}}$), leads to a higher uncertainty in those surface displacements estimated with pairs containing a shifted satellite image.

Table 2. Static shift applied to the SkyMap50 images to minimize the measured displacement of stationary terrain in each image pair.

Rock Glacier	Image 1	Image 2	Δx (m E)	Δy (m N)
Galena Creek	10 July 2021 ^{S*}	8 August 2022 ^{U}	−0.99	−1.20
Sulphur Creek	10 July 2021 ^{S}	7 August 2022 ^{S*}	0.84	−1.25
Sulphur Creek	25 August 2020 ^{A}	7 August 2022 ^{S*}	0.81	0.68

^{U} UAS image; ^{A} Piloted airborne image; ^{S} SkyMap50 satellite image; * Denotes that the image was shifted relative to the other image in the pair.

2.3. Surface Elevation Change

For the UAS and airborne datasets, the photogrammetric processing workflow produces DEMs with pixel widths twice those of their corresponding orthomosaics (Supplementary Figure S1); thus, the DEMs used in this study range from approximately 10–50 $\frac{\text{cm}}{\text{pixel}}$. Using these elevation maps, we calculated the surface elevation changes for Galena Creek, Sourdough, and McCarthy Creek over the time intervals between the earliest and latest photogrammetric acquisitions in order to observe any detectable signatures of horizontal flow or vertical thinning. For Galena Creek, the elevation change was calculated over the

August 2020/August 2022 interval; for Sourdough, the May 2014/July 2022 interval was used; and for McCarthy Creek, the May 2020/June 2021 interval was used. All of these elevation differences were calculated by subtracting the earlier DEM from the later DEM in each pair using the *Raster Calculator* tool in QGIS with the coarsest-resolution DEM as the reference.

To examine the surface change at Sulphur Creek, we calculated the difference between the DEM produced by the August 2020 airborne photogrammetry flight and the $\frac{1}{3}$ arcsecond resolution (approximately 10 m) DEM tile from the USGS 3D Elevation Program (3DEP) [41]. The spatial metadata for this tile indicate that the 3DEP data at Sulphur Creek was sourced from topographic information measured in 1985. Because this location in the 3DEP dataset is mostly barren land, the error of this $\frac{1}{3}$ arcsecond DEM is estimated to have a mean bias of −0.85 m with a standard deviation of 2.42 m [42]. We consider this bias in the data when interpreting the elevation change results discussed below.

3. Results

3.1. Wyoming

3.1.1. Galena Creek

At Galena Creek, the August 2020/August 2022 image pair resampled to 8.0 $\frac{\text{cm}}{\text{pixel}}$ resolution provided the best change detection signal in the set of image pairs. Surface displacement is detectable along the main trunk of the rock glacier, in contrast to the adjacent stationary terrain (Figure 3b). The displacements have a strong correlation with the surface slope, indicating a direct relationship between driving stress and flow velocity. The change detection algorithm cannot measure displacements on surface regions when there is snow in one or both of the images due to the low contrast and lack of pixel correlation within the reference block. These regions cause "noisy" results, which are identifiable in the mapped displacement vectors as regions with random displacement vector magnitude and direction that are associated with the snow patches when the displacement vectors are mapped over the base images. The effect of these noisy regions on the analysis of the data is mitigated by ignoring the displacement vectors greater than a noise threshold, which is determined by visually evaluating the flow field to find the maximum displacement with a direction that agrees with the local topography. For the case of Galena Creek, this threshold is approximately 1.6 $\frac{\text{m}}{\text{yr}}$. The noise may be further filtered by ignoring vectors where the displacement direction differs from the slope azimuth by more than 45°. The minimum displacement error for each image pair is taken to be the greater value of either the minimum displacement over a region interpreted to be stationary or the pixel size of the images. Using these metrics, a pair of low-altitude UAS images acquired with the same camera and positioning system for both surveys returned the best displacement measurements out of all the datasets presented here.

The photogrammetry data collected at Galena Creek allowed us an opportunity to directly compare the efficacy of the change detection method using imagery from homogeneous and heterogeneous platforms. We performed a hybrid change detection experiment using the 42 $\frac{\text{cm}}{\text{pixel}}$ SkyMap50 satellite image from July 2021 and the August 2022 UAS image resampled to 42 $\frac{\text{cm}}{\text{pixel}}$ with the cubic spline method. The displacement patterns in the upper two-thirds of the rock glacier are similar between the homogeneous change detection results and the hybrid results. The magnitude of the displacement is directly correlated with the surface slope (Figure 3c); however, the coarser resolution of the satellite image in the lower third of the rock glacier means that the proportion of mismatches and undetected movements increases due to the increased difference in pixel size between the original UAS and satellite images [43].

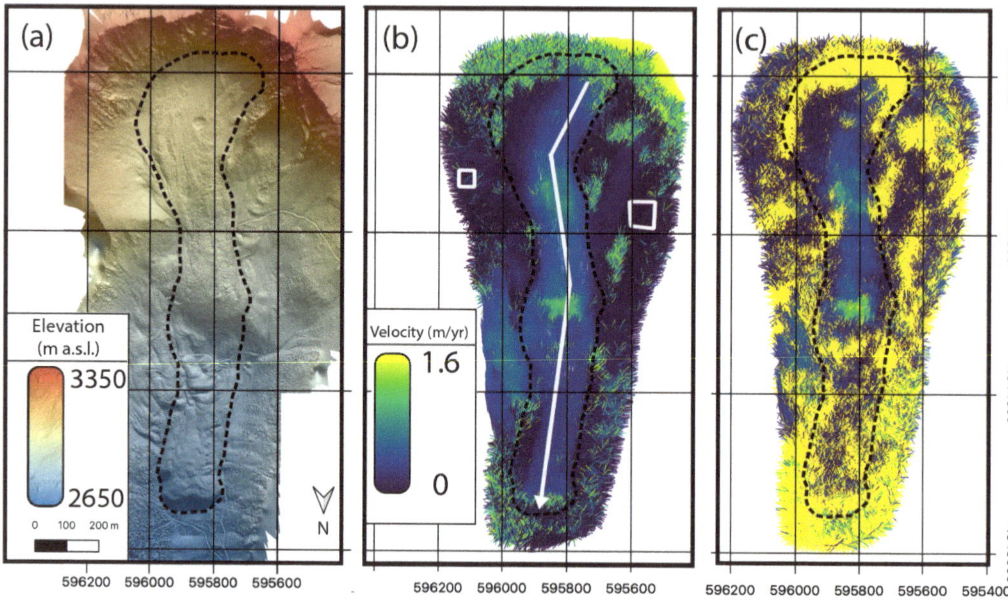

Figure 3. Galena Creek photogrammetry results: (**a**) DEM and hillshade produced from the August 2022 UAS flight; (**b**) surface velocity field derived from change detection between the August 2020 and August 2022 UAS-derived orthomosaics; (**c**) surface velocity field derived from change detection between the shifted July 2021 satellite image and the August 2022 orthomosaic. For reference, the extended rock glacier outline is delineated as a dashed line in each panel, the white line in (**b**) marks the profile that was sampled for the age analysis discussed below, and the white boxes in (**b**) indicate the stable points used in the uncertainty analysis.

3.1.2. Sulphur Creek

We detected a flow signal on the Sulphur Creek surface (Figure 4) using the August 2020 airborne imagery combined with the August 2022 SkyMap50 image, which was linearly shifted to account for the co-registration error between the two images (Table 2). Although the displacement values on the stable surfaces of the rock glacier surface indicate a relatively high baseline uncertainty, there is a signal of increased flow velocity on the lowest lobe of the rock glacier. In contrast, there does not appear to be substantial downslope movement in the middle portion of the rock glacier, where GPR and geomorphic observations indicate ice thicknesses of less than 10 m and stagnation of the ice [31,37].

These change detection results support the hypothesis of ice stagnation on a deflating debris-covered glacier that is transitioning to dead ice. The directions of the displacement vectors on this central portion of the glacier agree with the slope aspect (approximately 150°), suggesting movement toward the middle line of the glacier. This movement may be an effect of rapid recent thinning by incision of a supraglacial stream and subsequent ice flow from the thicker ice at the glacier margins to the thinned ice in the middle (see the elevation change results in Section 3.3). Alternatively, if the thinning is concentrated along a longitudinal line associated with a stream in the center of the glacier, this could lead to a reduction in the cross-flow buttressing force, allowing the lateral portions of the glacier to cohesively slide along the base towards the central trough. With either mechanism, this downwasting appears to preserve the surface debris structure, as the change detection algorithm successfully tracks features in the region over a spatial scale of a few hundred meters.

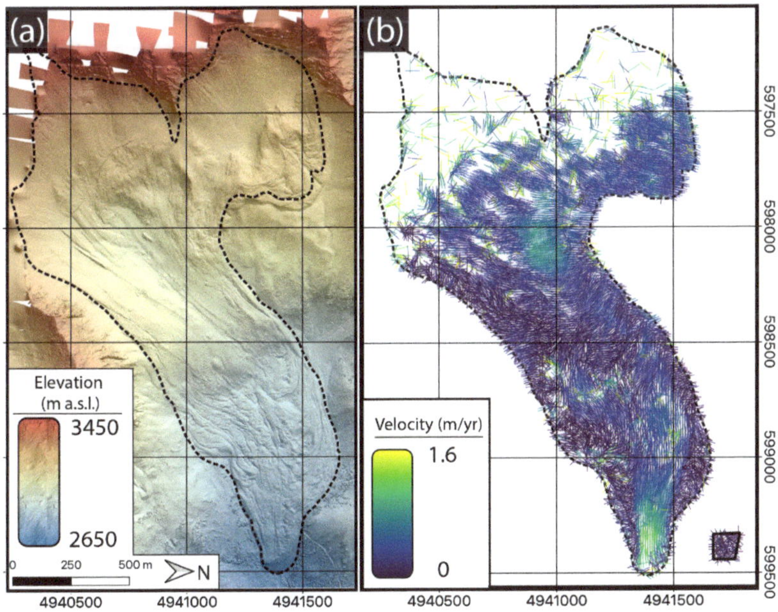

Figure 4. Sulphur Creek photogrammetry results: (**a**) DEM and hillshade produced from the August 2020 piloted overflight and (**b**) surface velocity field derived from change detection between the August 2020 airborne orthomosaic and the shifted August 2022 satellite image. For reference, the extended rock glacier outline is delineated as a dashed line in each panel. The box in the lower right corner of (**b**) indicates the stable area used for the uncertainty analysis.

3.2. Alaska

3.2.1. Sourdough

The data at Sourdough represent the longest monitoring period of the sites presented here, spanning from May 2014 to July 2022 (Figure 5, Table 1). To characterize the temporal variations in the flow field, we calculated surface displacements for image pairs from adjacent years (Figure 6) and the progressive displacement for subsequent images with respect to the May 2014 initial image. All of the annual image pairs show a consistently fast-moving region with distinct shear margins on the lower trunk of the rock glacier flowing at rates greater than 1 $\frac{m}{yr}$ (Figure 6), indicating a high level of rock glacier activity. We ignore velocity vectors with magnitudes greater than 1.8 $\frac{m}{yr}$, as visual examination indicates that all greater values are qualitative outliers with no directional correlation to the surrounding data points. These spurious velocity vectors are considered noise due to mismatched pixel blocks in the change detection routine. The resulting velocity maps distinguish an active secondary lobe that branches southeast from the main trunk after flowing around a bedrock pinning point. Below this pinning point, the flow of the main trunk and secondary lobe diverges and slows as the slope flattens, creating the characteristic tongue-shaped lobes of the lower rock glacier. Stagnant overflow lobes with low displacements are observed along the west edge of the feature (Figure 1a). While these flow patterns are evident when observing the full set of image pairs with intervals of one year or greater; the results of individual image pairs vary in quality, making it difficult to asses possible seasonal signals in the flow field.

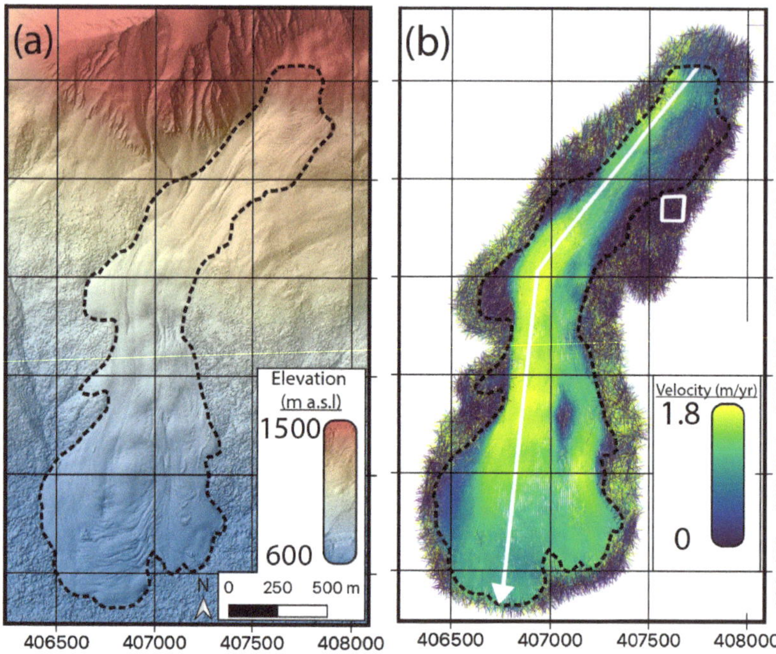

Figure 5. Sourdough photogrammetry results: (**a**) DEM and hillshade produced from the May 2014 piloted overflight and (**b**) surface velocity field derived from change detection between the May 2014 and September 2019 airborne orthomosaics. For reference, the extended rock glacier outline is delineated as a dashed line in each panel, the white line in (**b**) marks the profile that was sampled for the age analysis detailed below, and stable area used for the uncertainty analysis is marked by the white box in (**b**).

The displacement signal is generally stronger in the image pairs with longer time intervals. This indicates that larger displacements are detected more readily and consistently as long as the search window is large enough to contain the range of realistic displacements. The image pair with the greatest amount of noise is the September 2019/October 2020 interval; this noise is largely correlated with the presence of snow on the upper two-thirds of the rock glacier in the October 2020 image. This snow obscures surface features, which leads to inconsistencies in pixel intensity patterns, causing the normalized cross-correlation algorithm to fail. The remaining image pairs with annual time intervals exhibit a consistent pattern of increased surface velocity in the trunk of the rock glacier, although these results have varying degrees of signal and noise. To estimate the total displacement and average velocity of the rock glacier surface over the entire measurement period, we measured the displacement for all of the images as referenced to the May 2014 image (Supplementary Figure S2). This method successfully detects a peak velocity of approximately 1.5 $\frac{\text{m}}{\text{yr}}$ in the central trunk of the rock glacier; however, comparing the results from images acquired at different times of year does not reveal any surges or seasonal signals in the velocity field. Shorter time intervals between acquisitions and imagery with increased spatial resolution paired with permanent GNSS stations on the rock glacier's surface could shed further light on its seasonal flow patterns [44].

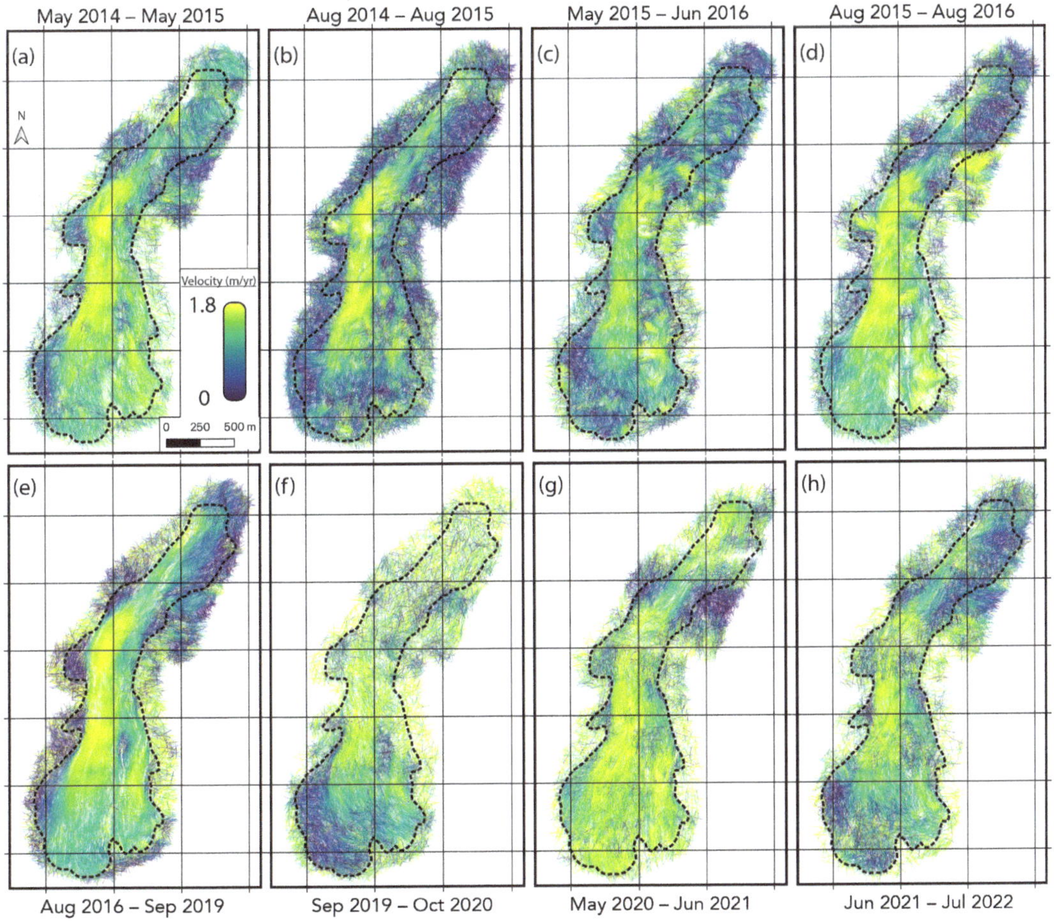

Figure 6. Surface velocity results for Sourdough using image pairs with time intervals of one year or more, demonstrating the range in quality of the change detection results for different image pairs and time intervals. The grid used for these maps is equivalent to the grid used in Figure 5, where the grid lines are drawn at 500 m intervals in the x and y directions in the projected coordinate system. Each panel (**a–h**) shows the results from image pairs, progressing in chronological order.

3.2.2. McCarthy Creek

The results at McCarthy Creek are generally noisier than at Sourdough, and the change detection results for the entire measurement period at McCarthy Creek show a flow pattern with a maximum velocity approximately half that of Sourdough (Figure 7). We chose to use the August 2014 image as the base image for the McCarthy Creek analysis because there was lingering snow on the upper portion of the rock glacier in the May 2014 image, meaning that the August 2014 image was able to detect a flow signal at higher reaches of the rock glacier. The fastest section of the rock glacier surface is the southern/upstream portion of its trunk, moving about 50 $\frac{cm}{yr}$, before slowing as the flow of the lower lobe diverges. At approximately 1200 m elevation the rock glacier branches into a fast southern lobe and a more stagnant northern lobe.

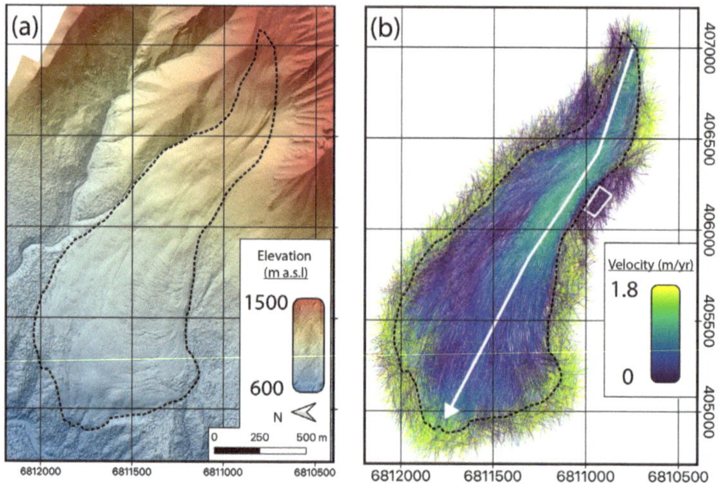

Figure 7. McCarthy Creek photogrammetry results: (**a**) DEM and hillshade produced from the May 2014 piloted overflight and (**b**) surface velocity field derived from change detection between the August 2014 and June 2021 airborne orthomosaics. For reference, the extended rock glacier outline is delineated as a dashed line in each panel, the white line in (**b**) marks the profile that was sampled for the age analysis below, and the stable area used for the uncertainty analysis is marked by the white box in (**b**).

There appears to be an increase in flow speed at the toe of the rock glacier, which could indicate a recent frontal advance or an increase in wasting and potential collapse of the rock glacier toe near the river channel of the McCarthy Creek drainage. However, this apparent signal could alternatively be caused by the combined geometric effects of the photogrammetric data acquisition, the surface slope at this location, and/or uncertainty due to vegetation on the surface. Similar to Sourdough, the change detection at McCarthy Creek performs the best for time intervals greater than one year, and the signal is generally stronger for longer time intervals (Figure 8). In Section 4.2, we discuss the estimation of the baseline uncertainty in these change detection results and how this affects further analysis and interpretation of the data.

3.3. Surface Elevation Change

At Galena Creek, subtracting the earlier DEM from the later DEM reveals indicators of both vertical thinning and surface-parallel motion (Figure 9a). There is an apparent bias of approximately −20 cm between the two DEMs, as shown by differencing the elevations of stable terrain. The elevation difference measured along a longitudinal profile on the rock glacier surface indicates a mean DEM difference of −40 cm with a standard deviation of 19 cm. By comparison, a sample of DEM differences on the stable ground provides a mean value of −18 cm with a standard deviation of 6 cm. A cross-flow profile of the elevation differences supports this observation as well (Supplementary Figure S3). This suggests that the rock glacier surface has lowered by 22 ± 13 cm over the two-year time interval. This 10 $\frac{\text{cm}}{\text{yr}}$ thinning rate for the upper two-thirds of the rock glacier agrees with previous estimates [12,38]. This thinning rate measurement further agrees with a thermal conduction model using air temperature data from the *Evening Star* Snowpack Telemetry (SNOTEL) meteorological station located <1 km east of Galena Creek at a similar elevation as the rock glacier's terminus (station ID = 472). This model uses an observed supraglacial debris thickness of 1.5 m [31], and the measured thinning rate fits a plausible range of thermal conductivities for the debris (Appendix A). The ice in the cirque of Galena Creek has a

GPR-measured thickness of >50 m; thus, assuming that this interpreted thinning rate of 10 $\frac{\text{cm}}{\text{yr}}$ remains constant, the glacial ice will be preserved here past the year 2500.

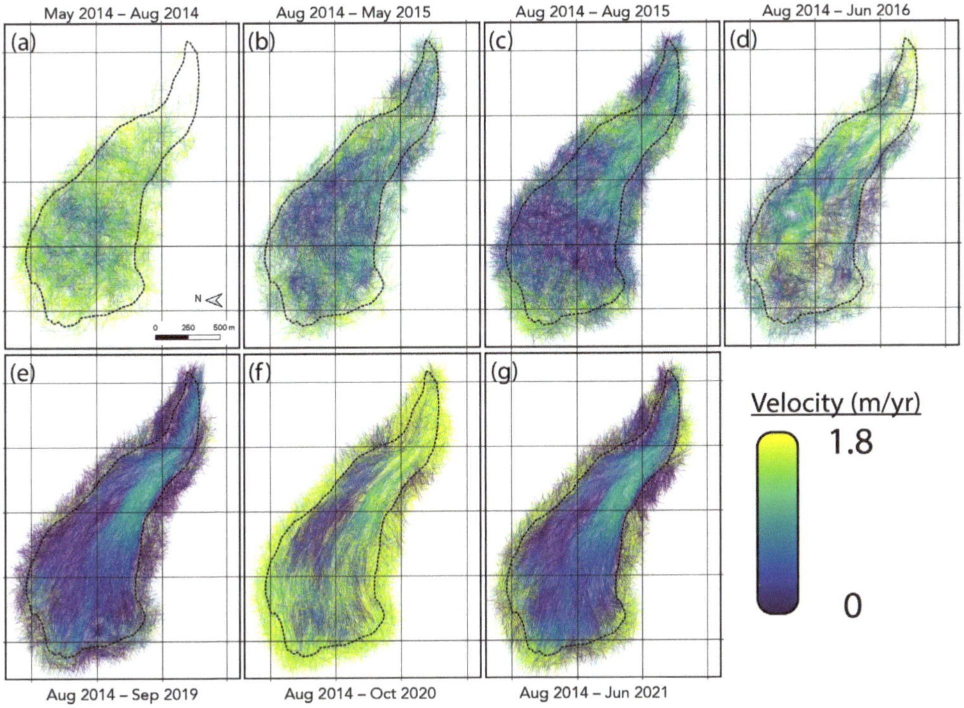

Figure 8. Surface velocity results for McCarthy Creek using image pairs, where each labeled panel was derived in reference to the August 2014 base image, demonstrating the general increase in quality of the change detection results at McCarthy Creek with increasing surface displacement while showing the impact of snow in the October 2020 image. The grid used for these maps is equivalent to the grid used in Figure 7, where the grid lines are drawn at 500 m intervals in the x and y directions in the projected coordinate system. Each panel (**a**–**g**) shows the results where images are chronologically compared to the August 2014 base image.

In addition to the overall thinning of the upper two-thirds of the rock glacier, flow-parallel oscillations in the DEM difference rasters are indicative of the translational motion of surface ridges/furrows, creating a positive value where a ridge has occupied previously void space and a negative value where a furrow has replaced a ridge. The distances between these troughs and crests in the oscillations of the DEM difference data are comparable with the surface displacement measured over the same period. Furthermore, strong negative values within the rock glacier boundaries appear to correlate with a sub-debris supraglacial creek that has been observed to expose ice to the surface. This observation suggests that ablation is concentrated in regions where ice has been exposed to the atmosphere due to the mass wasting of debris by supraglacial melt. These results exemplify the utility of high-resolution photogrammetry data in resolving cm-scale elevation changes on rock glacier surfaces from year to year.

To estimate the surface elevation change at Sulphur Creek, we calculated the difference between the $\frac{1}{3}$ arcsecond resolution (about 10 m/pixel) USGS 3DEP DEM surveyed in 1985 [41] and the 2020 airborne-derived DEM. The surface elevation change appears to be biased towards surface lowering, with a 0.315 km^2 section interpreted to be stable ground showing a mean difference of -4.9 ± 3.5 m, with the error here represented as one standard deviation. However, three regions in the Sulphur Creek image display surface lowering that exceeds the bias in the difference calculated between the 1985 and 2020 DEMs. The first region (labeled "1" in Figure 9b) has a surface area of 0.072 km^2 and displays a mean surface elevation change of -18.3 ± 4.1 m. Region 1 in this map corresponds with the region exhibiting inward flow in the change detection results, and the combination of these observations suggests rapid stagnation and collapse of the middle section of Sulphur Creek, supporting the interpretation of [37]. Region 2 has a surface area of 0.085 km^2 with -23.9 ± 3.5 m of surface change, while Region 3 has a surface area of 0.141 km^2 with -27.9 ± 5.9 m of surface change.

Regions 2 and 3 correspond with the two small cirque glaciers occupying the two forks of the upper Sulphur Creek basin. Accounting for the bias in elevation change estimated from the off-glacier terrain, Regions 1, 2, and 3 of Sulphur Creek have experienced mean thinning rates of $38 \pm 12\ \frac{\text{cm}}{\text{yr}}$, $54 \pm 10\ \frac{\text{cm}}{\text{yr}}$, and $66 \pm 17\ \frac{\text{cm}}{\text{yr}}$, respectively, over the 35-year DEM interval. These thinning rates indicate significant recent negative mass balance for the higher-elevation components of the Sulphur Creek system. We applied our simple thermal model here using a debris thickness of 0.5 along with SNOTEL data spanning the years 1990–2020; this model supports the result that Sulphur Creek has lost upwards of 20 m of ice to cumulative melt under a reasonable range of thermal conductivities for the debris (Appendix A). At these melt rates, the Sulphur Creek basin may lose the entirety of its glacial ice before 2100, and the only remaining subsurface ice in this basin will be preserved in an ice-cemented rock glacier. Future Wyoming fieldwork should aim to collect a UAS-derived DEM at Sulphur Creek to measure surface elevation change after the acquisition of the 2020 dataset and compare the results with the 1985–2020 surface elevation change rates as well as with the results from the neighboring Galena Creek.

In contrast to the Wyoming sites, the surface change between the earliest and latest datasets for the Alaska sites does not suggest broad patterns of elevation increase or decrease across the entirety of each rock glacier. Stationary regions near the rock glacier margins show a mean systematic bias of about +40 cm for the May 2014/July 2022 DEM pair for Sourdough (Figure 9c) and about +30 cm for the May 2014/June 2021 DEM pair for McCarthy Creek (Figure 9d). The mean systematic biases have corresponding standard deviations of about 50 cm for both rock glaciers. This estimate is complicated by the dense vegetation surrounding much of the rock glaciers' perimeters, meaning that stable bedrock estimations must be taken from locations with steep slopes, where the DEM error is likely the highest. There is not a clear change in the mean surface elevation change on either rock glacier surface compared to the surrounding stable terrain in the photogrammetric DEMs when compared with one another or when compared with the corresponding USGS 3DEP product. However, these elevation change results are similar to the Galena Creek results in that the variability of the surface change increases on the surface of the rock glaciers as opposed to off-glacier locations. This variability appears to be an effect of the translational motion of surface ridges and furrows, as the topographic oscillations are oriented perpendicular to flow while their wavelengths and velocities are generally out of phase with the timing of the data acquisition.

The translational motion of ridges can be observed by plotting elevation profiles from multiple flights. These profiles show that any thickness changes of the rock glacier are less than the vertical uncertainty in the elevation data, which is on the order of a few decimeters (Supplementary Figures S4 and S5). At Sourdough, there is a region of apparent thinning in its uppermost reaches corresponding with debris and avalanche cones, as well as a broad region of negative elevation change about 200 m wide on the lower lobe. The elevation change variability in the trunks of the Alaskan sites may include localized thinning that falls within the uncertainty of the elevation data; however, neither of the Alaskan sites indicate broadly consistent thinning across the surface. This result is supported by the thermal conduction model described in Appendix A. Localized elevation gain is observed near the toe of Sourdough as a result of its terminus advancing, which is corroborated by field observations of "bulldozed" trees.

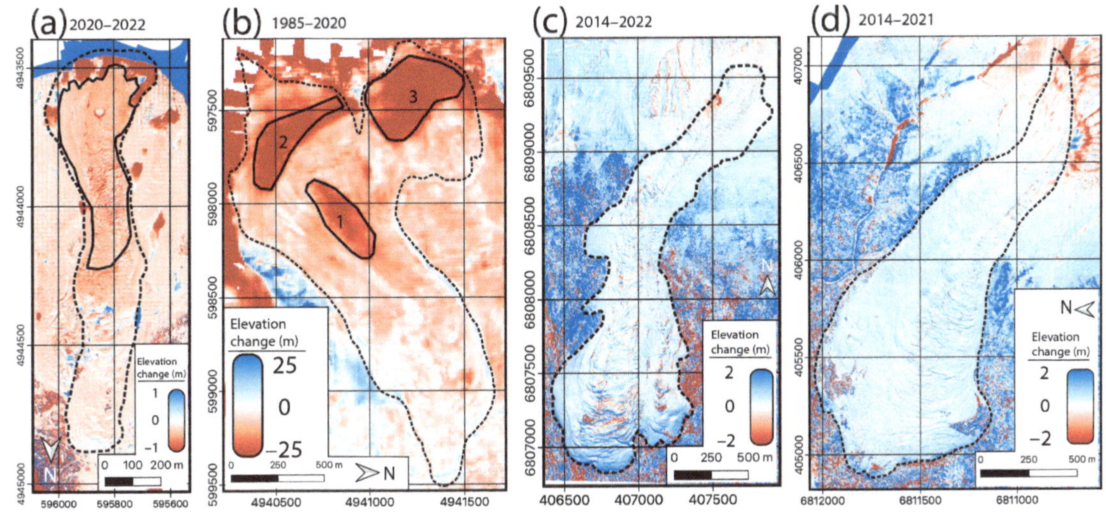

Figure 9. Surface elevation change results: (**a**) the August 2020/August 2022 UAS DEM pair for Galena Creek, with the region of increased ablation outlined in the solid black line; (**b**) the 1985 DEM from USGS 3DEP paired with the August 2020 airborne DEM for Sulphur Creek, where the regions labeled 1, 2, and 3 indicate areas of high thinning rates; (**c**) the May 2014/July 2022 airborne DEM pair for Sourdough; (**d**) the May 2014/June 2021 airborne DEM pair for McCarthy Creek.

In general, all of our rock glacier surface change maps exhibit indicators of longitudinal flow in agreement with the optical change detection results. Further, the Wyoming sites demonstrate clear signals of vertical thinning due to ice melt, with Sulphur Creek experiencing the fastest melt rate. The Alaska sites do not exhibit the same thinning signals. These trends are consistent with a thermal conduction model (Appendix A) using GPR-derived debris thickness measurements that show the debris to be thinnest at the upper part of Sulphur Creek and thickest at Sourdough [31]. Assuming similar mean annual air temperatures and constant thermal conductivities for the debris at all four sites, it is expected that the thinner debris at Sulphur Creek would lead to the highest melt rate, while the thick debris at the Alaska sites would inhibit melt to a greater degree. In the following section, we discuss the implications of these surface change results for the accumulation and evolution of each of these field sites; in addition, we further consider the sources of uncertainty in these results by defining criteria for assessing the accuracy of the horizontal and vertical surface change products.

4. Discussion

4.1. Validation and Uncertainty Analysis

The image pair collected with the UAS in both August 2020 and August 2022 shows a baseline velocity uncertainty of 6.8 $\frac{\text{cm}}{\text{yr}}$ (Table 3). This is the average value of the displacements returned from a subset of the CIAS results consisting of 320 points where bedrock is assumed to be motionless (Figure 3b). Examining the means of the vector components and their standard deviations provides information about the sources of uncertainty [45]. The mean x and y components of the UAS-derived velocity measurements at Galena Creek show that the systematic error is less than 1 cm. The standard deviations indicate that the random error is uniform in both directions and is comparable to the pixel size of the image. These values are similar to the results of photogrammetric change detection surveys in the Swiss Alps [46]. In comparison, the image pair using the coarser satellite image returned much higher uncertainty values due to the increase in mismatches leading to more noise in the results. Using repeated surface-based position measurements of marked boulders at Galena Creek, we validate our change detection results with independent velocity measurements in consideration of the uncertainty of our remote sensing results. We compare the velocities measured from boulder positions in 2015 and 2022 with the four nearest UAS-derived velocity values over the 2020 to 2022 interval [38].

Table 3. List of image pairs used for velocity measurements and the associated velocity uncertainties measured using regions of stable terrain at each field site. All means and standard deviations reported here are provided in units of $\frac{m}{yr}$.

Site	Stable Terrain Area (m^2)	# of Points	Image Pair	Mean ($\lvert v \rvert$)	Mean (v_x)	Mean (v_y)	$\sigma(v_x)$	$\sigma(v_y)$
Galena Creek	8400	320	August 20 August 22	0.068	0.009	−0.006	0.109	0.103
	8400	320	July 21 August 22	5.65	−0.290	0.018	6.43	7.11
Sulphur Creek	11,550	111	August 20 August 22	0.104	0.010	−0.010	0.112	0.086
Sourdough	13,600	141	May 14 August 14	1.11	0.948	−0.011	0.877	0.519
	13,600	141	May 14 May 15	0.246	−0.147	0.029	0.229	0.106
	13,600	141	May 14 August 15	0.288	0.163	0.017	0.336	0.349
	13,600	141	May 14 June 16	0.208	−0.171	0.070	0.118	0.059
	13,600	141	May 14 August 16	0.466	−0.345	−0.250	0.114	0.198
	13,600	141	May 14 September 19	0.057	0.014	0.003	0.063	0.134
	13,600	141	May 14 May 20	0.105	−0.034	0.001	0.234	0.235
	13,600	141	May 14 October 20	0.259	−0.091	−0.112	0.356	0.373
	13,600	141	May 14 June 21	0.042	−0.004	−0.038	0.014	0.019
	13,600	141	May 14 July 22	0.091	−0.014	−0.051	0.135	0.154
McCarthy Creek	11,030	111	May 14 August 14	6.64	2.02	0.293	8.30	6.17
	11,030	111	August 14 May 15	0.470	0.006	0.018	0.612	0.721
	11,030	111	August 14 August 15	0.403	0.046	0.067	0.407	0.327
	11,030	111	August 14 June 16	1.31	−0.231	0.201	1.41	1.21
	11,030	111	August 14 September 19	0.086	0.033	0.004	0.235	0.155
	11,030	111	August 14 October 20	1.06	−0.043	−0.140	0.952	0.855
	11,030	111	August 14 June 21	0.078	0−0.021	−0.018	0.086	0.146

In the upper section of Galena Creek, the change detection and boulder position results are in good agreement, with a maximum velocity magnitude difference of about 5 $\frac{cm}{yr}$, similar to the baseline uncertainty in the August 2020/August 2022 CIAS results (Figure 10). On the lower third of the rock glacier, four measurements show a discrepancy of 10 $\frac{cm}{yr}$ or greater between the 2015–2022 boulder measurements and the 2020–2022 change detection measurements. The largest of these discrepancies (0.36 $\frac{m}{yr}$) can be explained by noise in the change detection data, where a patch of trees created a zone of mismatched pixel clusters at the location of the surface measurement. The three other points with discrepancies greater than 5 cm exhibit an anisotropic bias, where the change detection measurements are about 10 $\frac{cm}{yr}$ faster than the boulder point measurements; most of the variation occurs along the y-axis. This could indicate a rapid acceleration of the lower rock glacier lobe by 10 $\frac{cm}{yr}$ between 2015 and 2020, or could be an effect of errors in measurement and the coordinate system transformation of the boulder positions on this lower lobe between the 2015 and 2022 surveys. A coordinate rotation was applied to the 2015 points, and the boulders on the lower lobe are the most distant from the pole of rotation, making them the most susceptible to an error in the rotation angle between coordinate systems. The deviation between boulder displacements and CIAS results generally increases when using boulder position measurements from 1997 to 1999 (Supplementary Figure S6), which may be an effect of either a changing rock glacier surface velocity field or a decreased measurement error with newer global positioning technology.

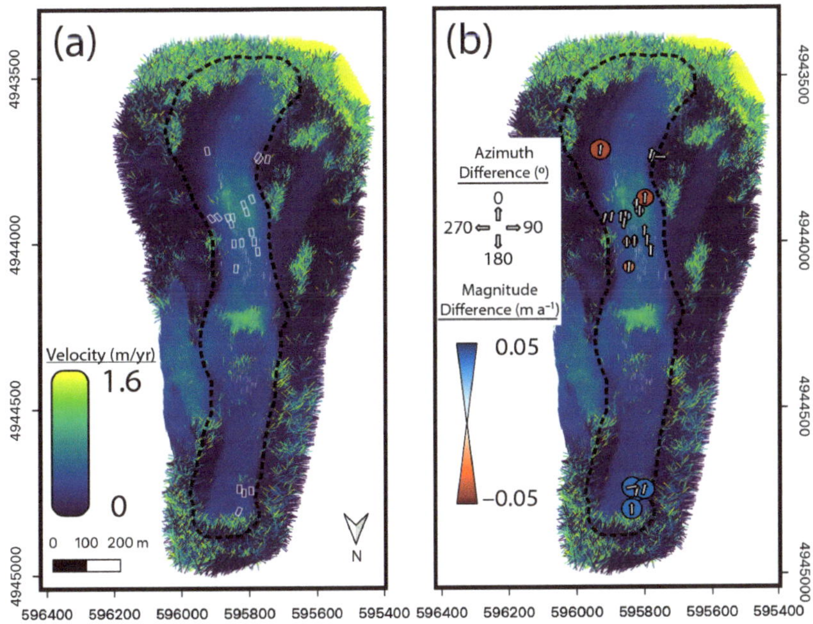

Figure 10. Comparison of photogrammetric change detection results between August 2020 and August 2022 UAS imagery with boulder velocities between August 2015 and August 2022 at Galena Creek. (**a**) The boulder velocity vectors are shown as rectangles with white borders, color-coded according to velocity magnitude and oriented according to the vector's direction; these boulder velocities are plotted over the automated change detection results using the same color scale. (**b**) The difference in magnitude and azimuth between the measured boulder velocities and the four nearest grid points in the change detection measurements. The size and color of each dot in (**b**) corresponds to its magnitude and sign and the direction of the arrow indicates the difference in vector azimuth, meaning that vectors with no change in azimuth display an arrow that faces directly upwards.

Although the airborne photogrammetry at the Alaska sites provides a clear signal of flow on both rock glaciers, uncertainty estimation using stable bedrock points is complicated by forest cover surrounding most of the perimeters of Sourdough and McCarthy Creek. The seasonally changing tree canopy provides poor references for the change detection algorithm, and this leads to mismatched and noisy results immediately surrounding the rock glacier. As we have no repeated surface-based boulder measurements for the Alaska sites, we examine the uncertainty in our change detection measurements here using two metrics: the range of peak velocities across a transverse profile, and the average minimum velocity on stable ground.

The first measure used to characterize the velocity uncertainty for the change detection results is the range in the magnitude of the peak velocities measured along a common transverse profile on the rock glacier surface for all of the time intervals examined. This range is about 0.4 $\frac{\text{m}}{\text{yr}}$, though this variability estimation may contain variations in the rock glacier's true velocity field during the measurement period. The second quantification of the uncertainty in measured velocity for each image pair is calculated by averaging a subset of low-magnitude displacement vectors selected at regions of the image interpreted to be bare stable ground. This value represents the minimum apparent velocity between stationary points in two images; therefore, we take these values as a representation of the uncertainty for the on-glacier velocity.

The averaged minimum velocity magnitudes of stable terrain are variable for the Sourdough image pairs, generally varying around a value of approximately 20 $\frac{\text{cm}}{\text{yr}}$. There is a trend of decreasing uncertainty with increasing time interval between image, though other factors such as a warped orthomosaic (August 2016) or a snow-covered surface (October 2020) contribute to these uncertainty values. Because the standard deviations of the vector components are generally random random, showing that there is no directional bias to these minimum velocity values, the value of ±20 $\frac{cm}{yr}$ (total range of 40 $\frac{cm}{yr}$) agrees with the range of 40 cm measured from the difference between the velocity peaks along a common profile for all image pairs. Therefore, we assume the random error in the Alaskan airborne datasets to be approximately ±20 $\frac{cm}{yr}$, while the systemic error approaches values less than 5 $\frac{cm}{yr}$, especially for longer time intervals.

Systematic biases likely exist in local regions of the individual image pairs. For example, in the May 2020/June 2021 image pair, the lower lobe of the rock glacier exhibits a velocity magnitude much greater than the magnitudes at the same area in the other image pairs, where the velocity decreases with proximity to the terminus (Figure 6). We suggest that this systematic bias in the May 2020/June 2021 image pair results from effects in the photogrammetric processing step, which could have led to geometric warping of the orthomosaics. The processing reports generated by the Agisoft Metashape software indicate that these two images had the highest percentage of regions on the rock glacier where the image overlap was less than nine (see the supplementary materials for the processing reports), which supports the hypothesis that the systematic error in this image pair is due to artifacts from the photogrammetric processing. Additionally, nonzero means of stable terrain velocity magnitudes combined with unequal standard deviations are indicative of systematic and/or nonuniform uncertainty distributions, such as that of the May 2014/June 2016 image pair at Sourdough (Table 3). Future studies should visually identify regions that may contain warping of the orthomosaic and the associated increase in uncertainty in order to avoid misinterpretation of the physical implications of the velocity field.

To further understand the relative quality of the change detection results for different image pairs, we examined the distributions of output velocities and maximum correlation coefficients for Sourdough and Galena Creek. We used the eight image pairs at Sourdough to examine the characteristics of velocity and correlation coefficient distributions. Qualitatively, one way to compare velocity measurements for different image pairs is to compare the tails of the distribution, where the output velocity is greater than the highest expected real velocity (about 2 $\frac{m}{yr}$ for Sourdough). For example, the August 14/August 2015 results for Sourdough have fewer outlying velocity values than those for May 14/May 15 (Figure 11). In Figure 6, it can be observed that the velocity field for August 2014/August 2015 has a lower baseline uncertainty than for May 14/May 15, leading to the conclusion that the quality of the results can partially be characterized by the number of outlying velocity magnitudes. These observations can be used to weight the velocity vector fields of specific image pairs during future kinematic analyses.

The distributions of the maximum correlation coefficients output by the CIAS algorithm provide another assessment of the reliability of the results for each image pair. These histograms (Figure 12) show a relationship between the width of the distribution of the maximum correlation coefficients and the quality of the change detection results. Two of the highest quality velocity fields as assessed by baseline uncertainty and flow signal continuity are August 2014/August 2015 and August 2016/September 2019. These two image pairs return the narrowest distribution of the maximum correlation coefficients. On the other hand, May 2014/May 2015 and September 2019/October 2020 have relatively low quality results, and broader distributions of the maximum correlation coefficients are observed. This effect is especially apparent for the September 2019/October 2020 pair.

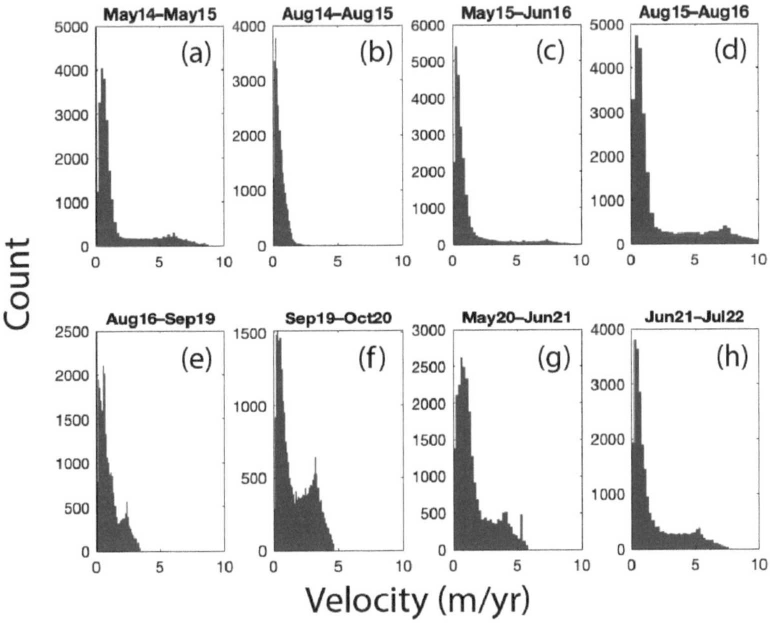

Figure 11. Histograms showing the distribution of velocity magnitudes for each image pair at Sourdough with a time interval of one year or longer. Each panel (**a**–**h**) corresponds with the results presented in the same panel labeled in Figure 6.

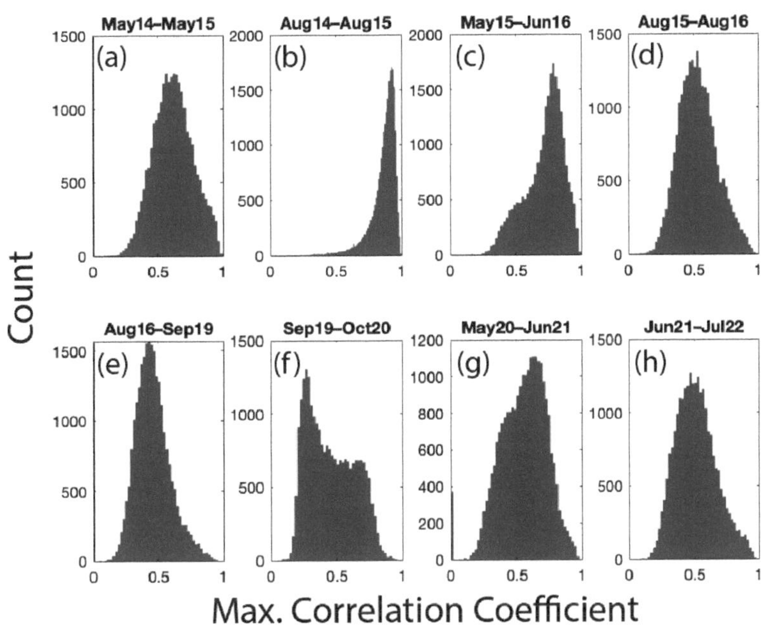

Figure 12. Histograms showing the distribution of the maximum correlation coefficients for each image pair at Sourdough with a time interval of one year or longer. Each panel (**a**–**h**) corresponds with results presented in the same panel labeled in Figure 6.

We tested these hypotheses with regard to the relationships between the number of outlying data points, the maximum correlation coefficient distribution, and the quality of change detection results by plotting the histograms for the August 2020/August 2022 image pair and the July 2021/August 2022 image pair at Galena Creek. Because the both the August 2020 and Augusut 2022 images were acquired using the UAS, the velocity field derived from this image pair has a low baseline uncertainty and a low amount of noisy regions in the velocity field. In comparison, the July 2021 image is a lower resolution and lower precision satellite product, leading to a higher baseline uncertainty and noise value for the July 2021/August 2022 velocity field. The distributions for Galena Creek (Figure 13) support our hypothesis that the velocity field quality can be characterized by both the size of the tail of outlying velocity magnitudes and the width of the distribution of the maximum correlation coefficients. In the case of the higher quality image pair at Galena Creek (August 2020/August 2022), the distributions contain a lower number of velocity magnitudes greater than 2 $\frac{\text{m}}{\text{yr}}$ and the peak of the maximum correlation coefficient distribution is narrower in comparison with the July 2021/August 2022 pair. Our observations of uncertainty patterns in the change detection results can be used to assess the propagation of error for future analyses using the velocity fields presented here. Evaluating the benefits and limitations of UAS, airborne, and satellite imaging platforms in regard to the measurement of rock glacier surface motion will contribute to the planning requirements of ongoing and future data acquisition campaigns [46,47].

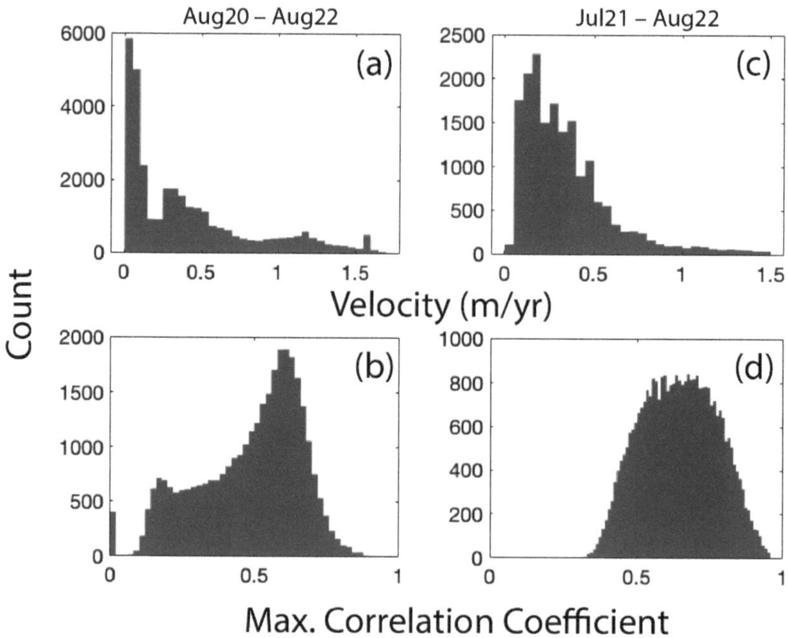

Figure 13. Histograms showing the distributions of velocity magnitude (**a**) and maximum correlation coefficient (**b**) for the August 2020/August 2022 image pair at Galena Creek and the velocity magnitude (**c**) and maximum correlation coefficient (**d**) for the July 2021/August 2022 hybrid image pair at Galena Creek.

4.2. Interpreted Flow History

The timescale required to transport a debris clast from the head of the glacier to the toe can estimated using the measured velocity fields, providing an estimate for the age of initial rock glacier accumulation. To obtain the age A_x at distance x from the rock glacier headwall, we integrate a smoothed profile of the inverse of calculated velocity magnitudes v_x (moving average window width = 5 pixels) along the central flowline for each rock glacier:

$$A_x = \int_0^x v_x^{-1}\, dx. \quad (1)$$

This method assumes a time-invariant velocity field along an interpreted flow path. The velocity field for Galena Creek has likely been dependent upon surface slope throughout its history, as it is today (Figure 14a); if the rock glacier was previously thicker and flowed faster due to higher driving stress, then our assumption of a constant velocity field would provide an upper bound for the age of the ice along the profile. Integrating the surface velocity profile of Galena Creek produces a terminus age of 3070 years (Figure 14a). This age falls between the estimated early neoglacial advance in Wyoming about 4000 years ago and the Audubon advance approximately 2000 to 1000 years ago [28]. Assuming that our age estimate is an upper bound, our results are most consistent with the rock glacier terminus of Galena Creek originating from the Audubon advance, while the debris-covered glacier comprising the upper two-thirds of the Galena Creek system contains ice that accumulated during the Little Ice Age (LIA), which spanned approximately the last half-millennium in the American Cordillera [48]. It is likely that this LIA advance interacted with the pre-existing rock glacier system, similarly to other landforms observed in the Swiss Alps and Chilean Andes [32–34]. This interaction of glacier ice and permafrost resulted in the complex topography and variable ice distribution found at the inflection in topography where they presently meet [31].

Our Galena Creek age profile is consistent with two calibrated radiocarbon measurements in leaf fragments at locations along the center flowline of Galena Creek (Figure 14a) [12]. The radiocarbon age acquired 100 m from the cirque headwall is 0–310 calendar years before present, and our velocity-derived age at this location is 180 years. Similarly, the radiocarbon age 800 m along the flow profile is 1410–1730 years before present, while the velocity-derived age is 1450 years. These independent measurements suggest that our method of age estimation is suitable for Galena Creek, where the low width/thickness ratio has allowed the velocity field to remain relatively consistent throughout the recent history of the rock glacier. Although Sulphur Creek is a close neighbor to Galena Creek, we did not perform a velocity profile analysis here, as the mid-glacier displacement vectors and elevation change since 1985 indicate substantial stagnation and surface subsidence in the past few decades, increasing the width/thickness ratio and likely invalidating the assumption of a time-invariant velocity field over the past few centuries.

We used the velocity profiles for three of the highest quality change detection results for Sourdough (August 2014/August 2015, August 2016/September 2019, and May 2014/July 2022) to better understand the propagation of uncertainty in the change detection results to the rock glacier age estimates (Figure 14b). The three profiles show rock glacier terminus ages ranging between 3,310 and 3,540 years. This range of ages is roughly consistent with the oldest of four late Holocene advances inferred from radiocarbon and tree ring dates [49], and although this indicates that these sites are older than the LIA, this result does not support the hypothesis that the ice in this population of rock glaciers in the Wrangell Mountains is related to the advance of the last glacial maximum (LGM), generally considered to be much more than 10,000 years ago.

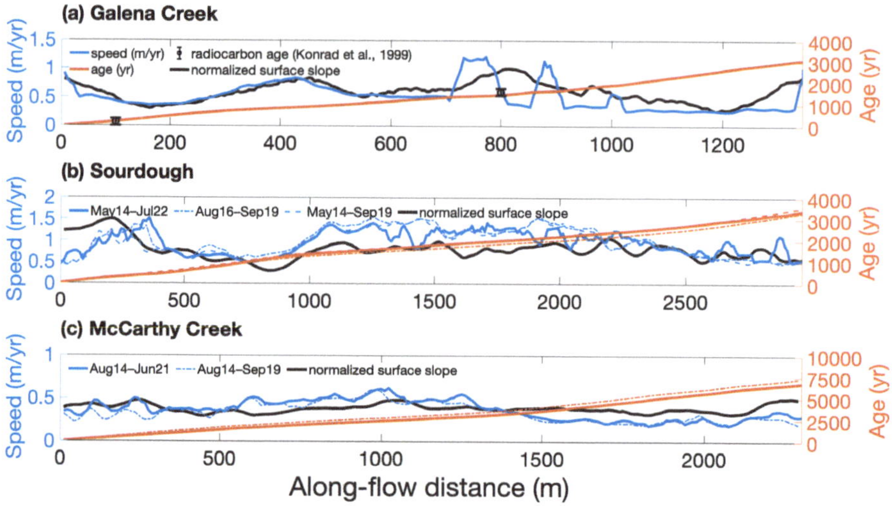

Figure 14. Velocity magnitudes (blue), age profiles (red), and normalized surface slopes (black) for (**a**) Galena Creek, including calibrated radiocarbon ages from [12] along the flow profile, (**b**) Sourdough, and (**c**) McCarthy Creek.

McCarthy Creek, just 2 km north of Sourdough, has a maximum velocity that is a factor of two less than that of Sourdough while covering a relatively similar length (Figure 14c). Using the August 2014/June 2021 velocity results for McCarthy Creek, which is the longest time interval without surface obfuscation due to snow, the estimated terminus age is approximately 6680 years. The August 2014/September 2019 image pair yields another high quality velocity field, and returns an age estimate of 7330 years. This range is about double the estimated age for Sourdough. Although this is an older age range, it is not consistent with an advance related to the LGM. While there is uncertainty in the exact path of a surface particle in comparison with our estimated flowline profiles, we do not expect this potential source of error to be the primary cause of the factor-of-two difference in the age calculations. Instead, this difference in estimated age for the two neighboring rock glaciers suggests the existence of local heterogeneities in rock glacier evolution, even if we assume that they are both related to the documented late Holocene ice advances in the Wrangell Mountains.

Heterogeneities in rock glacier evolution could influence variance in the surface velocity fields over time, refuting this age estimation's assumption of a time-invariant velocity field. Possible sources of differing velocity field evolution between the northward-flowing McCarthy Creek and southward-flowing Sourdough include the effect of slope aspect on accumulation and surface temperature as well as different series of rock glacier surges overriding less active older lobes. For example, if McCarthy Creek is in the process of stagnating while Sourdough's activity remains constant, our method would estimate an older age for McCarthy Creek due to the implication that slower surface velocities take longer to transport surface material along the length of the rock glacier. One possible line of evidence for a velocity field that has changed over time is the morphology and velocity distributions of the different lobes at Sourdough and McCarthy Creek. Sourdough consists of one major active lobe, where the peak velocity occurs within the steep trunk of the rock glacier, and one smaller active lobe to the east, where the velocity appears to be correlated to the surface slope as well. The western flank of Sourdough has two small steep lobes which are presently inactive (Figure 1a). These minor lobes are interpreted to be remnant overflow deposits from a past rock glacier advance when its thickness was greater than at present, indicating that this zone of maximum velocity has likely been correlated with the steep trunk of the rock glacier throughout its history. Even though the age estimate for Sourdough is taken to be an upper bound, we assume the relative velocities along its longitudinal profile have maintained similar trends in correlation with surface slope.

By contrast, McCarthy Creek consists of a large stagnant lobe to the north of the main active lobe, where the peak velocity is half that of Sourdough and which is found at the upper reaches of the rock glacier as opposed to the mid-glacier trunk, as it is at Sourdough. Sourdough is not substantially steeper than McCarthy Creek. In addition, McCarthy Creek displays no overflow lobes, suggesting that its movement may be more limited by its bedrock geometry and that the velocity field may be more susceptible to decreases in ice accumulation. Because the northern lobe of McCarthy Creek is presently stagnant and the peak velocity appears to have a lower correlation with surface slope (Figure 14c), it may be reasonable to assume that the northern lobe was previously more active and that the peak velocity would have been greater at the high slope regions on the rock glacier when the ice unit was thicker. All of these observations support the inference that McCarthy Creek has slowed over time, implying an overestimate of its total age; however, further work is needed to determine whether these differences in evolution between Sourdough and McCarthy Creek stem from heterogeneities in ice accumulation, debris input, valley geometry, the effects of slope aspect on insolation, or a combination of these processes.

In both Alaska and Wyoming, neighboring rock glaciers exhibit differences in flow rate distribution, suggesting that certain local controls may be influencing each rock glacier's evolution. As discussed above, the Alaskan rock glaciers differ significantly in their maximum flow speed, with Sourdough approaching 1.5 $\frac{\mathrm{m}}{\mathrm{yr}}$ and McCarthy Creek never exceeding 0.6 $\frac{\mathrm{m}}{\mathrm{yr}}$. Additionally, in Wyoming, Galena Creek's flow direction correlates with the down-valley topographical gradient and the flow velocity magnitude correlates with the magnitude of the longitudinal surface slope. Conversely, neighboring Sulphur Creek's velocity magnitude is greatest at the toe, and does not appear to be correlated with the longitudinal surface slope. The middle section of the glacier appears to be flowing perpendicular to the down-valley topography rather than parallel to it. This observation, combined with significant surface subsidence and a measured ice thicknesses of only about 10 m at this same location [31], reveals a recent rapid destabilization of the debris-covered ice in the upper Sulphur Creek basin. Although Galena Creek exhibits surface lowering as well, a comparison of its velocity-derived age with radiocarbon ages suggests no major deviations between its past and present velocity fields.

5. Conclusions

Our analysis demonstrates the capabilities and limitations of using multiple combinations of repeated imagery acquisition methods to perform photogrammetric change detection as a means of measuring rock glacier surface motion. The imagery acquired with the Phantom 4 RTK UAS had the highest success rate for detecting rock glacier surface flow due to this method's high image resolution and positioning accuracy. This effect is most apparent for the August 2020/August 2022 image pair at Galena Creek, Wyoming, where both images were acquired with the UAS. With this image pair, a strong flow signal was detected with a baseline measurement uncertainty of approximately 5 $\frac{\mathrm{cm}}{\mathrm{yr}}$. Airborne photogrammetry successfully detects rock glacier surface motion over annual time intervals, although the slightly diminished resolution and positioning accuracy due to the increased flight speed and altitude can propagate to the change detection results in comparison with the UAS-derived datasets. At each Alaska site (Sourdough and McCarthy Creek), strong flow signals were detected across the rock glacier surface with airborne imagery, especially for imaging intervals of three years or longer. The baseline uncertainty (20 $\frac{\mathrm{cm}}{\mathrm{yr}}$) is higher with the airborne method than that for image pairs where both images were acquired with the UAS.

While high-resolution satellite imagery provides the potential for consistent monitoring of rock glaciers, there is a tradeoff between this logistical convenience and the limitations of resolution and positioning for civilian data acquired from an orbital platform.

All of the orbital imagery used in this study required manual repositioning to improve co-registration with the images they were paired with, and the relatively coarse resolution limited the lower bound of the annual displacement magnitudes detectable by this method in comparison with the UAS and airborne platforms. The satellite imagery for both Wyoming field sites detected flow signals when paired with UAS and airborne data; however, the baseline uncertainty (40 cm) and the number of spurious displacement vectors caused by feature mismatches in the change detection were both the highest for all of the methods used.

A preliminary analysis of the rock glacier velocity fields shows that all of the study sites in Alaska and Wyoming likely originated during the early to middle Holocene, after the LGM and before the LIA. In addition to horizontal change detection, we used a combination of digital elevation models that are publicly available or generated with our photogrammetric processing method to estimate current thinning rates for each site. We found that both Wyoming sites exhibit a thinning signal that is consistent with the meteorological data and measured debris thickness, while neither of the the Alaska sites exhibits surface elevation change consistent with rock glacier thinning.

Due to the slow flow of rock glaciers in comparison with glaciers, all of the imagery applied to change detection experiments must have high-precision positioning information to reduce measurement error and detect a flow signal, especially when the time sampling interval is one year or less. This level of precision is vital for future studies focusing on remote sensing of seasonal patterns in rock glacier flow. With current technological resources, UAS imagery has the best resolution and positioning; however, it is limited to small spatial footprints at locations and times where the target is directly accessible. While airborne imagery provides a broader spatial extent and greater ease of data acquisition at regular time intervals for less accessible sites, it sacrifices resolution and precision. Satellite imagery offers a solution for regularly sampling a wide spatial area multiple times per year; however, the presently available data lack the resolution and precise positioning needed to achieve the lower uncertainty levels of UAS or airborne imagery. Increasing the number of high-resolution satellite imaging constellations for environmental studies would further improve the spatial and temporal capability to measure and monitor rock glaciers. We intend for the results of this study to be analyzed further using established glaciological principles, and our observations relating to the application of multi-platform change detection should be considered when planning campaigns to measure rock glacier surface motion.

Supplementary Materials: The following supporting information can be downloaded at: https://www.mdpi.com/article/10.3390/rs15194779/s1. Figure S1: Workflow diagram showing the methodology for combining UAS, airborne, and satellite data to measure each rock glacier's horizontal displacement and elevation change. Figure S2: Sourdough change detection using May 2014 base image. Figure S3: Galena Creek elevation change profile. Figure S4: Sourdough elevation profiles. Figure S5: McCarthy Creek elevation profiles. Figure S6: Surface boulder velocity measurements from 1997–2022 and comparison with remote change detection results.

Author Contributions: Conceptualization, all authors; methodology, all authors; software, T.M.M., R.A. and E.I.P.; validation, T.M.M. and R.A.; formal analysis, T.M.M., R.A. and E.I.P.; investigation, C.F.L., M.S.C., R.A., J.W.H., E.I.P. and T.M.M.; resources, J.W.H. and C.F.L.; data curation, C.F.L., M.S.C., R.A. and T.M.M.; writing—original draft preparation, T.M.M., R.A., E.I.P. and M.S.C.; writing—review and editing, all authors; visualization, T.M.M., E.I.P. and R.A.; supervision, J.W.H., J.S.L. and C.F.L.; project administration, J.W.H., J.S.L. and C.F.L.; funding acquisition, J.W.H., J.S.L. and C.F.L. All authors have read and agreed to the published version of the manuscript.

Funding: This research was funded by the NASA Solar System Workings program, grant number 80NSSC19K0561.

Data Availability Statement: The photogrammetry data, processed orthomosaics, DEMs, photogrammetric processing reports, horizontal change detection results, vertical elevation difference products, surface-based boulder position measurements, and the associated analysis code are openly available from the University of Arizona Research Data Repository (ReDATA) with the following DOI: https://doi.org/10.25422/azu.data.23272220.

Acknowledgments: We thank Kestrel Aerial Services for collecting the Wyoming airborne data presented here. This work utilized SkyMap50 satellite data made available through Soar. Extreme gratitude is felt for the field assistance provided by Brandon Tober, Stefano Nerozzi, Michael Daniel, and Rishi Chandra at the Wyoming sites and by Stefano Nerozzi, Tyler Kuehn, Victor Devaux-Chupin, and Bruno Belotti at Sourdough, Alaska. Many thanks to Noel Potter, Jr. and Noel Lewis Potter for sharing their boulder position data and for their valuable discussion regarding the history of exploration and glaciological research at the Wyoming field sites. Eric Yould provided essential logistical support throughout the years at the Alaska field sites. Thank you to three peer reviewers, whose comments and suggestions greatly improved this manuscript.

Conflicts of Interest: The authors declare no conflict of interest.

Abbreviations

The following abbreviations are used in this manuscript:

CIAS	Correlation Image Analysis Software
GNSS	Global Navigation Satellite System
GPR	Ground-Penetrating Radar
GDAL	Geospatial Data Abstraction Library
UAS	Uncrewed Aerial System
DEM	Digital Elevation Model
GCP	Ground Control Point
CP	Check Point
RMS	Root Mean Square
RTK	Real-Time Kinematic
LIA	Little Ice Age
LGM	Last Glacial Maximum
SNOTEL	Snowpack Telemetry

Appendix A. Thermal Conduction Model

To validate the remotely measured values of surface elevation change, we applied a simple melt model via 1D thermal conduction through the supraglacial debris layer. The objective of this model is to calibrate a realistic value for the thermal conductivity of the debris using the thinning rate measured with differenced DEMs at Galena Creek alongside temperature and snow depth data from the *Evening Star* meteorological station, which belongs to the SNOTEL network operated by the United States Department of Agriculture (station ID = 472). This station is located in the adjacent valley to the east of Galena Creek.

Using the SNOTEL data between the dates of 23 August 2020, and 8 August 2022, the melt rate was fixed to zero for dates with nonzero snow depth. For dates with zero snow, the melt rate ($\frac{\text{m}}{\text{day}}$) is provided by the following:

$$M = k\frac{dT}{dz}\frac{86{,}400}{\rho_{ice}L_f}, \tag{A1}$$

where k is the thermal conductivity of the debris layer, dT is simplistically assumed to be the SNOTEL air temperature measurement (T_{ice} = 0 °C), dz is the thickness of the debris layer, ρ_{ice} is the density of the ice (900 $\frac{\text{kg}}{\text{m}^3}$), and L_f is the latent heat of fusion (334,000 $\frac{\text{J}}{\text{kg}}$) [50]. For Galena Creek, we assume the debris thickness to be 1.5 m based on previous ice exposure observations and GPR measurements [1,3,31]. We test thermal conductivity values spanning the range of 0.3–1.8 $\frac{\text{W}}{\text{mK}}$, which is a feasible range for supraglacial debris, although it is likely that rock glacier debris resides on the lower end of this range due to its relatively high porosity [51,52]. The interpreted result of 20 cm cumulative melt during the

2020–2022 interval for Galena Creek is consistent with a thermal conductivity of 0.42 $\frac{W}{mK}$, approaching the lower bound of the plausible range (Figure A1b).

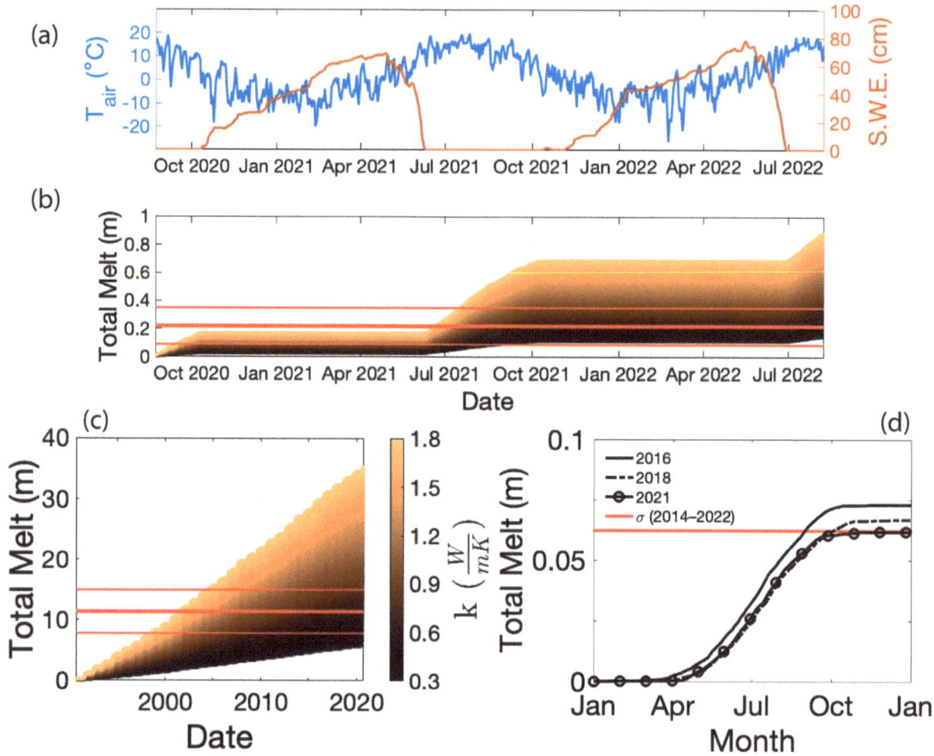

Figure A1. (**a**) Air temperature and snow water equivalent (SWE) data reported for the *Evening Star* SNOTEL site between 23 August 2020, and 8 August 2020. (**b**) Modeled melt over that same time interval for a range of possible thermal conductivity values for the debris layer and a debris thickness of 1.5 m. The horizontal red lines indicate the mean (thick line) and standard deviation (thin lines) of the photogrammetrically measured elevation change at upper Galena Creek. (**c**) Modeled melt using the *Evening Star* SNOTEL data between 1 September 1990, and 1 September 2020 for a debris thickness of 0.5 m as measured at upper Sulphur Creek in August 2020. The horizontal red lines indicate the expected mean (thick line) and standard deviation (thin lines) of the cumulative melt for that time period using melt rates measured between 1985 and 2020. (**d**) Modeled melt for Sourdough using air temperatures measured by an automated weather station near the rock glacier's toe, assuming a value of 3 m for the debris thickness and a thermal conductivity of 0.42 $\frac{W}{mK}$. The red line shows the standard deviation of the annual elevation change rate measured photogrammetrically between 2014 and 2022.

We applied the same thermal conduction model to the *Evening Star* SNOTEL data for the dates between 1 September 1990 and 1 September 2020 using a debris thickness of 0.5 m in order to test the remotely sensed cumulative surface lowering of between 18–28 m for upper Sulphur Creek over the 1985–2020 time interval. Although the SNOTEL data lack five years of data in comparison with the DEM interval, the thermal conduction results support the observation that upper Sulphur Creek thinned at a rate of tens of cm per year over three decades for plausible conductivities of the debris (Figure A1c), leading to cumulative melt well over 10 m for that period, further validating our photogrammetric results for Sulphur Creek.

We additionally tested the result of negligible surface elevation change for Sourdough and McCarthy Creek with the thermal conduction model using air temperature data acquired from an automated weather station on the lower lobe of Sourdough. Due to gaps in field campaigns, only the years 2016, 2018, and 2021 contained a complete and continuous calendar year of temperature data. We calculated the melt at Sourdough during each of these years using dz = 3 m and k = 0.42 $\frac{\mathrm{W}}{\mathrm{mK}}$ (Figure A1d). Each of these years returns melt rates less than 10 $\frac{\mathrm{cm}}{\mathrm{yr}}$. Further, these melt rates are likely overestimates, as this weather station does not measure the presence of snow; thus, this conditional step is removed from the melt rate calculation. Removing the condition that the melt rate equals zero when the snow depth is greater than zero leads to an increase in the estimated melt rates for both Galena Creek and Sulphur Creek by a factor of approximately 1.3. After correcting the Sourdough melt rates for this factor, the resulting these rates are within about 1 $\frac{\mathrm{cm}}{\mathrm{yr}}$ of the standard deviation of the 2014–2022 elevation difference product.

Additionally, the location of the weather station at the toe of the rock glacier may bias this calculation towards higher melt rates, as its elevation at the bottom of the rock glacier implies that it has the highest temperature on the surface of the rock glacier assuming normal atmospheric lapse rates. In addition to the uncertainty in the temperature change with increasing elevation and the relationship of the air temperature to the debris surface temperature, this thermal conduction model is subject to uncertainties in the debris thickness. Although the meteorological data indicate that melt may be occurring at Sourdough and McCarthy Creek, the surface change due to this melt falls within the measurement uncertainty of the DEMs used to calculate the surface change over the 2014–2021 interval for McCarthy Creek and the 2014-2022 interval for Sourdough. Future refinement of rock glacier melt rate estimates should consider the effects of nonconductive heat fluxes [52].

References

1. Potter, N. Ice-Cored Rock Glacier, Galena Creek, Northern Absaroka Mountains, Wyoming. *Geol. Soc. Am. Bull.* **1972**, *83*, 3025–3057. [CrossRef]
2. Anderson, R.S.; Anderson, L.S.; Armstrong, W.H.; Rossi, M.W.; Crump, S.E. Glaciation of alpine valleys: The glacier–debris-covered glacier–rock glacier continuum. *Geomorphology* **2018**, *311*, 127–142. [CrossRef]
3. Petersen, E.I.; Levy, J.S.; Holt, J.W.; Stuurman, C.M. New insights into ice accumulation at Galena Creek Rock Glacier from radar imaging of its internal structure. *J. Glaciol.* **2019**, *66*, 1–10. [CrossRef]
4. Wahrhaftig, C.; Cox, A. Rock glaciers in the Alaska Range. *Geol. Soc. Am. Bull.* **1959**, *70*, 383–436. [CrossRef]
5. Haeberli, W.; Hallet, B.; Arenson, L.; Elconin, R.; Humlum, O.; Kääb, A.; Kaufmann, V.; Ladanyi, B.; Matsuoka, N.; Springman, S.; et al. Permafrost creep and rock glacier dynamics. *Permafr. Periglac. Process.* **2006**, *17*, 189–214. [CrossRef]
6. IPA Action Group; Rock Glacier Inventories and Kinematics (RGIK). Towards Standard Guidelines for Inventorying Rock Glaciers: Baseline Concepts (Version 4.2.2). Available online: www.rgik.org (accessed on 20 April 2023).
7. Kääb, A.; Vollmer, M. Surface Geometry, Thickness Changes and Flow Fields on Creeping Mountain Permafrost: Automatic Extraction by Digital Image Analysis. *Permafr. Periglac. Process.* **2000**, *11*, 315–326. [CrossRef]
8. Outcalt, S.I.; Benedict, J.B. Photo-Interpretation of two Types of Rock Glacier in the Colorado Front Range, U.S.A. *J. Glaciol.* **1965**, *5*, 849–856. [CrossRef]
9. White, S.E. Rock Glacier Studies in the Colorado Front Range, 1961 to 1968. *Arct. Alp. Res.* **1971**, *3*, 43–64. [CrossRef]
10. Francou, B.; Reynaud, L. 10 year surficial velocities on a rock glacier (Laurichard, French Alps). *Permafr. Periglac. Process.* **1992**, *3*, 209–213. [CrossRef]
11. Berthling, I.; Etzelmüller, B.; Eiken, T.; Sollid, J.L. Rock glaciers on Prins Karls Forland, Svalbard. I: Internal structure, flow velocity and morphology. *Permafr. Periglac. Process.* **1998**, *9*, 135–145. [CrossRef]
12. Konrad, S.K.; Humphrey, N.F.; Steig, E.J.; Clark, D.H.; Potter, N.; Pfeffer, W.T. Rock glacier dynamics and paleoclimatic implications. *Geology* **1999**, *27*, 1131–1134. [CrossRef]
13. Bucki, A.K.; Echelmeyer, K.A. The flow of Fireweed rock glacier, Alaska, U.S.A. *J. Glaciol.* **2004**, *50*, 76–86. [CrossRef]
14. Bertone, A.; Barboux, C.; Bodin, X.; Bolch, T.; Brardinoni, F.; Rafael Caduff, R.; Christiansen, H.H.; Darrow, M.M.; Delaloye, R.; Etzelmüller, B.; et al. Incorporating InSAR kinematics into rock glacier inventories: Insights from 11 regions worldwide. *Cryosphere* **2022**, *16*, 2769–2792. [CrossRef]
15. Brencher, G.; Handwerger, A.L.; Munroe, J.S. InSAR-based characterization of rock glacier movement in the Uinta Mountains, Utah, USA. *Cryosphere* **2021**, *15*, 4823–4844. [CrossRef]
16. Liu, L.; Millar, C.I.; Westfall, R.D.; Zebker, H.A. Surface motion of active rock glaciers in the Sierra Nevada, California, USA: Inventory and a case study using InSAR. *Cryosphere* **2013**, *7*, 1109–1119. [CrossRef]

17. Kääb, A. Remote Sensing of Mountain Glaciers and Permafrost Creep. *Schriftenreihe Physische Geographie Glaziologie und Geomorphodynamik*; Geographisches Institut der Universität Zürich: Zürich, Switzerland, 2005; Volume 48.
18. Robson, B.A.; MacDonell, S.; Ayala, Á.; Bolch, T.; Nielsen, P.R.; Vivero, S. Glacier and rock glacier changes since the 1950s in the La Laguna catchment, Chile. *Cryosphere* **2022**, *16*, 647–665. [CrossRef]
19. Gärtner-Roer, I.; Brunner, N.; Delaloye, R.; Haeberli, W.; Kääb, A.; Thee, R. Glacier–permafrost relations in a high-mountain environment: 5 decades of kinematic monitoring at the Gruben site, Swiss Alps. *Cryosphere* **2022**, *16*, 2083–2101. [CrossRef]
20. Vivero, S.; Bodin, X.; Farías-Barahona, D.; MacDonell, S.; Schaffer, N.; Robson, B.A.; Lambiel, C. Combination of Aerial, Satellite, and UAV Photogrammetry for Quantifying Rock Glacier Kinematics in the Dry Andes of Chile (30° S) Since the 1950s. *Front. Remote Sens.* **2021**, *2*, 784015. [CrossRef]
21. Lei, Y.; Gardner, A.; Agram, P. Autonomous Repeat Image Feature Tracking (autoRIFT) and Its Application for Tracking Ice Displacement. *Remote Sens.* **2021**, *13*, 749. [CrossRef]
22. Kenner, R.; Phillips, M.; Beutel, J.; Hiller, M.; Limpach, P.; Pointner, E.; Volken, M. Factors Controlling Velocity Variations at Short-Term, Seasonal and Multiyear Time Scales, Ritigraben Rock Glacier, Western Swiss Alps. *Permafr. Periglac. Process.* **2017**, *28*, 675–684. [CrossRef]
23. Heid, T.; Kääb, A. Repeat optical satellite images reveal widespread and long term decrease in land-terminating glacier speeds. *Cryosphere* **2012**, *6*, 467–478. [CrossRef]
24. Delaloye, R.; Lambiel, C.; Gärtner-Roer, I. Overview of rock glacier kinematics research in the Swiss Alps. *Geogr. Helv.* **2010**, *65*, 135–145. [CrossRef]
25. Jones, D.B.; Harrison, S.; Anderson, K.; Betts, R.A. Mountain rock glaciers contain globally significant water stores. *Sci. Rep.* **2018**, *8*, 2834. [CrossRef] [PubMed]
26. Barsch, D. The problem of the ice-cored rock glacier. In *Rock Glaciers*; Giardino, J.R., Shroder, J.E., Jr., Vitek, J.D., Eds.; Allen & Unwin: Boston, MA, USA, 1987; pp. 45–53.
27. Clark, D.H.; Steig, E.J.; Potter, N.; Gillespie, A.R. Genetic variability of rock glaciers. *Geogr. Ann. Ser. A Phys. Geogr.* **1998**, *80*, 175–182. [CrossRef]
28. Ackert, R.P., Jr. A Rock Glacier/Debris-Covered Glacier System at Galena Creek, Absaroka Mountains, Wyoming. *Geogr. Ann. Ser. A Phys. Geogr.* **1998**, *80*, 267–276. [CrossRef]
29. Potter, N.; Steig, E.J.; Clark, D.H.; Speece, M.A.; Clark, G.M.; Updike, A.B. Galena Creek rock glacier revisited—New observations on an old controversy. *Geogr. Ann. Ser. A Phys. Geogr.* **1998**, *80*, 251–265. [CrossRef]
30. Steig, E.J.; Fitzpatrick, J.J.; Potter, N.; Clark, D.H. The geochemical record in rock glaciers. *Geogr. Ann. Ser. A Phys. Geogr.* **1998**, *80*, 277–286. [CrossRef]
31. Meng, T.M.; Petersen, E.I.; Holt, J.W. Rock glacier composition and structure from radio wave speed analysis with dipping reflector correction. *J. Glaciol.* **2023**, *69*, 639–657. [CrossRef]
32. Kunz, J.; Kneisel, C. Glacier–permafrost interaction at a thrust moraine complex in the glacier forefield Muragl, Swiss Alps. *Geosciences* **2020**, *10*, 205. [CrossRef]
33. Bodin, X.; Rojas, F.; Brenning, A. Status and evolution of the cryosphere in the Andes of Santiago (Chile, 33.5° S.). *Geomorphology* **2010**, *118*, 453–464. [CrossRef]
34. Wee, J.; Delaloye, R. Post-glacial dynamics of an alpine Little Ice Age glacitectonized frozen landform (Aget, western Swiss Alps). *Permafr. Periglac. Process.* **2022**, *33*, 370–385. [CrossRef]
35. Potter, N. (Dickinson College, Carlisle, PA, USA). Personal communication, 2022.
36. Östrem, G. Ice Melting under a Thin Layer of Moraine, and the Existence of Ice Cores in Moraine Ridges. *Geogr. Ann.* **1959**, *41*, 228–230. [CrossRef]
37. Petersen, E.I.; Holt, J.W.; Levy, J.S.; Meng, T.M.; Tober, B.S.; Christoffersen, M.; Stuurman, C.M.; Cardenas, B. The transition from Alpine Glacier to Rock Glacier at Sulphur Creek, Wyoming. In Proceedings of the American Geophysical Union Fall Meeting, San Francisco, CA, USA, 9–13 December 2019; Abstract Number C41E–1506.
38. Potter, N.L.; Potter, N.; Retelle, M.J. Continued Movement and Ablation Monitoring, Galena Creek Rock Glacier, Absaroka Mountains, WY. In Proceedings of the Geological Society of America Annual Meeting, Denver, CO, USA, 25 September 2016; Abstract Number 81-14.
39. Nolan, M.; Larsen, C.; Sturm, M. Mapping snow depth from manned aircraft on landscape scales at centimeter resolution using structure-from-motion photogrammetry. *Cryosphere* **2015**, *9*, 1445–1463. [CrossRef]
40. Heid, T.; Kääb, A. Evaluation of existing image matching methods for deriving glacier surface displacements globally from optical satellite imagery. *Remote Sens. Environ.* **2012**, *118*, 339–355. [CrossRef]
41. U.S. Geological Survey. USGS 1/3 Arc Second n45w110: U.S. Geological Survey. 2023. Available online: www.sciencebase.gov/catalog/item/63c78c29d34e06fef14edbd2 (accessed on 13 February 2023).
42. Gesch, D.B.; Oimoen, M.J.; Evans, G.A. Accuracy assessment of the U.S. Geological Survey National Elevation Dataset, and comparison with other large-area elevation datasets—SRTM and ASTER. In *U.S. Geological Survey Open-File Report 2014–1008*; U.S. Geological Survey: Wriston, WV, USA, 2014
43. Debella-Gilo, M.; Kääb, A. Sub-pixel precision image matching for measuring surface displacements on mass movements using normalized cross-correlation. *Remote Sens. Environ.* **2011**, *115*, 130–142. [CrossRef]

44. Wirz, V.; Gruber, S.; Purves, R.S.; Beutel, J.; Gärtner-Roer, I.; Gubler, S.; Vieli, A. Short-term velocity variations at three rock glaciers and their relationship with meteorological conditions. *Earth Surf. Dyn.* **2016**, *4*, 103–123. [CrossRef]
45. Taylor, J.R. *An Introduction to Error Analysis: The Study of Uncertainties in Physical Measurements*, 2nd ed.; University Science Books: Sausalito, CA, USA, 1997.
46. Vivero, S.; Hendrickx, H.; Frankl, A.; Delaloye, R.; Lambiel, C. Kinematics and geomorphological changes of a destabilising rock glacier captured from close-range sensing techniques (Tsarmine rock glacier, Western Swiss Alps). *Front. Earth Sci.* **2022**, *10* , 1017949. [CrossRef]
47. Kaufmann, V.; Kellerer-Pirklbauer, A.; Seier, G. Conventional and UAV-Based Aerial Surveys for Long-Term Monitoring (1954–2020) of a Highly Active Rock Glacier in Austria. *Front. Remote Sens.* **2021**, *2*, 732744. [CrossRef]
48. Davis, P. Holocene glacier fluctuations in the American Cordillera. *Quat. Sci. Rev.* **1988**, *7*, 129–157. [CrossRef]
49. Wiles, G.C.; Jacoby, G.C.; Davi, N.K.; McAllister, R.P. Late Holocene glacier fluctuations in the Wrangell Mountains, Alaska. *Geol. Soc. Am. Bull.* **2002**, *114*, 896–908. [CrossRef]
50. Cuffey, K.M.; Paterson, W.S.B. *The Physics of Glaciers*, 4th ed.; Elsevier: New York, NY, USA, 2010.
51. Monnier, S.; Kinnard, C. Reconsidering the glacier to rock glacier transformation problem: New insights from the central Andes of Chile. *Geomorphology* **2015**, *238*, 47–55. [CrossRef]
52. Petersen, E.I.; Hock, R.; Fochesatto, G.J.; Anderson, L.S. The Significance of Convection in Supraglacial Debris Revealed Through Novel Analysis of Thermistor Profiles. *JGR Earth Surf.* **2022**, *127*, 47–55. [CrossRef]

remote sensing

MDPI

Article

Surface Displacement of Hurd Rock Glacier from 1956 to 2019 from Historical Aerial Frames and Satellite Imagery (Livingston Island, Antarctic Peninsula)

Gonçalo Prates [1,2,3,*] and Gonçalo Vieira [1]

1 Centro de Estudos Geográficos, Associate Laboratory TERRA, Instituto de Geografia e Ordenamento do Território, Universidade de Lisboa, 1600-276 Lisboa, Portugal
2 Instituto Superior de Engenharia, Universidade do Algarve, 8005-139 Faro, Portugal
3 Laboratorio de Astronomía, Geodesia y Cartografía, Universidad de Cádiz, 11510 Cádiz, Spain
* Correspondence: gprates@ualg.pt; Tel.: +351-289-800-100

Abstract: In the second half of the 20th century, the western Antarctic Peninsula recorded the highest mean annual air temperature rise in the Antarctic. The South Shetland Islands are located about 100 km northwest of the Antarctic Peninsula. The mean annual air temperature at sea level in this Maritime Antarctic region is close to −2 °C and, therefore, very sensitive to permafrost degradation following atmospheric warming. Among geomorphological indicators of permafrost are rock glaciers found below steep slopes as a consequence of permafrost creep, but with surficial movement also generated by solifluction and shallow landslides of rock debris and finer sediments. Rock glacier surface velocity is a new essential climate variable parameter by the Global Climate Observing System, and its historical analysis allows insight into past permafrost behavior. Recovery of 1950s aerial image stereo-pairs and structure-from-motion processing, together with the analysis of QuickBird 2007 and Pleiades 2019 high-resolution satellite imagery, allowed inferring displacements of the Hurd rock glacier using compression ridge-and-furrow morphology analysis over 60 years. Displacements measured on the rock glacier surface from 1956 until 2019 were from 7.5 m to 22.5 m and surface velocity of 12 cm/year to 36 cm/year, measured on orthographic images, with combined deviation root-mean-square of 2.5 m and 2.4 m in easting and northing. The inferred surface velocity also provides a baseline reference to assess today's displacements. The results show patterns of the Hurd rock glacier displacement velocity, which are analogous to those reported within the last decade, without being possible to assess any displacement acceleration.

Keywords: historical photogrammetry; structure-from-motion; rock glaciers; South Shetland Islands; Antarctic Peninsula

1. Introduction

Rock glaciers are debris landforms generated by the former or current creep of frozen ground (permafrost), detectable in the landscape due to front and lateral margins and generally ridge-and-furrow surface topography [1]. They result mainly from the deformation of permafrost under the influence of gravity via permafrost creep, with solifluction playing a role near the surface in the active layer, but also with the possibility of land-sliding. Rock glaciers have been widely used as a geomorphic indicator of permafrost and are landforms typical for the lower limit of permafrost in mountain regions, such as the European Alps [2]. Rock glacier surface velocities typically vary from centimeters up to a few meters per year. It depends on slope steepness, material properties, hydrology, and ground thermal conditions [3]. Rock glacier kinematics is forced by climate, with most rock glaciers accelerating following the warming of permafrost and resulting changes in rheology and hydrology in the debris masses [4]. Rock glacier velocity is hence a good proxy for the effects of climate change on permafrost, and its significance [5,6] is leading to its possible integration as an associated parameter to the essential climate variable of permafrost in the Global Climate Observing System—World Meteorological Organization (WMO).

Contrary to mountainous regions worldwide, where rock glaciers have been a focus of research because they are significant for hazard assessment, hydrology (water supply), ecosystem niches, climate reconstruction, and climate change impacts [7–9], very little is known about rock glacier dynamics in the Antarctic. However, several rock glaciers have been identified in the South Shetland Islands (SSI), mainly on ice-free peninsulas of King George and Livingston islands, below 300 m above sea level (asl) [10]. Antarctic Peninsula rock glaciers with lengths of 55–470 m and widths of 100–360 m are small when compared to others in mountain environments worldwide. All those rock glaciers are active, three of them showing signs of recently reduced activity [10].

Rates of surface displacement of rock glaciers have barely been reported for the Antarctic Peninsula and the South Shetland Islands. One exception is the Hurd rock glacier in Livingston Island, where this study focuses. Studied by Interferometric Synthetic Aperture Radar (InSAR), its surface velocity was measured using phase-differential Global Navigation Satellite Systems (GNSS) with reported velocities of up to about 30 cm/year [11]. Rock glaciers can also be monitored using ground- or aerial-based photogrammetry, laser scanning, and visible-band satellite imagery [12]. Most data feeding these techniques may only be used for recent years. For the SSI, the historical aerial photogrammetric frames of the Falkland Islands and Dependencies Aerial Survey Expedition (FIDASE) from the 1950s [13,14] are possibly the oldest remote sensing data for the archipelago and are comparable to the decimeter-level pixel resolution of nowadays satellite imagery.

This study assesses the capability of extending monitoring back in time [15,16] to infer the Hurd rock glacier mean surface velocity with cm/year precision over 60 years, from 1956 to 2019. For this purpose, we used historical aerial frames and recent high-resolution satellite imagery.

2. Study Area

Since the 1950s, the western Antarctic Peninsula has suffered one of the highest maximum mean annual air temperature (MAAT) increases of the Antarctic, with as much as 3.4 °C, or about 0.5 °C/decade based on the Faraday/Vernadsky station record [17], placing the region as one of the world's climate warming hotspots [18–20]. Higher temperatures have been accompanied by increased precipitation [21,22] and higher snow accumulation, particularly in the western Antarctic Peninsula [23]. However, a regional cooling has been reported for the northwest Antarctica Peninsula from 1999 to 2015. Since then, the warming trend has resumed, with the summer of 2020 breaking record highs with maxima of 18.3 °C recorded at Esperanza Station [24]. In the SSI, about 100 km off the northwest Antarctic Peninsula (Figure 1a), the MAAT increased nearly 2 °C from 1968 until 2022 (Figure 1b). The MAAT is close to −2 °C at sea level [25], which makes the cryosphere very sensitive to small air temperature changes [26–28].

Permafrost is widespread in the Antarctic Peninsula, except close to sea level in its northwestern sector [29]. Modeling of the temperature at the top of permafrost shows that the SSI have permafrost temperatures above −2 °C, but frequently, just below freezing [28,30]. The comparison between the first measurements of the active layer thickness in the 1950s–60s and recent monitoring data has revealed an increase in the active layer depth in some areas of the SSI and Anvers Island [26,27]. Starting in 2000 and extending along the International Polar Year 2007–2008, a borehole network was installed, first in the SSI and later in the Antarctic Peninsula, measuring primarily permafrost temperature and active-layer thickness but also geomorphological changes [26–28,31–33]. Accompanying the reported atmospheric cooling, in some regions of the SSI, from 2006 to 2015, there has been a reduction of the active layer thickness at a rate of about 1.5–2.0 cm/year [31,32] and soil cooling as inferred from annual freezing indexes [34]. Lower summer temperatures have promoted a longer snow cover period [33,35], cooling the ground [25].

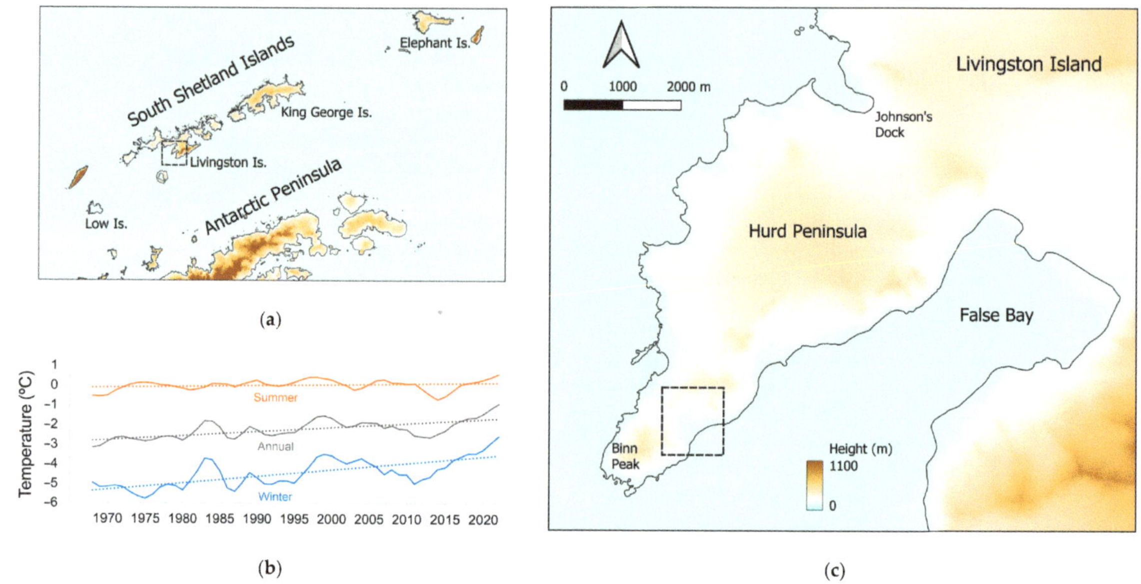

Figure 1. Location of Hurd Peninsula, within the dashed box, in Livingston Island of the South Shetland Islands about 100 km north of the Antarctic Peninsula (**a**). Air temperature at Bellingshausen station, King George Island, retrieved from Reference Antarctic Data for Environmental Research (legacy.bas.ac.uk/met/reader/) (accessed on 19 April 2023). Annual (gray), winter (blue), and summer (orange) means in solid line for weighted moving average and dotted line for trend (**b**). Location of Hurd rock glacier, inside the dashed box, in Hurd Peninsula with digital elevation model from historical aerial frames and REMA outside the SfM computed area (**c**).

No data has yet been published on the effects of the warming recorded since 2015 on permafrost temperatures and the active layer, but these are expected to warm and increase in thickness. In this paper, we study the Hurd rock glacier, which is located in a small glacial valley in the south of Hurd Peninsula in Livingston Island (Figure 1c). The bedrock is composed of sandstones, shales, and greywackes of the Myers Bluff Formation [10]. Binn Peak (392 m) is the highest summit on the ridge, with over 200 m elevation that bounds the valley. The rock glacier is formed by a debris accumulation about 470 m long and 360 m wide, with the surface showing longitudinal and transversal pressure ridges and furrows (Figure 2). The latter are especially present in the lower sector close to the front, which shows a maximum slope of 45° and is 15 to 20 m high [10]. The rock glacier forms in front of a small retreating cirque glacier and extends from about 110 m to 20 m asl, terminating over a fluvioglacial infill lying in contact with a raised marine terrace [36] about 150 m from the shoreline of False Bay. The glacial cirque headwalls the rock glacier with debris feeding the rock glacier, especially on the eastern side.

Permafrost in the Hurd Peninsula is continuous above about 150 m asl and absent in the raised beach terraces up to 30 m asl but is known to occur in relict bodies in ice-cored moraines, as well as in rock glaciers down to sea level [37]. Permafrost temperatures range from −0.4 to −1.8 °C, showing that it is very susceptible to thawing and generating thermokarst features in ice-rich terrain, but also debris flows and active-layer detachment slides [26–28]. The MAAT at sea level is about −1.2 °C and, in the short summer, rainfall events are frequent, and mean summer temperatures are 1.9 °C [38]. In Hurd rock glacier, the snow cover can melt completely in warm summers, but it is frequent that it prevails in furrows and sheltered areas.

Figure 2. Oblique photograph of Hurd rock glacier depicting the eastern valley slope and the ridge and furrows topography, as well as the frontal zone over the Holocene raised beach.

The first detailed research about Hurd rock glacier presented a geomorphological map and several vertical electrical soundings [36] that showed the presence of a high resistivity unit below a 1 to 3 m unfrozen superficial layer, having about 2 m thickness, and being interpreted as permafrost. The fluvioglacial plain in front of the rock glacier showed no permafrost, but the lobate debris features present in the slopes at the same elevation as the top of the rock glacier body suggested that its roots were in the continuous permafrost sector. The study of impacts of the atmosphere on phase delay for using InSAR to detect surface displacement of the rock glaciers in Livingston Island [11] refer to annual phase-differential GNSS data for the period of 2011 to 2015, which showed surface velocities at the Hurd rock glacier body of up to 30 cm/year, with faster sectors in the rock glacier front and central body and slower sectors in its eastern and western margins.

3. Materials and Methods

3.1. Structure from Motion and Rational Polynomial Processing

Classical stereo-photogrammetry is a mapping tool applied for almost one century, where from several overlapping images of the Earth's surface, both surface models and orthographic images are derived. The geometric consistency of the obtained information is appropriate for studying topographical changes and geomorphological dynamics [12].

Recently, structure from motion (SfM) was developed, being a near-automated compilation of digital imagery processing strategies that solves, together, camera position and surface geometry [39]. SfM's complete solution is based on the geometry of the photography and a highly redundant number of automatically detected matching features proportional to surface texture and image resolution identified in several images from diverse perspectives and preferably with a high degree of overlap [40]. Together with the ongoing increase in processing power, SfM made digital stereo-photogrammetry cost-effective, which gave a significant advance to the field [39,41,42].

Furthermore, by independently detecting and computing additional matching features on handy subsets of the overlapping images by multi-view stereo (MvS) processing, massive geometric data on dense point clouds allow for complete surface reconstruction [43]. The SfM-MvS together involves the detection of matching features in individual images, their coordinates being measured in the camera reference system, and the computation of camera and features' relative positions in a non-scaled arbitrary coordinate system. Therefore, ground control points or known camera centers' positions in an appropriate

Earth reference system are required to generate a dense point cloud and, from it, a surface mesh and surface reconstruction or digital elevation model.

The camera positions and digital elevation model are then needed to correct the displacement of each pixel in the images to form the orthographic image. The displacement of each pixel position is due to the projection of the feature by a non-vertical line-of-sight between the camera center and the feature, thus radial function of feature height [44]. Hence, the complete imagery processing can be performed with human intervention almost limited to the ground control points' identification [45].

In recent years, satellite imagery has also been acquired in stereo-pairs, but typically only a single image is acquired in each passage [44]. Similarly, orthographic images from satellite imagery require the sensor geometry, position, and digital elevation model to correct the displacement of each pixel due to feature height. The rational polynomial coefficients (RPC) models are commonly applied in satellite imagery to describe the acquisition process of its sensors without a camera geometry model. Handily, the coefficients of the RPC model of each satellite image are provided with them and enhanced using ground control points to relate surface coordinates to image coordinates by cubic polynomials [46,47]. To detect topographical changes and geomorphological dynamics, digital elevation models and orthographic images can be compared. The surface reconstruction is not obtainable from satellite single images; however, their orthographic images can be applied to quantify geomorphological dynamics [45].

3.2. Dataset Characteristics and Selection of Ground Control Points

Two sets of stereo-pairs of aerial frames of the Hurd Peninsula and two satellite single images were analyzed. The stereo-pairs made with a Fairchild metric camera were from 17 December 1956 and 26 December 1957, with a lens focal distance of 152.88 mm on the first and 153.19 mm on the second date. The flight altitudes were about 3960 m on the first and about 4115 m on the second date, with the subsequent scales of 1:26,000 and 1:27,000. The digital softcopies' pixel length is about 0.02 mm on the hardcopy frame (1016 dpi), standard panchromatic-film grain resolution [44], and about 0.6 m on the Earth's surface.

Two satellite single images from 7 February 2007 and 28 February 2019, respectively taken by QuickBird and Pleiades sensors, with pixel length resolution better than 0.6 m in panchromatic mode, were also used. Hence, the initial estimated resolutions of all digital images involved were similar (Table 1).

Table 1. Applied imagery material information.

Proprietary	Type and Details	Number of Frames/Images	Altitude and Scale	Date
Falkland Islands and Dependencies Aerial Survey Expedition (FIDASE)	f: 152.9 mm f: 153.2 mm	4 4	3960 m 1:26,000 4115 m 1:27,000	17 December 1956 26 December 1957
DigitalGlobe (QuickBird)	panchromatic 4 bands f: 8800.0 mm	1	450,000 m	7 February 2007
CNES/Airbus (Pleiades)	panchromatic 4 bands f: 12,905.0 mm	1	694,000 m	28 February 2019

Additional topographic data was accessed on the Reference Elevation Model of Antarctica (REMA) 2 m resolution digital elevation model. REMA was constructed from several individual stereoscopic digital elevation models extracted from pairs of submeter resolution satellite imagery acquired between 2009 and 2021 and vertically registered to satellite altimetry measurements, resulting in absolute uncertainties of less than 1 m and relative uncertainties of decimeters [48].

Apart from the selection of a few parameters, the identification of the ground control points was the main human intervention in the image processing and the main potential error source. Because of mobility and security reasons, it was not possible to survey these points with phase-differential GNSS, except on the rock glacier surface. Further, the lack of artificial ground control points was a strong limitation, as the interpretation of natural ground control points can be somewhat ambiguous, making it less crucial to survey with phase-differential GNSS precision. Nonetheless, phase-differential GNSS accuracy is expected to impact the overall precision of the orthographic images [45].

Scale and reference frame are the main purposes of ground control points, and hence, the same was used in all orthographic images to minimize the effect of a possible reference frame bias when differentiating them to quantify displacements and also accounting for the maximum height difference to constrain vertical scale. Their precision influence on the horizontal scale and reference frame rotation is inversely proportional to the distance between them, as for larger distances, lesser inaccuracy in reference frame rotation and scale is imposed. The required lens and frame centering, optional lens distortion, frame aspect, and skew calibration are adjusted to the scaled surface model and hence also affected by the ground control points precision.

While satellite images were processed by the ENVI 5.2 software with their rational polynomial coefficients and REMA digital elevation model, stereo-pairs were processed by the PhotoScan 1.0 software with SfM-MvS strategies.

The 2007 QuickBird satellite image was orthorectified with their RCP and the REMA model only and pan-sharpened (Figure 3a). This orthographic image was chosen as the reference for co-registration, from where coordinates in the World Geodetic System 84 reference system and projected in Universal Transverse Mercator at fuse 20 South were retrieved. Height was retrieved from the REMA model. Nine well-identified features in the historical aerial frames and satellite images, mostly at rock outcrops and dispersed locations (Figure 3a), were selected as ground control points in the common area to the aerial frames and satellite images. The 2019 Pleiades satellite image was then orthorectified with their RCP, the REMA model, and nine ground control points to be co-registered to the QuickBird orthographic image and pan-sharpened (Figure 3b).

Historical aerial frames were processed together: three stereo-pairs from 17 December 1956 and three stereo-pairs from 26 December 1957, totaling eight frames, and the nine ground control points, to generate a common dense point cloud and digital elevation model with 2.6 m resolution (Figure 1c). All stereo-pairs were processed together to increase the degree of overlap and, therefore, increase the common digital elevation model precision, also increasing the constraint in the co-registration of both orthographic images. The 1956 and 1957 orthographic images were generated from each date single frame with its center closer to the rock glacier (Figure 4a,b), which is the area where features are less affected by radial displacement due to ground height.

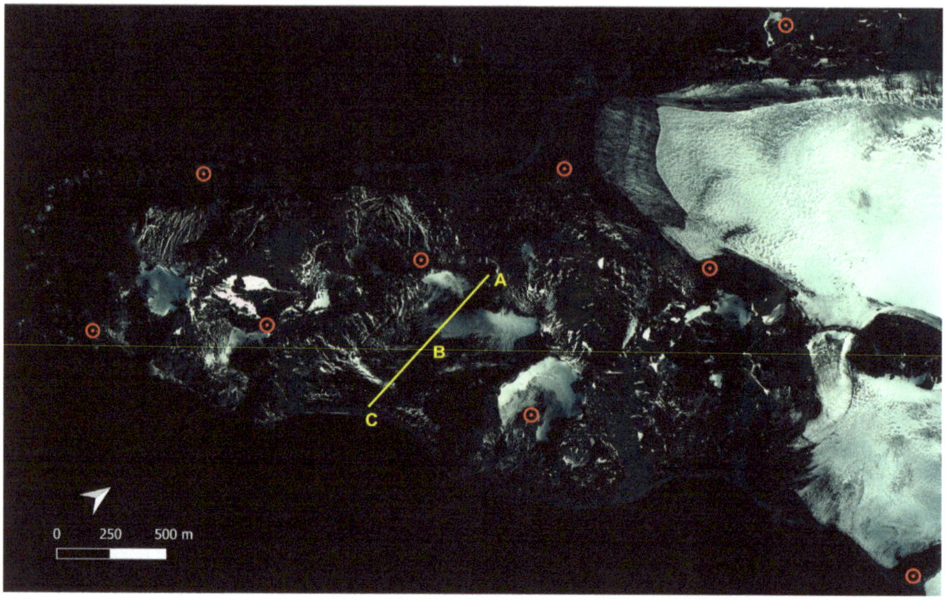

(a)

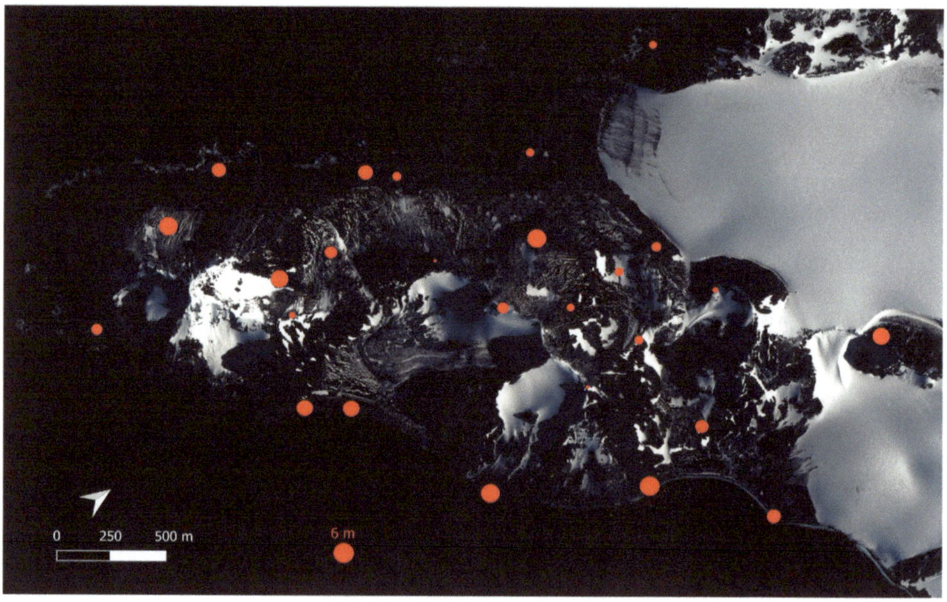

(b)

Figure 3. Orthographic image of 2007 by QuickBird sensor (©DigitalGlobe Inc., Boulder, Colorado, USA, 2007) with the location of the ground control points used shown in red circles (see Appendix A) and profile (A-B-C) of measured heights comparison (**a**). Orthographic image of 2019 by Pleiades sensor (©CNES, Paris, France, 2019, Airbus distribution service) with the location of the twenty-six assessed features and their orthographic images (1956, 1957, 2007, and 2019) coordinates' common deviation root-mean-squared (rms), varying from 0.9 m to 6.2 m (see Appendix B) (**b**).

(a)

(b)

Figure 4. Orthographic image of 1956 and location of the frame center in a red box (**a**). Orthographic image of 1957 and location of the frame center in a red box (**b**).

4. Results

Orthographic Images Generation and Combined Accuracy Assessment

The orthographic 2019 Pleiades satellite image was co-registered to the 2007 QuickBird orthographic image with a deviation root-mean-squared (rms) of 3.4 m and 2.4 m in easting and northing, respectively, at the nine ground control points. The 1956 and 1957 historical aerial frames processed together gave a deviation rms of 2.1 m, 3.1 m, and 5.6 m in easting, northing, and height, respectively, at the nine ground control points.

Furthermore, to assess the relative accuracy among all orthographic images, twenty-six features were selected based on their expected immobility (Figure 3b). The coordinates of each feature were measured in every orthographic image, and their deviation rms were computed. The identification of corresponding features can be ambiguous and was minimized by avoiding steep terrain, shadows, and snow. Still, to evenly disperse the analyzed features throughout the orthographic images, this criterium was not always achieved, particularly due to shadows in the 1957 and 2019 orthographic images covering most of the southeast coastline (Figures 3b and 4b). A common deviation rms of all orthographic images (1956, 1957, 2007, and 2019) of 2.5 m and 2.4 m in easting and northing, respectively, was computed. The deviation rms computed at each feature is shown in Figure 3b, where 0.3 m and 6.0 m (easting) and 0.6 m and 4.6 m (northing) were the extreme values.

5. Discussion

Hurd Rock Glacier Surface Velocity

The controls of climate on rock glacier displacement have been analyzed by several authors, and rock glacier acceleration has been shown to be a consequence of climate warming in many cases [4]. However, other environmental factors control rock glacier displacements, such as variable sediment and ice/water supply to the rock glacier body. All these require being interpreted carefully and framed within the geomorphological and climate setting of each rock glacier. In steep and thin rock glaciers, rapid impact from atmospheric warming on rock glacier temperatures and creep are expected, while thicker rock glaciers may need decades for deformation to occur [3].

From the orthographic images of 1956, 1957, 2007, and 2019, mean displacement velocities can be measured at the rock glacier and the moraine at its root over nearly 60 years. As the rock glacier develops from nearly 100 m asl, north, to nearly 10 m asl, south, its gravity-driven displacement is expected mainly from north to south, and the features that best serve to measure it are the pressure ridges not covered by snow and with a nearly east–west direction (Figures 5 and 6). Due to different conditions of snow cover and illumination, no automatic inference of displacements was possible. These different conditions were useful to visually select common features on the rock glacier, for which displacement was inferred, as the identification of individual ridges benefits from furrows covered by snow and by the lack of shadows over the rock glacier. In Figures 5 and 6, the displacement vectors can be depicted, where their initial points are at the measured features in 1956 and 1957 (Figure 5) while their terminal points are at these features in 2007 and 2019 (Figure 6), allowing for their identification in each orthographic image.

The southward displacement is measured with uncertainty mainly from the northing component. Over 60 years, the deviation rms of 2.4 m in northing implies a north-to-south displacement velocity uncertainty of 4 cm/year. For achieving a 95% confidence level, only north-to-south displacements above 4.7 m were considered. All displacements were measured via the difference of corresponding features' coordinates in each pair of orthographic images of 1956 with 2007 and 2019, and 1957 with 2007 and 2019. Therefore, the average displacement velocities have an uncertainty of about 2 cm/year.

Displacements measured on the rock glacier surface vary between 7.5 m and 22.5 m, and the corresponding surface velocity values were from 12 cm/year to 36 cm/year for the period of 1956 to 2019, with the fastest sectors being the eastern front sector, as well as the front moraine at the root of the rock glacier.

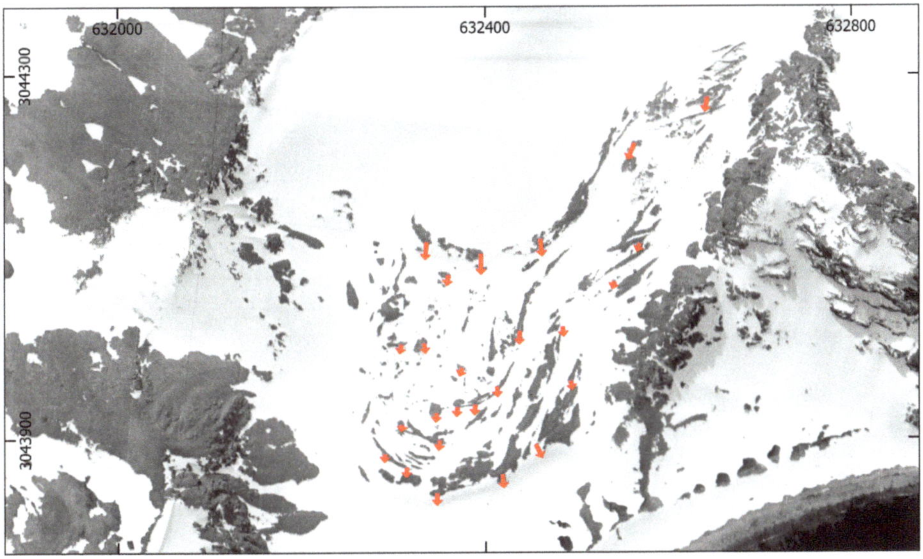

(a)

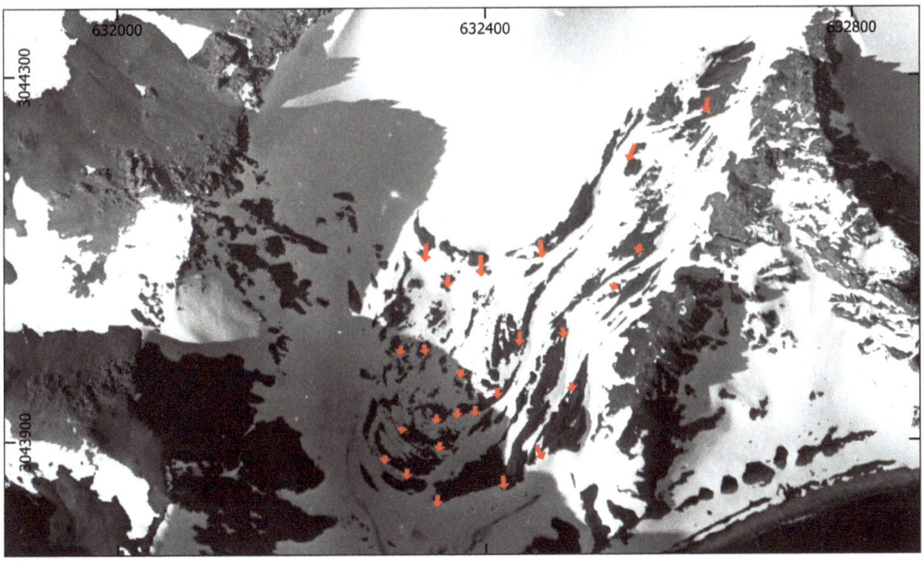

(b)

Figure 5. Initial point of displacement and displacement vectors (in red) at the measured features in 1956 (**a**) and 1957 (**b**). Displacement vectors in image scale range from 7.5 m to 22.5 m. Coordinates in WGS84–UTM20S.

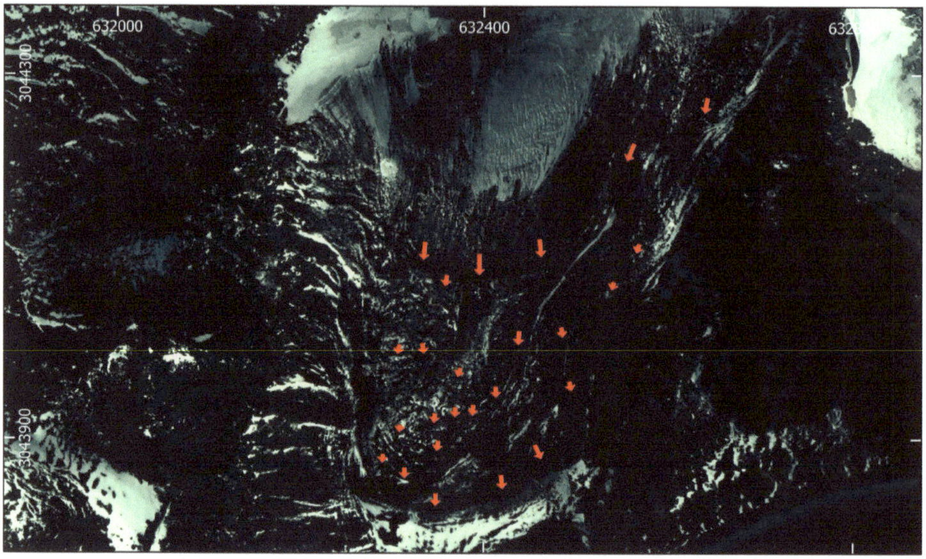

(a)

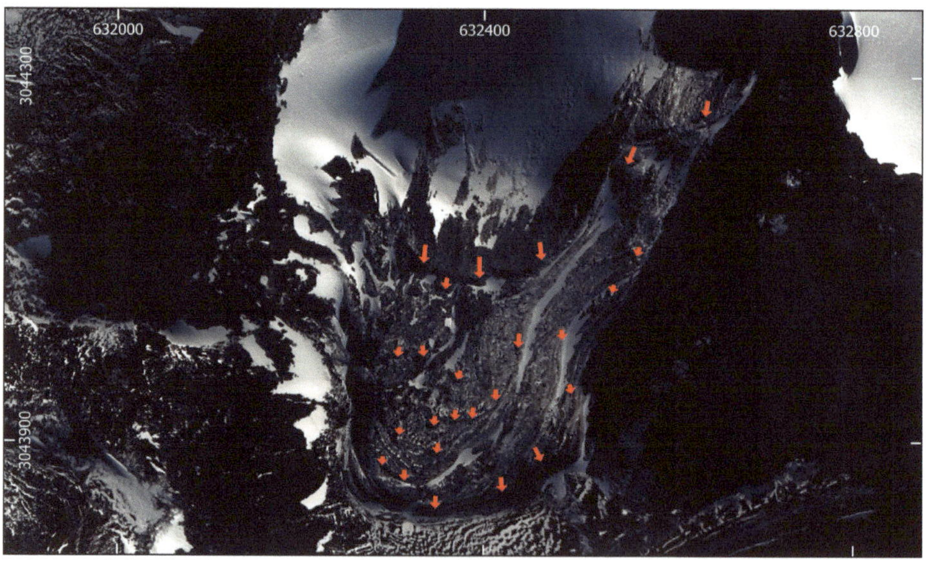

(b)

Figure 6. Terminal point of displacement and displacement vectors (in red) at the measured features in 2007 (©DigitalGlobe Inc., Boulder, Colorado, USA, 2007) (**a**) and 2019 (©CNES, Paris, France, 2019, Airbus distribution service) (**b**). Displacement vectors in image scale range from 7.5 m to 22.5 m. Coordinates in WGS84–UTM20S.

The highest displacement velocities were measured at the front moraine at the root of the rock glacier, ranging from 29 cm/year to 36 cm/year (Figure 7). The lower displacement velocity values were measured at the western sector of the rock glacier body, ranging from 12 cm/year to 18 cm/year. At the eastern sector of the rock glacier body and its front, the displacement velocity ranges from 21 cm/year to 26 cm/year. Any displacement could be detected on the rock glacier for the very short interval of 1956 to 1957. No displacement was detected greater than 4.7 m from 2007 to 2019 (Figure 6). This measured velocity field agrees well with that measured by phase-differential GNSS and reported within the last decade [6], not allowing for a clear assessment of any displacement acceleration.

Figure 7. Surface velocities of displaced features range from 12 cm/year to 36 cm/year. Initial point of velocity vectors (in red) at the measured features in 1956. Includes material ©CNES, Paris, France, 2019 (Airbus distribution service). Coordinates in WGS84–UTM20S.

Although the height precision obtained from SfM applied to the 1956 and 1957 photographic images was inferred at about 5.6 m, the height profile relative to the 2021 REMA digital model (Figure 8) reveal high agreement between them, both above the rock glacier (A–B), over the rock glacier and its adjacent raised beach (B–C). Horizontal displacements at the rock glacier root can be measured at about 10–20 m between profile inflections, although displacement at the rock glacier foot is less clear, perhaps smoothed by the snow accumulation at the 1956 and 1957 images. At the eastern sector of the rock glacier, a slight surface uplift may have taken place, extending to a small, elevated terrace at the raised beach. Yet, these differences may be associated to the limited vertical precision.

The Hurd rock glacier rheology and evolution of its permafrost thermal state are unknown, limiting its surface velocity modeling. The regional temperature evolution at the South Shetland Islands may be assessed by analyzing the MAAT time series of Bellingshausen station in King George Island (Figure 1b). MAAT shows an increase of about 1.5 °C from 1967 to 1999, followed by a decrease of about −1.0 °C until 2014, and again by an increase of about 1.5 °C in MAAT until 2022 (Figure 1b). Despite the significant atmospheric warming reported for the archipelago, since at least the 1950s, the measured displacement velocities do not show a clear acceleration, as could be expected following the warming in the climate series. This may be due to the very warm setting of the Hurd rock glacier in the permafrost zone of the South Shetland, below the continuous permafrost [23]. Furthermore, the complex interannual variability in snow cover conditions [30], with poorly understood impacts on ground hydrology, may add extra complexity to the understanding of Hurd rock glacier rheology as well as its reaction to climate change.

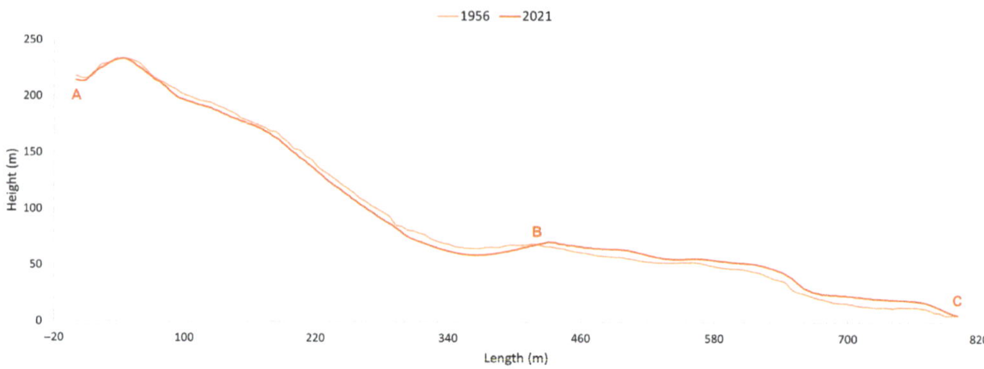

Figure 8. Height profile of the Hurd Rock Glacier in 1956 based on SfM from 1956 and 1957 FIDASE photographic data relative to the 2021 REMA from satellite imagery from 2009 to 2021. A-B-C location in the profile as in Figure 3a.

6. Conclusions

The analysis of co-registered high-resolution satellite orthographic images from 2007 and 2019 together and orthographic images generated from SfM-MvS processed stereo-pairs from 1956 and 1957 shows that historical photogrammetric frames allow developing detailed surface deformation monitoring studies in remote Antarctic areas. This is the first time that long-term displacement velocities for rock glaciers are presented for the Antarctic Peninsula and the South Shetland Islands region, which makes this data especially valuable and shows the potential applicability of the technique and datasets used.

Results depict that the analysis of historical aerial frames of the South Shetland Islands from the 1950s provides reliable information for the analysis of geomorphological changes with considerable resolution. Nevertheless, this type of analysis requires a sufficient time span to accommodate its uncertainty level. In this study, the 1950s stereo-pairs recovery and SfM-MvS processing allowed inferring displacements of the Hurd rock glacier compression ridges over 60 years. Orthographic images' precision can be enhanced by ground control points with accurate coordinates at least one order of magnitude above the images' resolution, measured by phase-differential GNSS. The uncertainty of the produced orthographic images was 2.5 m and 2.4 m in easting and northing, respectively.

At the Hurd rock glacier, displacement velocities were measured over about 60 years, ranging from 12 cm/year to 36 cm/year, depending on the rock glacier sector. These values reveal a similar spatial distribution to that reported from short phase-differential GNSS time series within the last decade. From 2007 to 2019, no displacement surpassing the displacement uncertainty on the orthographic images was detected. Jointly, these statements support the long-term strength and the short-term limitation of visible-band imagery to detect ground displacements. The implemented strategy, where no automatic co-registration with local adjustment of features and additional pixel displacements to those acknowledged for camera and central projection geometry, showed to be trustworthy at metric-level, avoiding spurious diminished to null displacements. The orthographic image data did not allow for detecting changes in the Hurd rock glacier displacement velocities above the inferred 4 cm/year uncertainty when compared to reported short time series of phase-differential GNSS measurements at this time, following the recorded atmospheric warming of the South Shetland region.

Author Contributions: Conceptualization, G.P. and G.V.; formal analysis, G.P.; funding acquisition, G.V.; methodology, G.P.; resources, G.P. and G.V.; validation, G.P.; visualization, G.P.; writing—original draft, G.P.; writing—review and editing, G.P. and G.V. All authors have read and agreed to the published version of the manuscript.

Funding: This research was funded by the Portuguese Science and Technology Foundation through the Portuguese Polar Program under the PERMANTAR projects (PROPOLAR/FCT) and THAWIMPACT/2022.06628.PTDC.

Data Availability Statement: Not applicable.

Acknowledgments: The access to FIDASE photogrammetric frames from the Hurd Peninsula, Livingston Island, downloaded from Earth Explorer, is acknowledged by the British Antarctic Survey (BAS) and United States Geological Survey (USGS). The access to the REMA digital elevation model is acknowledged by the University of Minnesota. The reviewers are acknowledged for their comments and suggestions.

Conflicts of Interest: The authors declare no conflict of interest. The founding sponsors had no role in the design of the study, in the collection, analysis, or interpretation of data, in the writing of the manuscript, or in the decision to publish the results.

Appendix A

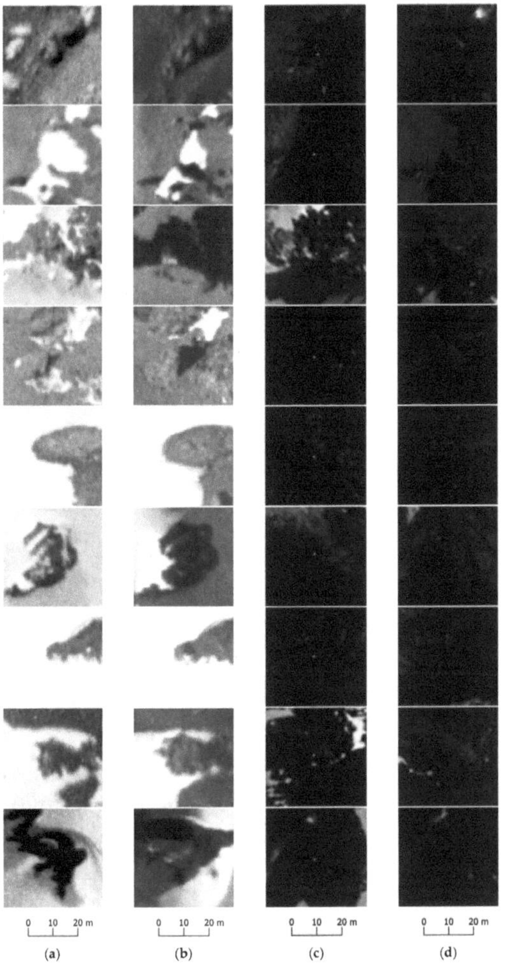

Figure A1. Georeferencing ground control points' (red dots) at orthographic images of 1956 (**a**), 1957 (**b**), 2007 (©DigitalGlobe Inc., Boulder, Colorado, USA, 2007) (**c**) and 2019 (©CNES, Paris, France, 2019, Airbus distribution service) (**d**) for assessment of visual identification and relative precision.

Appendix B

Table A1. Coordinates of ground control points for precision assessment in WGS84-UTM20S.

1956		1957		2007		2019	
X (m)	Y (m)	X (m)	Y (m)	X (m)	Y (m)	X (m)	Y (m)
631,523.445	3,042,909.857	631,525.561	3,042,907.211	631,526.818	3,042,912.039	631,528.406	3,042,913.098
631,253.305	3,043,775.044	631,253.570	3,043,774.780	631,253.702	3,043,777.227	631,249.733	3,043,782.783
631,980.248	3,043,675.958	631,981.306	3,043,674.899	631,981.835	3,043,675.098	631,981.571	3,043,679.066
632,141.908	3,044,346.610	632,142.437	3,044,346.081	632,142.173	3,044,346.081	632,142.702	3,044,347.801
633,005.111	3,044,584.801	633,006.170	3,044,585.066	633,005.773	3,044,586.323	633,006.831	3,044,584.206
631,986.598	3,044,978.171	631,989.508	3,044,979.626	631,986.333	3,044,977.377	631,984.216	3,044,979.494
632,971.906	3,045,301.359	632,971.377	3,045,303.740	632,971.245	3,045,304.137	632,969.393	3,045,303.873
633,962.506	3,044,934.382	633,962.241	3,044,935.176	633,965.549	3,044,927.900	633,961.051	3,044,929.223
632,358.205	3,043,488.633	632,358.999	3,043,486.384	632,364.158	3,043,479.174	632,359.925	3,043,484.730
631,730.481	3,044,419.834	631,731.539	3,044,419.305	631,733.656	3,044,420.032	631,730.481	3,044,424.001
632,490.894	3,044,479.630	632,490.629	3,044,478.571	632,489.174	3,044,480.754	632,496.053	3,044,476.851
632,475.812	3,043,658.892	632,478.194	3,043,658.495	632,480.839	3,043,650.161	632,480.046	3,043,653.071
631,336.120	3,043,439.685	631,336.649	3,043,439.949	631,331.952	3,043,442.331	631,327.984	3,043,447.887
631,848.618	3,043,978.906	631,846.765	3,043,978.376	631,848.948	3,043,978.906	631,844.979	3,043,984.991
631,639.861	3,044,313.074	631,640.390	3,044,311.222	631,641.515	3,044,312.016	631,634.900	3,044,318.366
631,807.607	3,043,727.287	631,806.813	3,043,725.435	631,815.875	3,043,720.672	631,813.229	3,043,725.699
632,669.487	3,044,727.213	632,671.869	3,044,726.949	632,669.950	3,044,725.824	632,666.775	3,044,728.206
633,441.938	3,044,906.270	633,441.872	3,044,902.037	633,446.106	3,044,899.920	633,444.518	3,044,898.862
631,900.079	3,045,711.331	631,899.021	3,045,710.273	631,896.441	3,045,711.662	631,895.383	3,045,710.868
633,577.802	3,045,792.426	633,573.833	3,045,795.138	633,569.600	3,045,798.048	633,567.483	3,045,796.395
632,323.677	3,044,774.706	632,322.619	3,044,776.294	632,316.798	3,044,770.870	632,321.560	3,044,781.982
632,663.402	3,045,197.378	632,662.608	3,045,197.180	632,662.608	3,045,197.709	632,658.110	3,045,200.288
632,657.449	3,045,001.454	632,661.021	3,044,998.742	632,660.095	3,044,997.022	632,657.713	3,044,997.882
632,959.074	3,044,894.364	632,960.926	3,044,895.422	632,964.762	3,044,892.313	632,961.984	3,044,894.033
633,146.680	3,043,955.199	633,150.120	3,043,956.787	633,152.501	3,043,947.791	633,153.824	3,043,948.585
633,530.591	3,044,556.862	633,528.739	3,044,554.216	633,539.322	3,044,548.660	633,534.057	3,044,554.102

References

1. RGIK. *Towards Standard Guidelines for Inventorying Rock Glaciers: Baseline Concepts (Version 4.2.2)*; IPA Action Group Rock Glacier Inventories and Kinematics: Troms and Finnmark, Norway, 2022; 13p.
2. Haeberli, W. *Creep of Mountain Permafrost: Internal Structure and Flow of Alpine Rock Glaciers*; Eidgenössische Technische Hochschule Zürich: Zürich, Switzerland, 1985; Volume 77, 142p.
3. Müller, J.; Vieli, A.; Gärtner-Roer, I. Rock Glaciers on the Run—Understanding Rock Glacier Landform Evolution and Recent Changes from Numerical Flow Modeling. *Cryosphere* **2016**, *10*, 2865–2886. [CrossRef]
4. Marcer, M.; Cicoira, A.; Cusicanqui, D.; Bodin, X.; Echelard, T.; Obregon, R.; Schoeneich, P. Rock Glaciers throughout the French Alps Accelerated and Destabilised since 1990 as Air Temperatures Increased. *Commun. Earth Environ.* **2021**, *2*, 81. [CrossRef]
5. Sorg, A.; Kääb, A.; Roesch, A.; Bigler, C.; Stoffel, M. Contrasting Responses of Central Asian Rock Glaciers to Global Warming. *Sci. Rep.* **2015**, *5*, 8228. [CrossRef]
6. Rangecroft, S.; Suggitt, A.J.; Anderson, K.; Harrison, S. Future Climate Warming and Changes to Mountain Permafrost in the Bolivian Andes. *Clim. Change* **2016**, *137*, 231–243. [CrossRef] [PubMed]
7. Reinosch, E.; Gerke, M.; Riedel, B.; Schwalb, A.; Ye, Q.; Buckel, J. Rock Glacier Inventory of the Western Nyainqêntanglha Range, Tibetan Plateau, Supported by InSAR Time Series and Automated Classification. *Permafr. Periglac. Process.* **2021**, *32*, 657–672. [CrossRef]
8. Tielidze, L.G.; Cicoira, A.; Nosenko, G.A.; Eaves, S.R. The First Rock Glacier Inventory for the Greater Caucasus. *Geosciences* **2023**, *13*, 117. [CrossRef]
9. Wagner, T.; Pleschberger, R.; Kainz, S.; Ribis, M.; Kellerer-Pirklbauer, A.; Krainer, K.; Philippitsch, R.; Winkler, G. The First Consistent Inventory of Rock Glaciers and Their Hydrological Catchments of the Austrian Alps. *Austrian J. Earth Sci.* **2020**, *113*, 1–23. [CrossRef]
10. Serrano, E.; López-Martínez, J. Rock Glaciers in the South Shetland Islands, Western Antarctica. *Geomorphology* **2000**, *35*, 145–162. [CrossRef]
11. Reis, A.R.; Catalao, J.; Vieira, G.; Nico, G. Mitigation of Atmospheric Phase Delay in InSAR Time Series Using ERA-Interim Model, GPS and MODIS Data: Application to the Permafrost Deformation in Hurd Peninsula, Antarctica. In Proceedings of the 2015 IEEE International Geoscience and Remote Sensing Symposium (IGARSS), Milan, Italy, 26–31 July 2015; pp. 3454–3457.

12. Kääb, A.; Vollmer, M. Surface Geometry, Thickness Changes and Flow Fields on Creeping Mountain Permafrost: Automatic Extraction by Digital Image Analysis. *Permafr. Periglac. Process.* **2000**, *11*, 315–326. [CrossRef]
13. Gavin-Robinson, C. The story of the Falkland Island dependencies aerial survey expeditions. *Flight* **1957**, 802–805.
14. Fox, A.J.; Cziferszky, A. Unlocking the Time Capsule of Historic Aerial Photography to Measure Changes in Antarctic Peninsula Glaciers. *Photogramm. Rec.* **2008**, *23*, 51–68. [CrossRef]
15. Tanguy, R.; Whalen, D.; Prates, G.; Vieira, G. Shoreline Change Rates and Land to Sea Sediment and Soil Organic Carbon Transfer in Eastern Parry Peninsula from 1965 to 2020 (Amundsen Gulf, Canada). *Arct. Sci.* **2023**. [CrossRef]
16. Tanguy, R.; Whalen, D.; Prates, G.; Pina, P.; Freitas, P.; Bergstedt, H.; Vieira, G. Permafrost Degradation in the Ice-Wedge Tundra Terrace of Paulatuk Peninsula (Darnley Bay, Canada). *Geomorphology* **2023**, *435*, 108754. [CrossRef]
17. Turner, J.; Barrand, N.E.; Bracegirdle, T.J.; Convey, P.; Hodgson, D.A.; Jarvis, M.; Jenkins, A.; Marshall, G.; Meredith, M.P.; Roscoe, H.; et al. Antarctic Climate Change and the Environment: An Update. *Polar Rec.* **2014**, *50*, 237–259. [CrossRef]
18. Vaughan, D.G.; Marshall, G.J.; Connolley, W.M.; Parkinson, C.; Mulvaney, R.; Hodgson, D.A.; King, J.C.; Pudsey, C.J.; Turner, J. Recent rapid regional climate warming on the Antarctic Peninsula. *Clim. Change* **2003**, *60*, 243–274. [CrossRef]
19. Turner, J.; Colwell, S.R.; Marshall, G.J.; Lachlan-Cope, T.A.; Carleton, A.M.; Jones, P.D.; Lagun, V.; Reid, P.A.; Iagovkina, S. Antarctic Climate Change during the Last 50 Years. *Int. J. Climatol.* **2005**, *25*, 279–294. [CrossRef]
20. Turner, J.; Bindschadler, R.; Convey, P.; di Prisco, G.; Fahrbach, E.; Gutt, J.; Hodgson, D.; Mayewski, P.; Summerhayes, C. *Antarctic Climate Change and the Environment*; Scientific Committee on Antarctic Research: Cambridge, UK, 2009.
21. Turner, J.; Lachlan-Cope, T.; Colwell, S.; Marshall, G.J. A Positive Trend in Western Antarctic Peninsula Precipitation over the Last 50 Years Reflecting Regional and Antarctic-Wide Atmospheric Circulation Changes. *Ann. Glaciol.* **2005**, *41*, 85–91. [CrossRef]
22. Miles, G.M.; Marshall, G.J.; McConnell, J.R.; Aristarain, A.J. Recent Accumulation Variability and Change on the Antarctic Peninsula from the ERA40 Reanalysis. *Int. J. Climatol.* **2008**, *28*, 1409–1422. [CrossRef]
23. Thomas, E.R.; Marshall, G.J.; McConnell, J.R. A Doubling in Snow Accumulation in the Western Antarctic Peninsula since 1850. *Geophys. Res. Lett.* **2008**, *35*, L01706. [CrossRef]
24. González-Herrero, S.; Barriopedro, D.; Trigo, R.M.; López-Bustins, J.A.; Oliva, M. Climate Warming Amplified the 2020 Record-Breaking Heatwave in the Antarctic Peninsula. *Commun. Earth Environ.* **2022**, *3*, 122. [CrossRef]
25. Oliva, M.; Navarro, F.; Hrbáček, F.; Hernández, A.; Nývlt, D.; Pereira, P.; Ruiz-Fernández, J.; Trigo, R. Recent Regional Climate Cooling on the Antarctic Peninsula and Associated Impacts on the Cryosphere. *Sci. Total Environ.* **2017**, *580*, 210–223. [CrossRef] [PubMed]
26. Vieira, G.; Bockheim, J.; Guglielmin, M.; Balks, M.; Abramov, A.A.; Boelhouwers, J.; Cannone, N.; Ganzert, L.; Gilichinsky, D.A.; Goryachkin, S.; et al. Thermal State of Permafrost and Active-Layer Monitoring in the Antarctic: Advances during the International Polar Year 2007-2009. *Permafr. Periglac. Process.* **2010**, *21*, 182–197. [CrossRef]
27. Bockheim, J.; Vieira, G.; Ramos, M.; López-Martínez, J.; Serrano, E.; Guglielmin, M.; Wilhelm, K.; Nieuwendam, A. Climate Warming and Permafrost Dynamics in the Antarctic Peninsula Region. *Glob. Planet. Change* **2013**, *100*, 215–223. [CrossRef]
28. Ferreira, A.; Vieira, G.; Ramos, M.; Nieuwendam, A. Ground Temperature and Permafrost Distribution in Hurd Peninsula (Livingston Island, Maritime Antarctic): An Assessment Using Freezing Indexes and TTOP Modelling. *Catena* **2017**, *149*, 560–571. [CrossRef]
29. López-Martínez, J.; Serrano, E.; Schmid, T.; Mink, S.; Linés, C. Periglacial Processes and Landforms in the South Shetland Islands (Northern Antarctic Peninsula Region). *Geomorphology* **2012**, *155–156*, 62–79. [CrossRef]
30. Obu, J.; Westermann, S.; Vieira, G.; Abramov, A.; Balks, M.R.; Bartsch, A.; Hrbáček, F.; Kääb, A.; Ramos, M. Pan-Antarctic Map of near-Surface Permafrost Temperatures at 1 Km^2 Scale. *Cryosph.* **2020**, *14*, 497–519. [CrossRef]
31. Goyanes, G.; Vieira, G.; Caselli, A.; Mora, C.; Ramos, M.; de Pablo, M.A.; Neves, M.; Santos, F.; Bernardo, I.; Gilichinsky, D.; et al. Régimen térmico y variabilidad espacial de la capa active en isla Decepción, Antártida. *Rev. Asoc. Geol. Argent.* **2014**, *71*, 112–124.
32. Ramos, M.; Vieira, G.; de Pablo, M.A.; Molina, A.; Abramov, A.; Goyanes, G. Recent Shallowing of the Thaw Depth at Crater Lake, Deception Island, Antarctica (2006–2014). *Catena* **2017**, *149*, 519–528. [CrossRef]
33. De Pablo, M.A.; Ramos, M.; Molina, A. Snow Cover Evolution, on 2009–2014, at the Limnopolar Lake CALM-S Site on Byers Peninsula, Livingston Island, Antarctica. *Catena* **2017**, *149*, 538–547. [CrossRef]
34. Ramos, M.; de Pablo, M.A.; Molina, A.; Vieira, G.; Trindade, A.; Correia, A. Study of the Active Layer Freezing and Thawing Cycles in the Surroundings of the Spanish Antarctic Station JCI 2000 to 2011. In Proceedings of the 7th Portuguese Spanish Assembly of Geodesy and Geophysics, San Sebastian, Spain, 25–29 June 2012; pp. 723–726.
35. Ramos, M.; Vieira, G.; de Pablo, M.A.; Molina, A.; Jimenez, J.J. Transition from a Subaerial to a Subnival Permafrost Temperature Regime Following Increased Snow Cover (Livingston Island, Maritime Antarctic). *Atmosphere* **2020**, *11*, 1332. [CrossRef]
36. Serrano, E.; Giner, J.; Gumiel, P.; López-Martínez, J. El glaciar rocoso de Hurd: Estructura e inserción en el sistema de transferencia de derrubios Antártico Marítimo (Isla Livington, Islas Shetland de Sur, Antártida). *Rev. Cuat. Geomorfol.* **2004**, *18*, 13–24.
37. Hauck, C.; Vieira, G.; Gruber, S.; Blanco, J.; Ramos, M. Geophysical Identification of Permafrost in Livingston Island, Maritime Antarctica. *J. Geophys. Res.* **2007**, *112*, F02S19. [CrossRef]
38. Recondo, C.; Corbea-Pérez, A.; Peón, J.; Pendás, E.; Ramos, M.; Calleja, J.F.; de Pablo, M.Á.; Fernández, S.; Corrales, J.A. Empirical Models for Estimating Air Temperature Using MODIS Land Surface Temperature (and Spatiotemporal Variables) in the Hurd Peninsula of Livingston Island, Antarctica, between 2000 and 2016. *Remote Sens.* **2022**, *14*, 3206. [CrossRef]

39. Westoby, M.J.; Brasington, J.; Glasser, N.F.; Hambrey, M.J.; Reynolds, J.M. "Structure-from-Motion" Photogrammetry: A Low-Cost, Effective Tool for Geoscience Applications. *Geomorphology* **2012**, *179*, 300–314. [CrossRef]
40. Snavely, N.; Seitz, S.M.; Szeliski, R. Modeling the World from Internet Photo Collections. *Int. J. Comput. Vis.* **2008**, *80*, 189–210. [CrossRef]
41. Javernick, L.; Brasington, J.; Caruso, B. Modeling the Topography of Shallow Braided Rivers Using Structure-from-Motion Photogrammetry. *Geomorphology* **2014**, *213*, 166–182. [CrossRef]
42. Cook, K.L. An Evaluation of the Effectiveness of Low-Cost UAVs and Structure from Motion for Geomorphic Change Detection. *Geomorphology* **2017**, *278*, 195–208. [CrossRef]
43. Furukawa, Y.; Ponce, J. Accurate, Dense, and Robust Multi-View Stereopsis. In Proceedings of the IEEE Computer Society Conference on Computer Vision and Pattern Recognition, Minneapolis, MN, USA, 17–22 June 2007.
44. Lillesand, T.; Kiefer, R.W.; Chipman, J. *Remote Sensing and Image Interpretation*, 7th ed.; Wiley: Hoboken, NJ, USA, 2015; ISBN 9781118919453.
45. Prates, G.; Torrecillas, C.; Berrocoso, M.; Goyanes, G.; Vieira, G. Deception Island 1967–1970 Volcano Eruptions from Historical Aerial Frames and Satellite Imagery (Antarctic Peninsula). *Remote Sens.* **2023**, *15*, 2052. [CrossRef]
46. Tao, C.V.; Hu, Y. A Comprehensive Study of the Rational Function Model for Photogrammetric Processing. *Photogramm. Eng. Remote Sens.* **2001**, *67*, 1347–1358.
47. Marí, R.; de Franchis, C.; Meinhardt-Llopis, E.; Anger, J.; Facciolo, G. A Generic Bundle Adjustment Methodology for Indirect RPC Model Refinement of Satellite Imagery. *Image Process. Line* **2021**, *11*, 344–373. [CrossRef]
48. Howat, I.M.; Porter, C.; Smith, B.E.; Noh, M.J.; Morin, P. The Reference Elevation Model of Antarctica. *Cryosphere* **2019**, *13*, 665–674. [CrossRef]

remote sensing

MDPI

Article

Snow Cover and Climate Change and Their Coupling Effects on Runoff in the Keriya River Basin during 2001–2020

Wei Yan [1,2], Yifan Wang [1,2,*], Xiaofei Ma [3], Minghua Liu [1,2], Junhui Yan [1,2], Yaogeng Tan [4] and Sutao Liu [1,2]

1 School of Geographic Sciences, Xinyang Normal University, Xinyang 464000, China; yanwei.jw@xynu.edu.cn (W.Y.); lmhhqs@126.com (M.L.); yanjh2015@126.com (J.Y.); liusutao2021@163.com (S.L.)
2 Xinyang Key Laboratory of Climate Change and Environmental Evolution, Xinyang Normal University, Xinyang 464000, China
3 State Key Laboratory of Desert and Oasis Ecology, Xinjiang Institute of Ecology and Geography, Chinese Academy of Sciences, Urumqi 830011, China; mxf@ms.xjb.ac.cn
4 Information Center (Hydrology Monitor and Forecast Center), Ministry of Water Resources, Beijing 100053, China; yaogengtan@mwr.gov.cn
* Correspondence: wangyf@xynu.edu.cn

Abstract: As a significant component of the cryosphere, snow cover plays a crucial role in modulating atmospheric circulation and regional hydrological equilibrium. Therefore, studying the dynamics of snow cover and its response to climate change is of great significance for regional water resource management and disaster prevention. In this study, reanalysis climate datasets and a new MODIS snow cover extent product over China were used to analyze the characteristics of climate change and spatiotemporal variations in snow cover in the Keriya River Basin (KRB). Furthermore, the effects of climate factors on snow cover and their coupling effects on runoff were quantitatively evaluated by adopting partial least squares regression (PLSR) method and structural equation modeling (SEM), respectively. Our findings demonstrated the following: (1) Air temperature and precipitation of KRB showed a significant increase at rates of 0.24 °C/decade and 14.21 mm/decade, respectively, while the wind speed did not change significantly. (2) The snow cover frequency (SCF) in the KRB presented the distribution characteristics of "low in the north and high in the south". The intra-annual variation of snow cover percentage (SCP) of KRB displayed a single peak (in winter), double peaks (in spring and autumn), and stability (SCP > 75%), whose boundary elevations were 4000 m and 6000 m, respectively. The annual, summer, and winter SCP in the KRB declined, while the spring and autumn SCP experienced a trend showing an insignificant increase during the hydrological years of 2001–2020. Additionally, both the annual and seasonal SCF (except autumn) will be further increased in more than 50% of the KRB, according to estimates. (3) Annual and winter SCF were controlled by precipitation, of which the former showed a mainly negative response, while the latter showed a mainly positive response, accounting for 43.1% and 76.16% of the KRB, respectively. Air temperature controlled SCF changes in 45% of regions in spring, summer, and autumn, mainly showing negative effects. Wind speed contributed to SCF changes in the range of 11.23% to 26.54% across annual and seasonal scales. (4) Climate factors and snow cover mainly affect annual runoff through direct influences, and the total effect was as follows: precipitation (0.609) > air temperature (−0.122) > SCP (0.09).

Keywords: snow cover; climate change; runoff; partial least squares regression; Keriya River Basin

1. Introduction

Snow cover, as an indispensable constituent of the cryosphere, plays a highly important role within the global and regional climate system, exerting an essential impact on surface radiation balance, energy balance, and hydrological partitioning [1–4]. Concurrent with the relentless warming of the global climate, the Northern Hemisphere is experiencing a pronounced decline in snow cover [5,6], which profoundly impacts the regional water cycle [7,8]. Inland rivers situated in the arid and semi-arid regions of northwest China predominantly rely on the runoff derived from the thawing of adjacent alpine mountain snowpacks [9]. Changes in the total and seasonal runoff distributions caused by snow cover changes will affect social and economic development in the middle and lower reaches [10–12]. Therefore, studying the influence of snow cover dynamics and climate change on runoff to inform decisions about the scientific management and planning of water resources in Northwest arid areas has become imperative [13,14].

The conventional method for monitoring snow cover involves the collection of snow cover data from meteorological stations or observation sites. However, this approach suffers from spatial inhomogeneity due to the uneven distribution of the stations, primarily concentrated in low-altitude regions [15–18]. In remote alpine areas, the complex terrain and challenging environmental conditions impede field monitoring efforts [19]. Nevertheless, recent advancements in remote sensing technology have unlocked new opportunities for snow cover research. Remote sensing data offer the advantages of wide broad coverage, frequent updating, and high spatial resolution, which compensates for the limitations of ground-based monitoring data [20,21]. Currently, various remote sensing data are leveraged for snow cover research, including the moderate resolution imaging spectroradiometer (MODIS), advanced very high resolution radiometer (AVHRR), scanning multichannel microwave radiometer (SMMR), and other related products [22–24]. Among these products, MODIS snow cover products have emerged as the mainstream data of remote sensing snow cover products due to their high spatiotemporal resolutions. Additionally, they have been widely adopted in the study of snow cover changes across regions of varying scales [25–27]. For instance, Zou et al. [28] employed MOD10A2 and MYD10A2 snow cover products to explore the variation of snow cover in Northern Xinjiang, Qinghai-Tibet Plateau, and Northeast China. Their study revealed an insignificant increasing trend in snow cover areas and snow depths from 2001 to 2020. Thapa et al. [29] combined three different 8 day composite snow products, including MOD10A2, MYD10A2, and MOYDGL06, to analyze the variation trend of snow cover in the Karakoram region. They discovered a negligible decline in snow cover areas from 2003 to 2018. However, despite the widespread utilization of MODIS data in snow cover research, the accuracy of MODIS data can be affected by factors such as cloud cover and land cover. Therefore, a daily cloud-gap-filled MODIS snow cover extent product produced by Hao et al. [30], which comprehensively considered the impact of land cover and cloud on original MODIS snow cover data, was selected as the main data source for the present study. A detailed explanation of these data can be seen in Section 2.2.1.

Snow cover, as an exceedingly responsive component of the cryosphere, is profoundly influenced by climate change and is regarded as a vital indicator of global climate change [31]. Unveiling the response of snow cover to climate change has emerged as a primary focus within snow cover research, garnering extensive attention from the scholarly community. For instance, Du et al. [32] conducted an analysis of the relationship between snow cover frequency (SCF) and climate factors in the Qilian Mountains from 2000 to 2020. They suggested that the SCF was dominated by precipitation rather than air temperature, with precipitation playing a positive role. However, Hussain et al. [33] investigated the impact of climate factors on snow cover area within the Gilgit River Basin from 2001 to 2015 and reported a negative correlation between snow cover area and air temperature, while precipitation exhibited no evident relationship. In recent years, the focus of scholars has predominantly centered around the influence of air temperature and precipitation on snow cover, often neglecting the potential effect of wind speed. Although certain researchers have discussed the potential impact of wind speed on snow cover [34–36], the majority of these discussions lie within the realm of qualitative studies, with few quantitative analyses conducted on the impact of wind speed on snow cover [37]. Moreover, regarding research methods, most researchers mainly applied the Pearson correlation method to investigate the influence of climate factors on snow cover. However, this kind of method ignores the interactions among various climate factors, which subsequently affects the accuracy of the results derived from the analysis. In contrast, partial least squares regression (PLSR) represents a new multivariate statistical data analysis method capable of eliminating the multiple correlations among independent variables [38], which enables one to clarify the degree of influence of different climate factors on snow cover change.

The Keriya River, a typical inland river in the arid region, stands as the largest river and primary water source for the Yutian Oasis. The alpine mountains contribute crucial meltwater, which serves as the principal supply for the Keriya River and represents a valuable resource for the survival and development of the downstream regions [39]. Although most of the glaciers within the Keriya River Basin (KRB) exhibited stability, there existed a slight trend of total area reduction [40]. In contrast, the snow cover area experienced a trend depicting a slight increase [41]. From 1957 to 2017, the KRB witnessed an increase in runoff depth at a rate of 4.27 mm/decade, which was mainly affected by air temperature and precipitation [41,42]. Recent studies concerning the KRB only focus on the changes in snow cover or runoff, with few conducting quantitative and systematic analyses regarding the relationship between climate, snow cover, and runoff. Structural equation modeling (SEM) can comprehensively analyze the relationship between various variables, allowing one to quantify the direct and indirect effects of climate and snow cover on runoff.

The research objectives of this study were as follows: (1) to reveal the variation characteristics of climate and snow cover in KRB based on reanalysis climate datasets and the new MODIS snow cover extent product over China, (2) to analyze the influence of different climate factors (air temperature, precipitation, and wind speed) on SCF at the pixel scale by adopting PLSR method, and (3) to discuss the interplay between snow cover, climate factors, and their collective influence on runoff through the application of structural equation modeling (SEM). The results of this paper could facilitate a better understanding the spatiotemporal variation of snow cover in the KRB and its influencing mechanism and clarify the regional water cycle process. Furthermore, this study is of great significance to the utilization and management of water resources in the context of climate change.

2. Materials and Methods

2.1. Study Area

The Keriya River, originating from the Guliya Ice Cap in the West Kunlun Mountains, holds the distinction of being the largest river coursing through Yutian County in Xinjiang Uygur Autonomous Region (Figure 1). The KRB exhibits a distinctive topographical gradient, with higher elevations situated in the southern region and lower elevations in the north. The KRB, controlled by Keriya Hydrological Station (36°45′N, 81°48′E), spans a geographical range from 35°11′ to 36°27′N and 81°27′ to 82°50′E with an altitude of 1972–6858 m and an expansive area of approximately 8350.25 km^2. The glacier-covered region in the source region of the KRB spans 682 km^2, approximately constituting 8.2% of the total basin area. Characterized by a warm temperate arid desert climate, the KRB experiences an average annual evaporation of 1922 mm (as observed by the Keriya Hydrological Station), an average annual air temperature of 9.6 °C, and an average annual precipitation of 129.7 mm. The average annual runoff depth of the KRB is 91.3 mm, with glacier and snow meltwater contributing to 47.3% of the total runoff, which is one of the important sources of replenishment for the KRB.

2.2. Data Sources

2.2.1. Snow Cover Dataset

The new MODIS snow cover extent product over China was obtained from the National Tibetan Plateau Data Center (http://data.tpdc.ac.cn/ (accessed on 1 October 2022)). This dataset offered a daily temporal resolution and a spatial resolution of 500 m for the period spanning from 2000 to 2020. The creation of this dataset involved the utilization of the MODIS reflectivity product MOD/MYD09GA, incorporating a snow-discriminant decision tree algorithm for diverse surface types. Furthermore, a vacancy-filling algorithm, such as a spatiotemporal interpolation algorithm for the hidden Markov random field model, was employed to complete the cloud removal within the dataset. The overall accuracy of the dataset exceeds 93%, with omission error and commission error values constrained within 10% [30,43], thus affirming the reliability and suitability of the dataset for snow cover research. In this study, the dataset was extracted according to the hydrological year (HY) scale (from 1 September to 31 August of the following year), enabling an analysis of the spatiotemporal variation characteristics of snow cover within the KRB from 2001 to 2020.

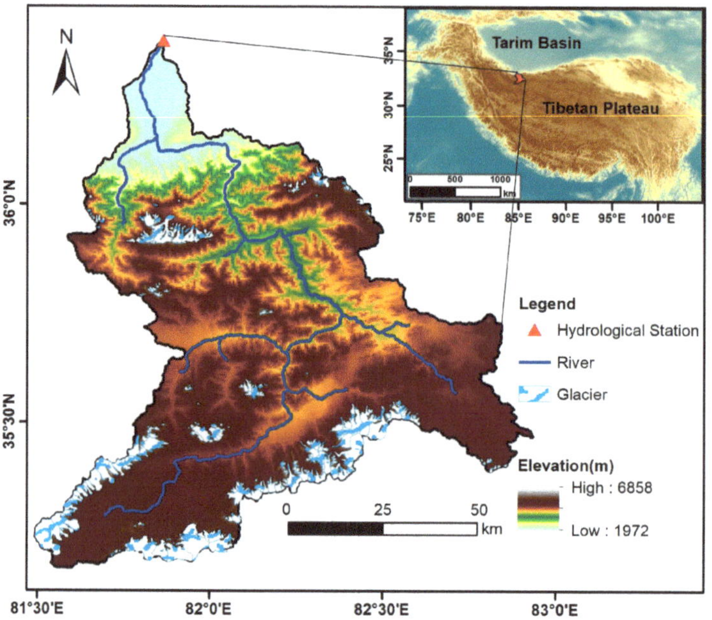

Figure 1. Location and topography of the KRB.

2.2.2. Climate and Runoff Data

The observation data regarding air temperature, precipitation, and wind speed from 2000 to 2020 were obtained from meteorological and hydrological stations positioned within the KRB and its surrounding areas. The monthly runoff data was sourced from the Keriya Hydrological Station.

The 1-km monthly mean temperature dataset for China was obtained from the National Tibetan Plateau Data Center (http://data.tpdc.ac.cn/ (accessed on 5 January 2023)). This dataset encompasses a temporal resolution of monthly intervals, a spatial resolution of 1 km, and a time series spanning from 1960 to 2020. The dataset underwent spatial downscaling from CRU TS v4.02 using WorldClim datasets, employing the Delta downscaling method. Its accuracy was verified against 496 national weather stations, confirming its high reliability [44–48]. We further evaluated its applicability to the KRB by using the observed air temperature data at stations and found that the correlation coefficients of both were above 0.97, which proved the reliability of the datasets.

The precipitation data utilized in this study was sourced from ERA5-Land monthly averaged data provided by European Centre for Medium-Range Weather Forecasts (https://cds.climate.copernicus.eu/ (accessed on 5 January 2023)). This dataset demonstrates a temporal resolution of monthly intervals, a spatial resolution of 0.1°, and a time series spanning from 1960 to 2020. ERA5-Land is derived by replaying the land component of the European Centre for Medium-Range Weather Forecasts ERA5 climate reanalysis, there providing an accurate description of the historical climate conditions [49]. Here, the accuracy of this dataset was analyzed based on the observed monthly precipitation data of different stations, and we found that the correlation coefficients were above 0.64 at the significance level of 0.05, indicating that this dataset could sufficiently reflect the real precipitation situation in the KRB.

The wind speed data adopted in this study were obtained from the High Asia Refined (HAR) analysis version 2 dataset (https://www.klima.tu-berlin.de/ (accessed on 8 January 2023)). This dataset offers a temporal resolution of monthly intervals, a spatial resolution of 10 km, and a time series spanning from 1980 to 2020. The HAR v2 dataset was generated through the application of the Weather Research and Forecasting model (WRF) version 4.1 to dynamically downscale the ERA5 reanalysis data, which is widely utilized for research purposes [50]. The accuracy of this dataset was analyzed based on the observed wind speed data from different stations, and the average correlation coefficient reached 0.50, meeting the needs of this study.

To align with the spatial resolution of the snow cover data, a statistical downscaling method was employed to reduce the spatial resolution of the three climate reanalysis datasets to 500 m. Subsequently, the data were divided according to the hydrological year scale.

2.2.3. Digital Elevation Model (DEM)

The DEM employed in this study possesses a spatial resolution of 90 m, which was provided by the Shuttle Radar Topography Mission (http://srtm.csi.cgiar.org (accessed on 1 October 2022)). To match the spatial resolution of the snow cover data, the DEM was resampled to 500 m by using a bilinear interpolation method. Subsequently, aspect and slope were calculated based on the DEM. Table 1 presents the zonal division results of elevation, aspect, and slope.

Table 1. Zonal features extracted from DEM of the KRB.

Elevation (m)	Area (km^2)	Aspect	Area (km^2)	Slope (°)	Area (km^2)
≤2500	145.75	North (N)	1427.75	≤5	2088.75
2500–3000	281.75	Northeast (NE)	1119.50	5–10	2039.00
3000–3500	324.5	East (E)	801.75	10–15	1713.00
3500–4000	357.75	Southeast (SE)	920.75	15–20	1263.25
4000–4500	692.75	South (S)	1050.00	20–25	772.00
4500–5000	2129.25	Southwest (SW)	818.00	25–30	348.50
5000–5500	2815.00	West (W)	891.75	>30	125.75
5500–6000	1271.50	Northwest (NW)	1320.75		
>6000	332.00				

2.3. Methods

2.3.1. Snow Cover Indices

Snow cover percentage (SCP) and snow cover frequency (SCF), regarded as two commonly used snow cover indices, hold substantial significance in investigating the spatiotemporal variation of snow cover. *SCP* is defined as the percentage of snow cover within the basin. The formula is as follows:

$$SCP = \frac{S}{A} \times 100\% \tag{1}$$

where S represents the snow cover area within the KRB, and A represents the total area of the KRB.

SCF is defined as the proportion of days with snow cover on a pixel within the basin to the total number of days. The formula is as follows:

$$SCF = \frac{D_s}{D} \times 100\% \tag{2}$$

where D_s represents the number of days with snow cover in a specific pixel, and D represents the total number of days in a year (or a season).

2.3.2. Time-Series Analysis

Sen's slope method, a nonparametric statistical approach, was adopted to estimate increasing or decreasing trend changes within time series [51]. The Mann–Kendall trend test, a quantitative analysis tool, was used to assess the significance of the changing trends in the time series, while the Mann–Kendall mutation test was applied to identify the position of mutation points within the time series [52,53]. Typically, the combination of Sen's slope method and the Mann–Kendall trend test allows for the evaluation of trend characteristics and the rates, which is why they are widely applied in hydro-climatic studies [54]. In this study, these methods were used to analyze changing trends and mutation points in snow cover and climate factors in time series.

The Hurst exponent, often determined via R/S analysis, is an effective method for the prediction of future change trends according to the long-term dependence or persistence of time series [55]. In this study, the Hurst exponent was adopted to analyze and predict the future change trends of SCF at each pixel.

2.3.3. Partial Least Squares Regression (PLSR)

PLSR is a comprehensive analysis method that combines principal component analysis, canonical correlation analysis, and multiple linear regression analysis. It possesses the advantages of these three methods and can eliminate multiple correlations among independent variables [38]. PLSR can be divided into univariate PLSR and multivariate PLSR. Since this study only involved a single dependent variable, univariate PLSR was adopted to explore the response of snow cover change to climate change.

Firstly, both the snow cover and climate data were standardized. F_0 represents the standardized variable of snow cover data (dependent variable), and E_0 represents the standardized matrix of the set of air temperature, precipitation, and wind speed data. A component u_1 was extracted from F_0, satisfying $u_1 = F_0 c_1$, where c_1 represents the first axis of F_0. Similarly, component t_1 was extracted from E_0, satisfying $t_1 = E_0 w_1$, where w_1 represents the first axis of E_0. Following the principles of principal component analysis and canonical correlation analysis, the following can be obtained:

$$w_1 = E_0^T F_0 / \left\| E_0^T F_0 \right\| \tag{3}$$

Next, the regression equations for E_0 and F_0 with respect to t_1 are obtained, respectively.

$$E_0 = t_1 p_1^T + E_1 \tag{4}$$

$$F_0 = t_1 r_1 + F_1 \tag{5}$$

where p_1 and r_1 represent regression coefficients, and E_1 and F_1 represent residual matrices of the regression equation. Subsequently, components $t_1, t_2, \ldots, t_m$ were successively extracted from component t_h using the same approach. The regression equation for F_0 with respect to t_h can be obtained using the formula:

$$F_0 = r_1 t_1 + r_2 t_2 + \cdots + r_m t_m + F_m \tag{6}$$

When $x_j^* = E_{0j}$ and $y^* = F_0$, the following formula can be obtained:

$$\hat{y}^* = \alpha_1 x_1^* + \alpha_2 x_2^* + \cdots + \alpha_p x_p^* \tag{7}$$

The regression coefficient of x_j^* is expressed as follows:

$$\alpha_j = \sum_{h=1}^{m} r_h w_{hj}^* \tag{8}$$

where w^*_{hj} represents the *j*th component of w^*_h. If x_j contributes significantly to the construction of t_h, the coefficient of x_j in the regression model will be larger. In this study, PLSR was applied to examine the influence of climate factors on SCF at each pixel.

2.3.4. Structural Equation Modeling (SEM)

SEM allows for the representation of complex direct and indirect causal relationships among multiple variables. It involves the establishment of a specific model based on existing theoretical knowledge, encompassing both the measurement model and the structural model for the hypothesized causal relationship among observed variables [56]. Path analysis, which is considered a special case in SEM, refers to the situation where all variables are directly available, resulting in the existence of only the structural model [57]. This approach not only enables the determination of the total effect of independent variables on the dependent variable but also observes the mutual effect between the independent variables and further divides the total influence into direct and indirect effects. In this study, SEM was employed to investigate the extent to which snow cover and climate elements affect runoff.

3. Results

3.1. Characteristics of Climate Change in the KRB

In this study, the average annual values of climate reanalysis data were extracted based on the hydrological year. Subsequently, Sen's slope method and the Mann–Kendall trend and mutation test were employed to analyze the trends and mutation characteristics of air temperature and precipitation in the KRB from 1961 to 2020, as well as wind speed from 1981 to 2020 (Figure 2).

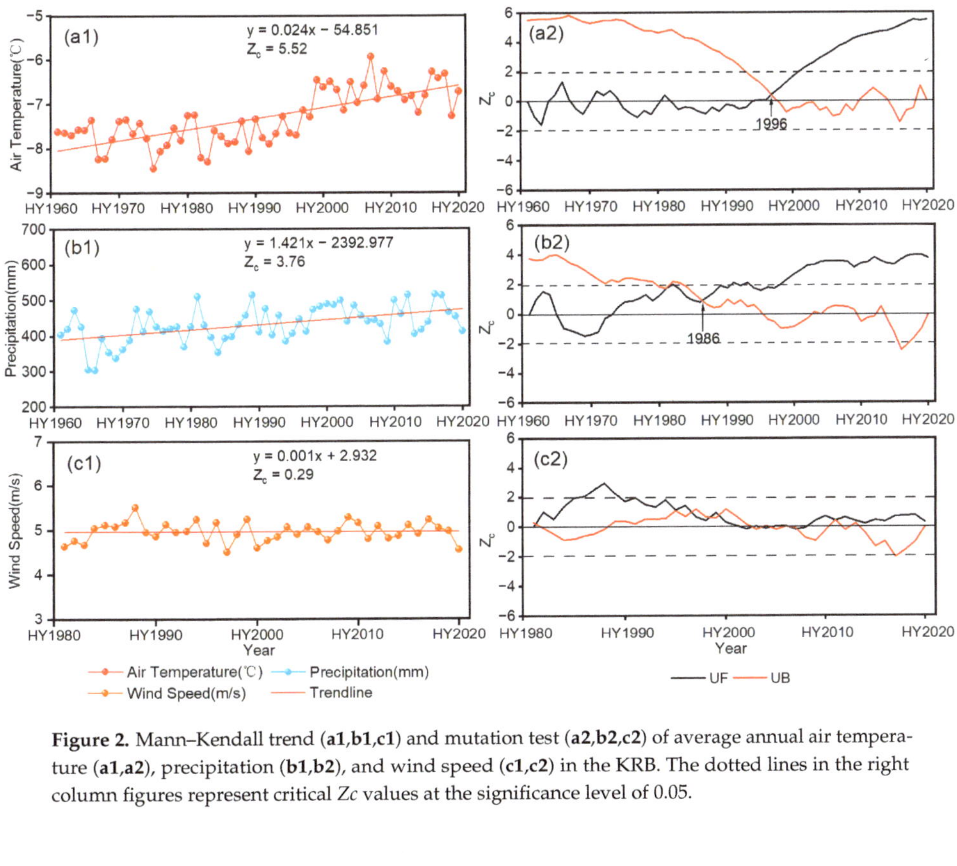

Figure 2. Mann–Kendall trend (**a1**,**b1**,**c1**) and mutation test (**a2**,**b2**,**c2**) of average annual air temperature (**a1**,**a2**), precipitation (**b1**,**b2**), and wind speed (**c1**,**c2**) in the KRB. The dotted lines in the right column figures represent critical *Zc* values at the significance level of 0.05.

From 1961 to 2020, there was a significant increase in air temperature and precipitation, with rates of 0.24 °C/decade and 14.21 mm/decade, respectively (both passing the significance level of 0.001). These findings aligned with the climate change trend observed in Xinjiang over the past 60 years [58]. The average annual air temperature was −7.3 °C, reaching a maximum value of −5.9 °C in 2007 (Figure 2(a1)). As shown in Figure 2(a2), the mutation year for air temperature occurred in 1996. Prior to the mutation year, the average annual air temperature demonstrated a fluctuating trend, which continued to rise after the mutation year and exceeded the significance level of 0.05 after 2001. The average annual precipitation was 431.9 mm, with a maximum value of 516.3 mm in 2016 (Figure 2(b1)). As depicted in Figure 2(b2), the variation trend of annual precipitation can be roughly divided into three stages: an increase during 1961–1965, a decrease during 1965–1973, and another increase during 1973–2020. There was a mutation year in 1986, followed by a significant increase after 1989. From 1981 to 2020, the average annual wind speed was 5.0 m/s, reaching a maximum value of 5.5 m/s in 1988 (Figure 2(c1)). As can be seen from Figure 2(c2), the average annual wind speed indicated a slight overall upward trend, with a significant increase from 1985 to 1990. However, the changing trend during the remaining periods was insignificant (without an abrupt change point).

3.2. Characteristics of Snow Cover Change in the KRB

3.2.1. Spatial Distribution Characteristics of Snow Cover

Figure 3 illustrates the spatial distribution characteristics of the average annual SCF in the KRB across 20 hydrological years. The SCF exhibited lower values within the low elevation area in the northern part of the KRB, while higher SCF values were observed within the high elevation area in the southern part. This spatial pattern demonstrated a positive correlation between SCF and elevation.

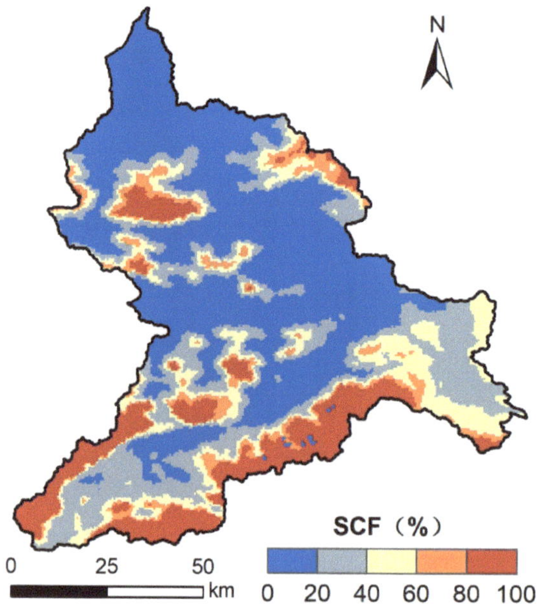

Figure 3. Spatial distribution of annual average SCF in the KRB from 2001 to 2020.

To study the distribution characteristics of SCP across different terrains in the KRB, regional statistics for SCP were conducted based on elevation, aspect, and slope (Figure 4).

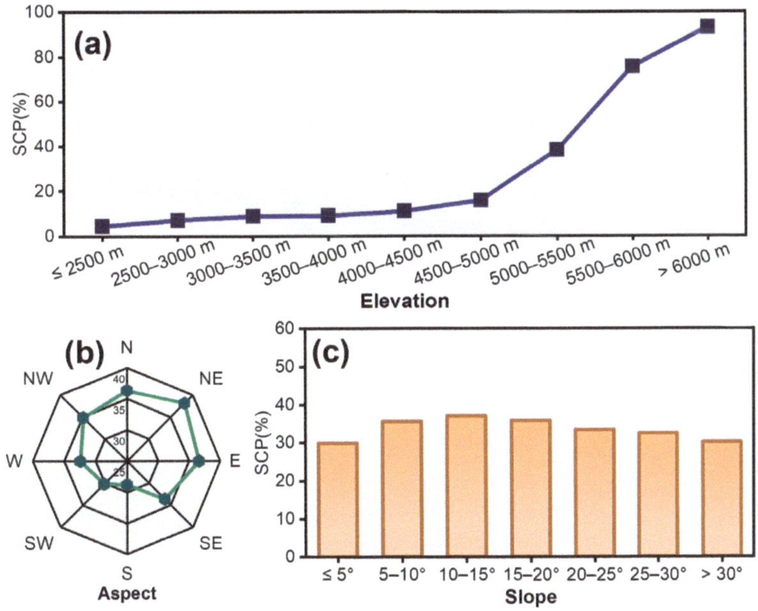

Figure 4. SCP distribution at different elevations (**a**), aspects (**b**), and slopes (**c**) in the KRB from 2001 to 2020.

The impact of elevation on SCP is presented in Figure 4a. With the increase in elevation, the SCP in the KRB increased from 4.46% to 93.22%. The SCP exhibited a gradual increase at an elevation below 5000 m, while a more rapid increase was observed above 5000 m. This may be due to the higher temperature in low-elevation areas compared to high-elevation areas, which hindered the formation of stable snow cover.

Figure 4b shows the distribution of SCP across various aspects. The SCP exhibited distinct patterns among different aspects. Specifically, the SCP values exceeded 35% in the north, northeast, and east aspects, with the highest SCP (38.12%) observed in the northeast aspect. In contrast, the SCP values were relatively low in the south, southwest, and west aspect, with the lowest SCP observed in the south aspect at only 28.78%. This difference can be attributed to the increased solar radiation received by the south-facing region, leading to more rapid snow melting compared to other regions.

Figure 4c depicts the relationship between SCP and slope. The SCP demonstrated an initial increase followed by a decrease as slope values increased, with a distinct boundary observed at 15 degrees. The highest SCP value, reaching 37.07%, was observed within the 10–15 degree zone. SCP values were relatively low at about 30% in the 0–5 degree and above 30 degree zones. This pattern can be attributed to the flatter terrain in the downstream areas of the KRB, particularly at lower elevations, where higher temperatures prevailed. When the slope exceeded 30 degrees, the terrain became steeper and more susceptible to avalanches [59].

3.2.2. Intra-Annual Variation of Snow Cover

This study extracted the annual average daily SCP for the KRB and different elevation zones over a period of 20 hydrological years. As shown in Figure 5a, the accumulation of snow cover in the KRB initiated in September, gradually increasing until it reached its peak in mid-October. Between November and December, the SCP experienced a slight decrease and was relatively low due to low precipitation. Subsequently, the SCP exhibited a fluctuating upward trend, with a maximum value of 49.68% observed in mid-April. As the air temperature increased, the SCP began to decline rapidly from June, reaching a minimum of 13% in early August. Overall, the SCP exhibited two peak periods, one in autumn and the other in spring. The inter-annual fluctuation of SCP was slight in summer but extensive in other seasons.

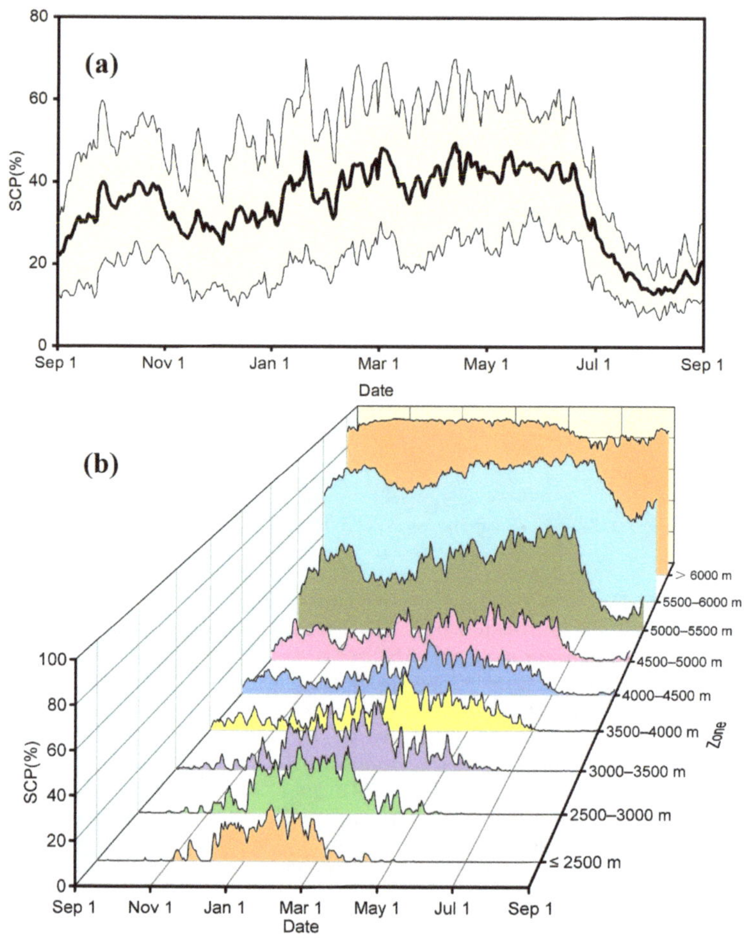

Figure 5. Intra-annual variation of SCP in KRB from 2001 to 2020 at basin scale (**a**) and different elevation zones (**b**).

The intra-annual variation of SCP at different elevation zones can be divided into three types (Figure 5b). (1) Below 4000 m, the SCP demonstrated a unimodal trend, with the peak value mainly concentrated in late-winter and early-spring. Below 3500 m, the peak value of SCP appeared in February, while it appeared in March for elevations between 3500 m and 4000 m. As the elevation increased, the peak value of SCP tended to be delayed. (2) The SCP between 4000 m and 6000 m exhibited bimodal variation characteristics, with the peaks occurring in autumn and spring, respectively. The autumn peak period, except for the 4500–5000 m zone in late-September, generally appeared in mid-October across other elevation zones. In the spring peak period, each elevation zone had a different peak period, ranging from early-March to early-June. With the increase in elevation, the peak time of SCP was further delayed, the peak value increased, and the duration of the snow accumulation period lengthened. (3) The SCP above 6000 m remained relatively stable, with the SCP values exceeding 75%. This zone was characterized by perennial snow and glaciers, which were less affected by seasonal changes.

3.2.3. Interannual Variation of Snow Cover

This study employed Sen's slope method and the Mann–Kendall methods to assess the interannual variation of SCP in the KRB across 20 hydrological years between 2001 and 2020 considering both annual and four-season time scales (Figure 6). Overall, the trends observed for SCP on both the annual and seasonal scales did not exhibit the statistical significance level of 0.05, and no evident mutation year was identified. In terms of annual scale, the SCP demonstrated a decrease at a rate of −1.17%/decade, with an average annual SCP of 34.09%. The SCP was higher in 2003 and 2019, reaching 43.05% and 45.02%, respectively. Prior to 2014, the UF values were greater than 0 in most years, whereas after 2014, the UF values were less than 0. Thus, based on a division at 2014, the average annual SCP displayed an increasing trend followed by a decreasing trend. Regarding the four-season time scale, the highest average seasonal SCP was observed in spring (42.95%), while the lowest was recorded in summer (24.72%). The change rate of SCP in summer was the highest, reaching −2.13%.

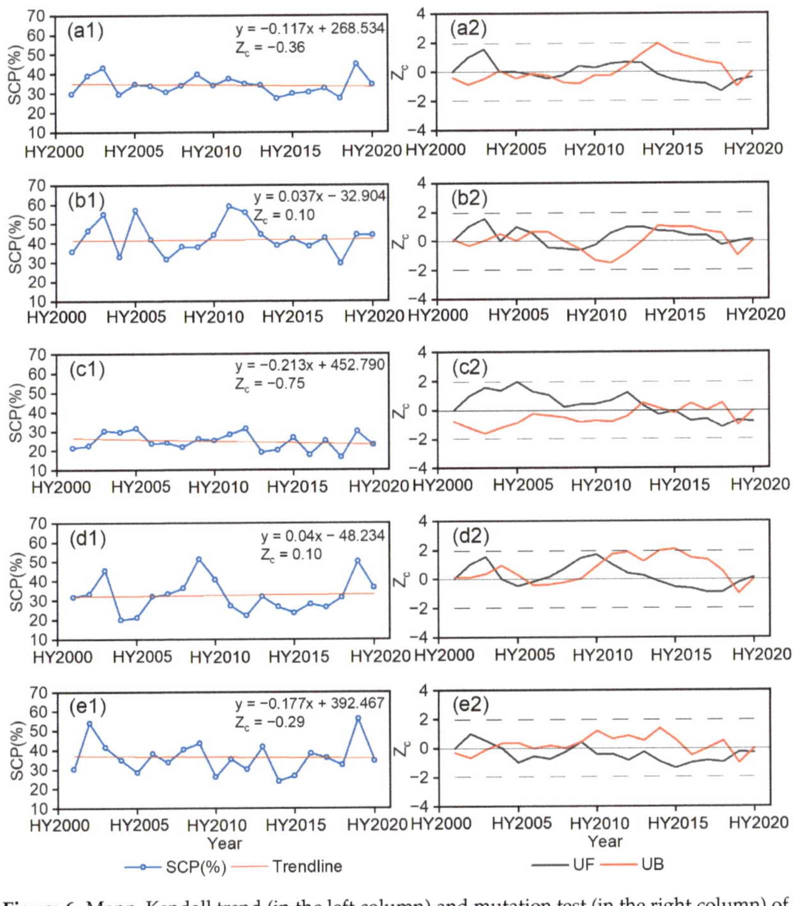

Figure 6. Mann–Kendall trend (in the left column) and mutation test (in the right column) of average SCP in the KRB from 2001 to 2020 at different time scales: annual (**a1**,**a2**), spring (**b1**,**b2**), summer (**c1**,**c2**), autumn (**d1**,**d2**), and winter (**e1**,**e2**). The dotted lines in the right column figures represent critical *Zc* values at a significance level of 0.05.

The spatial variation trend of the annual and seasonal SCF in the KRB during the study period was investigated by applying Sen's slope method (in the left column of Figure 7). The significance of the variation trend was analyzed in conjunction with the Mann–Kendall trend test (in the middle column of Figure 7). In addition, the Hurst exponent was adopted to predict the future variation trends (in the right column of Figure 7). The results indicated that the trends observed for the annual and seasonal SCF in the KRB from 2001 to 2020 were dominated by insignificant changes. Regions exhibiting a decreasing trend of SCF accounted for more than 50% of the total area in the annual, spring, summer, and winter seasons. Specifically, the areas with a decreasing trend of SCF in annual, spring, and winter were mainly concentrated in the northern, central, and southwestern regions of the KRB, while they were mainly located in the southern regions in summer. On the other hand, SCF in autumn displayed a primarily increasing trend, accounting for 38.24% of the total area of the basin, with 4.04% of the area showing a significant increase (significant at the 0.05 level), mainly distributed in the central area of the basin. In terms of future change trends, both the annual and seasonal SCF (except autumn) are expected to increase, with more than 45% of the regions displaying a reversal in the past change trend.

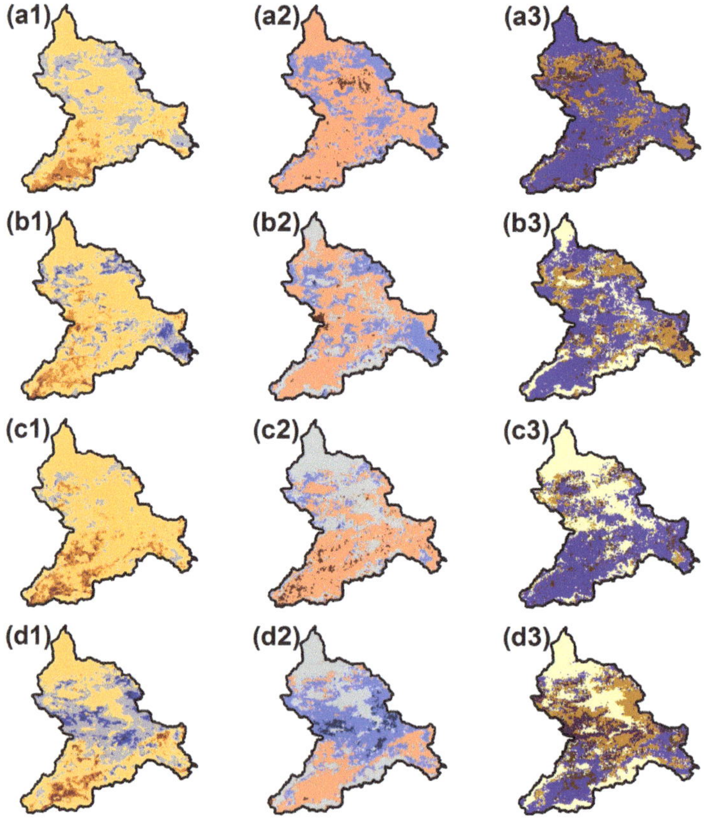

Figure 7. *Cont.*

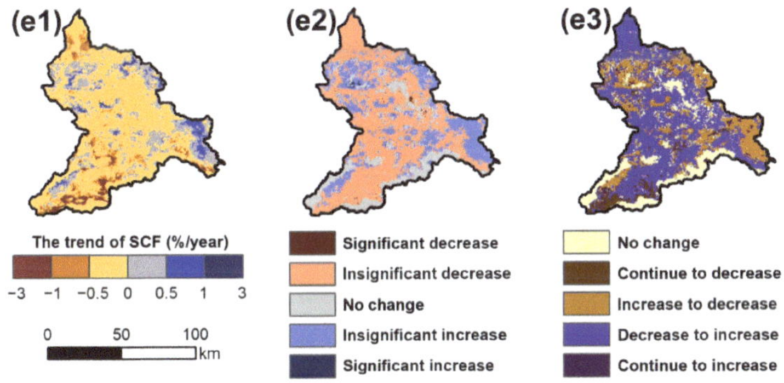

Figure 7. Spatial distribution of SCF variation trends (in the left column), significance (in the middle column), and future changes (in the right column) in the KRB from 2001 to 2020 at different time scales: annual (**a1**,**a2**,**a3**), spring (**b1**,**b2**,**b3**), summer (**c1**,**c2**,**c3**), autumn (**d1**,**d2**,**d3**), and winter (**e1**,**e2**,**e3**).

4. Discussion

4.1. Response of Snow Cover Change to Climate Change

Considering the multiple correlations among air temperature, precipitation, and wind speed, PLSR method was used to investigate the influence of different climate factors on SCF. By comparing the maximum of absolute value of PLSR coefficients on each pixel, the main controlling factors of SCF in each pixel of the KRB across 20 hydrological years (from 2001 to 2020) were obtained (Figure 8). On an annual scale, precipitation emerged as the main factor affecting the SCF, with a mainly negative influence, accounting for 43.10% of the total area. In contrast, SCF in winter was mainly positively affected by precipitation, accounting for 76.16% of the area. During spring, summer, and autumn, air temperature controlled SCF in over 45% of the KRB, with a dominant negative influence. Among these seasons, autumn accounted for the largest proportion (60.93%). The influence of wind speed on annual and seasonal SCF accounted for 11.23% to 26.54%, primarily exerting a positive effect on the annual and spring scales while having a negative influence during summer, autumn, and winter.

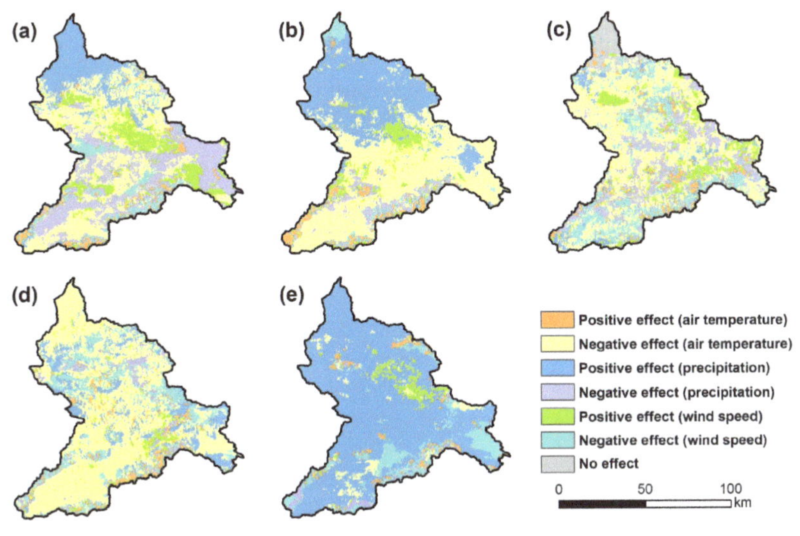

Figure 8. Influencing factors of SCF in the KRB from 2001 to 2020 at different time scales: annual (**a**), spring (**b**), summer (**c**), autumn (**d**), and winter (**e**).

In both annual and seasonal scales, air temperature exerted a predominantly negative influence on SCF due to the snow cover's susceptibility to air temperature changes. An increase in air temperature directly led to the snow melting. The influence of precipitation on SCF varied across the four seasons. In spring, autumn, and winter, precipitation mainly demonstrated a positive effect on SCF, possibly because most of the precipitation occurred as solid snow, facilitating snow accumulation. However, in annual and summer scales, precipitation had a negative effect on SCF, particularly in the high elevation areas in the southern part of the KRB. These regions contained a large amount of eternal snow, where additional snowfall did not contribute to increased SCF. Instead, excessive snow accumulation can lead to a decrease in the SCF due to snow avalanches [60]. In addition to air temperature and precipitation, wind speed also played a role in influencing SCF. In the annual and spring scales, wind speed exhibited a positive effect on SCF. It redistributed snow cover within the KRB, resulting in a more uniform spatial distribution. Conversely, wind speed had a negative impact on SCF in summer, autumn, and winter, mainly observed in the high elevation areas. This could be due to the persistently subzero air temperatures and arid conditions in these regions. Under the influence of wind speed, snow cover sublimation was significant, and the wind speed could also promote the occurrence of snow drifts and avalanches [61]. Combining three kinds of climate data, this study focuses on the effects of climate change on SCF. However, terrain also plays an important role in snow cover change. In the future, we will continue to examine the influencing factors of snow cover and attempt to analyze the elevation effect of snow cover change on climate response [62].

4.2. *Effect of Snow Cover and Climate on Runoff*

In this study, SEM was employed to analyze the impact of air temperature, precipitation, and SCP on runoff from 2001 to 2020. The results indicated the total effect, direct effect, and indirect effect of these three factors on runoff at the annual scale (Figure 9). In terms of the total effect, the standardized regression coefficients of the effect for air temperature, precipitation, and snow cover on runoff were −0.122, 0.609, and 0.09, respectively. Precipitation emerged as the major driver of runoff.

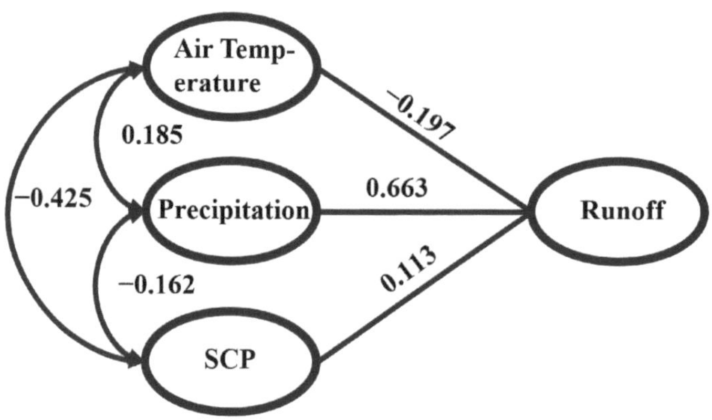

Figure 9. Effects of air temperature, precipitation, and SCP on runoff in the KRB from 2001 to 2020.

The direct effect of air temperature on runoff exhibited a magnitude of −0.197. This can be due to the consistently low average annual air temperature in the KRB, where the majority of areas experienced temperatures below 0 °C. The increase in air temperature had a limited effect on snow melt, but it resulted in higher evaporation rates, leading to a decrease in the runoff. Both precipitation and SCP had a direct positive effect on runoff, with the effect coefficients of 0.663 and 0.113, respectively. Wang et al. [63] found that the annual runoff of the Keriya River was negatively correlated with air temperature and positively correlated with precipitation, which was consistent with the findings of this study.

Regarding the indirect effects, the indirect effect coefficients of air temperature and SCP on runoff through precipitation were relatively significant, measuring 0.123 and −0.107, respectively. However, it is worth noting that the indirect effect coefficients of air temperature, precipitation, and SCP on runoff were all smaller than the direct effect coefficients, indicating that the three factors mainly exerted direct effects on runoff. Due to the temporal extent of the MODIS snow cover data, this study only analyzed the impact of SCP and climate factors on runoff across 20 hydrological years. In the future, it would be beneficial to incorporate observational data from meteorological stations to supplement the snow cover data. This would enable an analysis of the effect mechanism on runoff at longer time scales, facilitating a deeper understanding of the water cycle processes of the KRB.

5. Conclusions

In this study, reanalysis climate datasets were used to analyze the characteristics of climate change in the KRB. The spatiotemporal distributions of snow cover in the KRB during 20 hydrological years from 2001 to 2020 were analyzed utilizing a new MODIS snow cover extent product over China. The response of snow cover change to climate factors was evaluated using PLSR. Furthermore, the study delved into the impact of snow cover and climate factors on annual runoff by employing SEM. The findings from the present study can be summarized as follows:

(1) There was a significant increase in air temperature and precipitation, with rates of 0.24 °C/decade and 14.21 mm/decade, and the mutation year occurred in 1996 and 1986, respectively. However, wind speed did not change significantly.

(2) In terms of spatial distribution, the SCF in the KRB presented "low in the north and high in the south" distribution characteristics. The SCP in the KRB demonstrated an elevation-dependent increase, with the highest values observed in the north aspect and in the 10–15 degrees slope zone. Regarding the intra-annual variation, the SCP within the KRB demonstrated distinctive patterns, including a single peak in winter, double peaks in both spring and autumn, and a consistent high value (SCP > 75%) with turning elevations of 4000 m and 6000 m, respectively. Moreover, the peak SCP values showed a delayed trend with increasing elevation. In terms of temporal change, the SCP in the KRB decreased annually and in summer and winter; however, it increased in spring and autumn between 2001 and 2020. More than 50% of the KRB experienced a decreasing trend for annual, spring, summer, and winter SCF, whereas 38.24% of the areas showed an increasing trend in autumn. In addition to autumn, annual and seasonal SCF is estimated to show an upward trend in the future, accounting for more than 50% of the KRB.

(3) The annual SCF was mainly negatively affected by precipitation, while in winter, it was mainly positively affected by precipitation, accounting for 43.1% and 76.16% of the area, respectively. The spring, summer, and autumn SCF changes in more than 45% of KRB were controlled by air temperature, exerting a predominantly negative influence. Annually and during spring, the impact of wind speed on SCF was mainly positive; however, it negative in summer, autumn, and winter, with the area controlled by wind speed ranging from 11.23% to 26.54%.

(4) The total effect of climate factors and SCP on the annual runoff in the KRB was as follows: precipitation (0.609) > air temperature (−0.122) > SCP (0.09). Climate factors and SCP mainly exerted a direct effect on the changes in annual runoff.

Author Contributions: Conceptualization, W.Y. and Y.W.; methodology, W.Y. and X.M.; software, M.L.; validation, J.Y. and Y.T.; formal analysis, S.L.; data curation, Y.W.; writing—original draft preparation, W.Y. and Y.W.; writing—review and editing, W.Y.; supervision, X.M. All authors have read and agreed to the published version of the manuscript.

Funding: This research was funded by the Open Project of Key Laboratory, Xinjiang Uygur Autonomous Region (2023D04073), the Natural Science Foundation of Henan (232300421250, 222300420522), the Key Scientific Research Project in Colleges and Universities of Henan Province (22A170019), the Nanhu Scholars Program for Young Scholars of XYNU, and the Scientific Research Foundation of Graduate School of Xinyang Normal University (2022KYJJ017).

Data Availability Statement: Not applicable.

Acknowledgments: We would like to express our sincere thanks to the anonymous reviewers.

Conflicts of Interest: The authors declare no conflict of interest.

References

1. Li, Y.; Chen, Y.; Li, Z. Climate and topographic controls on snow phenology dynamics in the Tienshan Mountains, Central Asia. *Atmos. Res.* **2020**, *236*, 104813. [CrossRef]
2. Qin, D.; Liu, S.; Li, P. Snow cover distribution, variability, and response to climate change in western China. *J. Clim.* **2006**, *19*, 1820–1833. [CrossRef]
3. Qin, D.; Zhou, B.; Xiao, C. Progress in studies of cryospheric changes and their impacts on climate of China. *J. Meteorol. Res.* **2014**, *28*, 732–746. [CrossRef]
4. Yan, J.; Wei, R.; Wang, J.; Niu, J.; Ji, S.; Lu, S.; Li, X. Study on spatial-temporal varations of snow cover days and its indicating climatic significance in the Middle and Lower Reaches of Yangtze River for the psat 50 years. *J. Xinyang Norm. Univ. Nat. Sci. Ed.* **2022**, *35*, 92–96. [CrossRef]
5. Xiao, X.; Zhang, T.; Zhong, X.; Li, X. Spatiotemporal variation of snow depth in the Northern Hemisphere from 1992 to 2016. *Remote Sens.* **2020**, *12*, 2728. [CrossRef]
6. Wang, Y.; Huang, X.; Liang, H.; Sun, Y.; Feng, Q.; Liang, T. Tracking snow variations in the Northern Hemisphere using multi-source remote sensing data (2000–2015). *Remote Sens.* **2018**, *10*, 136. [CrossRef]
7. López-Moreno, J.I.; Beniston, M.; García-Ruiz, J.M. Environmental change and water management in the Pyrenees: Facts and future perspectives for Mediterranean mountains. *Glob. Planet. Chang.* **2008**, *61*, 300–312. [CrossRef]
8. Stone, R.S.; Dutton, E.G.; Harris, J.M.; Longenecker, D. Earlier spring snowmelt in northern Alaska as an indicator of climate change. *J. Geophys. Res. Atmos.* **2002**, *107*, ACL 10-1–ACL 10-13. [CrossRef]
9. Shao, Y.; Dong, Z.; Meng, J.; Wu, S.; Li, Y.; Zhu, S.; Zhang, Q.; Zheng, Z. Analysis of Runoff Variation and Future Trends in a Changing Environment: Case Study for Shiyanghe River Basin, Northwest China. *Sustainability* **2023**, *15*, 2173. [CrossRef]
10. Pulliainen, J.; Luojus, K.; Derksen, C.; Mudryk, L.; Lemmetyinen, J.; Salminen, M.; Ikonen, J.; Takala, M.; Cohen, J.; Smolander, T.; et al. Patterns and trends of Northern Hemisphere snow mass from 1980 to 2018. *Nature* **2020**, *581*, 294–298. [CrossRef]
11. Pulliainen, J. Mapping of snow water equivalent and snow depth in boreal and sub-arctic zones by assimilating space-borne microwave radiometer data and ground-based observations. *Remote Sens. Environ.* **2006**, *101*, 257–269. [CrossRef]
12. Zhang, Y.; Li, B.; Zheng, D. A discussion on the boundary and area of the Tibetan Plateau in China. *Geogr. Res.* **2002**, *21*, 1–8. [CrossRef]
13. Seidel, F.C.; Rittger, K.; Skiles, S.M.; Molotch, N.P.; Painter, T.H. Case study of spatial and temporal variability of snow cover, grain size, albedo and radiative forcing in the Sierra Nevada and Rocky Mountain snowpack derived from imaging spectroscopy. *Cryosphere* **2016**, *10*, 1229–1244. [CrossRef]
14. Zhao, G.; Han, Y.; Liu, W.; Zhang, R.; Meng, J.; Zeng, X. Spatial-temporal variations of extreme temperature events in Xinjiang from 1961 to 2016. *J. Xinyang Norm. Univ. (Nat. Sci. Ed.)* **2021**, *34*, 248–254. [CrossRef]
15. Tan, X.; Wu, Z.; Mu, X.; Gao, P.; Zhao, G.; Sun, W.; Gu, C. Spatiotemporal changes in snow cover over China during 1960–2013. *Atmos. Res.* **2019**, *218*, 183–194. [CrossRef]
16. Huang, X.; Liu, C.; Zheng, Z.; Wang, Y.; Li, X.; Liang, T. Snow cover variations across China from 1951–2018. *Cryosphere Discuss.* **2020**, *preprint*. [CrossRef]
17. Peng, S.; Piao, S.; Ciais, P.; Fang, J.; Wang, X. Change in winter snow depth and its impacts on vegetation in China. *Glob. Chang. Biol.* **2010**, *16*, 3004–3013. [CrossRef]
18. Liu, M.; Ma, J.; Wang, L.; Lu, S.; Yan, J.; Liu, J.; He, Z.; Cui, Y. Spatial and temporal varations of shallow ground temperature in the Huaihe River Source and its response to recent global warming hiatus. *J. Xinyang Norm. Univ. (Nat. Sci. Ed.)* **2023**, *36*, 180–185. [CrossRef]
19. Chen, W.; Ding, J.; Wang, J.; Zhang, J.; Zhang, Z. Temporal and spatial variability in snow cover over the Xinjiang Uygur Autonomous Region, China, from 2001 to 2015. *PeerJ* **2020**, *8*, e8861. [CrossRef]
20. Zhang, G.; Xie, H.; Yao, T.; Liang, T.; Kang, S. Snow cover dynamics of four lake basins over Tibetan Plateau using time series MODIS data (2001–2010). *Water Resour. Res.* **2012**, *48*, W10529. [CrossRef]
21. Chen, Y.; Zhou, S.; Zhu, M.; Wu, L. Correlation analysis of vegetation cover change and its driving factors in Henan province. *J. Xinyang Norm. Univ. (Nat. Sci. Ed.)* **2021**, *34*, 401–407. [CrossRef]
22. Hall, D.K.; Riggs, G.A.; Salomonson, V.V.; DiGirolamo, N.E.; Bayr, K.J. MODIS snow-cover products. *Remote Sens. Environ.* **2002**, *83*, 181–194. [CrossRef]

23. Kalluri, S.; Cao, C.; Heidinger, A.; Ignatov, A.; Key, J.; Smith, T. The advanced very high resolution radiometer: Contributing to earth observations for over 40 years. *Bull. Am. Meteorol. Soc.* **2021**, *102*, E351–E366. [CrossRef]
24. Venäläinen, P.; Luojus, K.; Lemmetyinen, J.; Pulliainen, J.; Moisander, M.; Takala, M. Impact of dynamic snow density on GlobSnow snow water equivalent retrieval accuracy. *Cryosphere* **2021**, *15*, 2969–2981. [CrossRef]
25. Saavedra, F.A.; Kampf, S.K.; Fassnacht, S.R.; Sibold, J.S. Changes in Andes snow cover from MODIS data, 2000–2016. *Cryosphere* **2018**, *12*, 1027–1046. [CrossRef]
26. Liang, T.G.; Huang, X.D.; Wu, C.X.; Liu, X.Y.; Li, W.L.; Guo, Z.G.; Ren, J.Z. An application of MODIS data to snow cover monitoring in a pastoral area: A case study in Northern Xinjiang, China. *Remote Sens. Environ.* **2008**, *112*, 1514–1526. [CrossRef]
27. Pu, Z.; Xu, L.; Salomonson, V.V. MODIS/Terra observed seasonal variations of snow cover over the Tibetan Plateau. *Geophys. Res. Lett.* **2007**, *34*, L06706. [CrossRef]
28. Zou, Y.; Sun, P.; Ma, Z.; Lv, Y.; Zhang, Q. Snow Cover in the Three Stable Snow Cover Areas of China and Spatio-Temporal Patterns of the Future. *Remote Sens.* **2022**, *14*, 3098. [CrossRef]
29. Thapa, A.; Muhammad, S. Contemporary snow changes in the karakoram region attributed to improved MODIS data between 2003 and 2018. *Water* **2020**, *12*, 2681. [CrossRef]
30. Hao, X.; Huang, G.; Zheng, Z.; Sun, X.; Ji, W.; Zhao, H.; Wang, J.; Li, H.; Wang, X. Development and validation of a new MODIS snow-cover-extent product over China. *Hydrol. Earth Syst. Sci.* **2022**, *26*, 1937–1952. [CrossRef]
31. Rupp, D.E.; Mote, P.W.; Bindoff, N.L.; Stott, P.A.; Robinson, D.A. Detection and Attribution of Observed Changes in Northern Hemisphere Spring Snow Cover. *J. Clim.* **2013**, *26*, 6904–6914. [CrossRef]
32. Du, W.; Kang, S.; Qian, L.; Jiang, Y.; Sun, W.; Chen, J.; Xu, Z.; Sun, W.; Qin, X.; Chai, X. Spatiotemporal variation of snow cover frequency in the Qilian Mountains (northwestern China) during 2000–2020 and associated circulation mechanisms. *Remote Sens.* **2022**, *14*, 2823. [CrossRef]
33. Hussain, D.; Kuo, C.-Y.; Hameed, A.; Tseng, K.-H.; Jan, B.; Abbas, N.; Kao, H.-C.; Lan, W.-H.; Imani, M. Spaceborne satellite for snow cover and hydrological characteristic of the Gilgit river basin, Hindukush–Karakoram mountains, Pakistan. *Sensors* **2019**, *19*, 531. [CrossRef] [PubMed]
34. VyshnevskiI, V.I.; Donich, O.A. Snow cover in the Ukrainian Carpathians. *Acta Hydrol. Slovaca* **2021**, *22*, 284–293. [CrossRef]
35. Shi, M.; Yuan, Z.; Hong, X.; Liu, S. Spatiotemporal Variation of Snow Cover and Its Response to Climate Change in the Source Region of the Yangtze River, China. *Atmosphere* **2022**, *13*, 1161. [CrossRef]
36. Ban, C.; Xu, Z.; Zuo, D.; Liu, X.; Zhang, R.; Wang, J. Vertical influence of temperature and precipitation on snow cover variability in the Yarlung Zangbo River basin, China. *Int. J. Climatol.* **2021**, *41*, 1148–1161. [CrossRef]
37. Dharpure, J.K.; Patel, A.; Goswami, A.; Kulkarni, A.V.; Snehmani. Spatiotemporal snow cover characterization and its linkage with climate change over the Chenab river basin, western Himalayas. *GIsci. Remote. Sens.* **2020**, *57*, 882–906. [CrossRef]
38. Geladi, P.; Kowalski, B.R. Partial least-squares regression: A tutorial. *Anal. Chim. Acta* **1986**, *185*, 1–17. [CrossRef]
39. Zubaida, M.; Xia, J.; Polat, M.; Zhang, R. Land use and landscape pattern changes in the middle reaches of the Keriya River. *Acta Ecol. Sin.* **2019**, *39*, 2322–2330. [CrossRef]
40. Xu, J.; Liu, S.; Zhang, S.; Shangguan, D. Glaciers fluctuations in the Karamilan-Keriya River Watershed in the past 30 years. *J. Glaciol. Geocryol.* **2006**, *28*, 312–318.
41. Ma, X.; Yan, W.; Zhao, C.; Kundzewicz, Z.W. Snow-cover area and runoff variation under climate change in the West Kunlun Mountains. *Water* **2019**, *11*, 2246. [CrossRef]
42. Ling, H.; Zhang, Q.; Shi, W.; Xu, H. Runoff variation law and its response to climate change in the headstream area of the Keriya River basin, Xinjiang. *J. Earth. Sci.* **2011**, *22*, 780–791. [CrossRef]
43. Hao, X. *A New MODIS Snow Cover Extent Product over China (2000–2020)*; National Tibetan Plateau Data Center: Lanzhou, China, 2021. [CrossRef]
44. Peng, S. *1-km Monthly Mean Temperature Dataset for China (1901–2021)*; National Tibetan Plateau Data Center: Lanzhou, China, 2020. [CrossRef]
45. Peng, S.; Gang, C.; Cao, Y.; Chen, Y. Assessment of climate change trends over the Loess Plateau in China from 1901 to 2100. *Int. J. Climatol.* **2018**, *38*, 2250–2264. [CrossRef]
46. Peng, S.; Ding, Y.; Wen, Z.; Chen, Y.; Cao, Y.; Ren, J. Spatiotemporal change and trend analysis of potential evapotranspiration over the Loess Plateau of China during 2011–2100. *Agric. For. Meteorol.* **2017**, *233*, 183–194. [CrossRef]
47. Ding, Y.; Peng, S. Spatiotemporal trends and attribution of drought across China from 1901–2100. *Sustainability* **2020**, *12*, 477. [CrossRef]
48. Peng, S.; Ding, Y.; Liu, W.; Li, Z. 1 km monthly temperature and precipitation dataset for China from 1901 to 2017. *Earth Syst. Sci. Data* **2019**, *11*, 1931–1946. [CrossRef]
49. Muñoz Sabater, J. ERA5-Land monthly data from 1950 to present: Copernicus Climate Change Service (C3S) Climate Data Store (CDS). 2019. Available online: https://cds.climate.copernicus.eu/cdsapp#!/dataset/reanalysis-era5-land-monthly-means?tab=overview (accessed on 21 May 2023).
50. Wang, X.; Tolksdorf, V.; Otto, M.; Scherer, D. WRF-based dynamical downscaling of ERA5 reanalysis data for High Mountain Asia: Towards a new version of the High Asia Refined analysis. *Int. J. Climatol.* **2021**, *41*, 743–762. [CrossRef]

51. Salmi, T.; Määttä, A.; Anttila, P.; Ruoho-Airola, T.; Amnell, T. Detecting trends of annual values of atmospheric pollutants by the Mann-Kendall test and Sen's slope estimates. In *The Excel Template Application MAKESENS*; Finnish Meteorological Institute: Helsinki, Finland, 2002.
52. Mann, H.B. Nonparametric tests against trend. *Econometrica* **1945**, *13*, 245–259. [CrossRef]
53. Kendall, M.G. *Rank Correlation Methods*, 4th ed.; Charles Griffin: London, UK, 1975; ISBN 0195208374.
54. Gocic, M.; Trajkovic, S. Analysis of changes in meteorological variables using Mann-Kendall and Sen's slope estimator statistical tests in Serbia. *Glob. Planet. Chang.* **2013**, *100*, 172–182. [CrossRef]
55. Hurst, H.E. Long-term storage capacity of reservoirs. *Trans. Am. Soc. Civ. Eng.* **1951**, *116*, 770–799. [CrossRef]
56. Blunch, N. *Introduction to Structural Equation Modeling Using IBM SPSS Statistics and AMOS*; Sage Publications Inc.: Thousand Oaks, CA, USA, 2012.
57. Lei, P.W.; Wu, Q. Introduction to structural equation modeling: Issues and practical considerations. *Educ. Meas. Issues Pract.* **2007**, *26*, 33–43. [CrossRef]
58. Chen, Y.; Li, Z.; Xu, J.; Shen, Y.; Xing, X.; Xie, T.; Li, Z.; Yang, L.; Xi, H.; Zhu, C.; et al. Changes and protection suggestions in water resources and ecological environment in arid region of Northwest China. *Bull. Chin. Acad. Sci.* **2023**, *38*, 385–393. [CrossRef]
59. Schweizer, J.; Jamieson, J.B. Snow cover properties for skier triggering of avalanches. *Cold Reg. Sci. Technol.* **2001**, *33*, 207–221. [CrossRef]
60. Wu, S.; Zhang, X.; Du, J.; Zhou, X.; Tuo, Y.; Li, R.; Duan, Z. The vertical influence of temperature and precipitation on snow cover variability in the Central Tianshan Mountains, Northwest China. *Hydrol. Process.* **2019**, *33*, 1686–1697. [CrossRef]
61. Zhang, C.; Mou, N.; Niu, J.; Zhang, L.; Liu, F. Spatio-Temporal Variation Characteristics of Snow Depth and Snow Cover Days over the Tibetan Plateau. *Water* **2021**, *13*, 307. [CrossRef]
62. Morán-Tejeda, E.; López-Moreno, J.I.; Beniston, M. The changing roles of temperature and precipitation on snowpack variability in Switzerland as a function of altitude. *Geophys. Res. Lett.* **2013**, *40*, 2131–2136. [CrossRef]
63. Wang, D.; Shi, Q.; Dong, D.; Chen, C. Response of runoff volume change to climate in the Keriya River in Xinjiang. *Arid Zone Res.* **2018**, *35*, 1271–1279. [CrossRef]

remote sensing

MDPI

Article

Unmanned Aerial Vehicle-Based Structure from Motion Technique for Precise Snow Depth Retrieval—Implication for Optimal Ground Control Point Deployment Strategy

Song Shu [1,*], Ok-Youn Yu [2], Chris Schoonover [2], Hongxing Liu [3] and Bo Yang [4]

1 Department of Geography and Planning, Appalachian State University, Boone, NC 28608, USA
2 Department of Sustainable Technology and the Built Environment, Appalachian State University, Boone, NC 28608, USA; yuo@appstate.edu (O.-Y.Y.)
3 Department of Geography, The University of Alabama, Tuscaloosa, AL 35487, USA
4 Department of Urban and Regional Planning, San Jose State University, San Jose, CA 95192, USA
* Correspondence: shus@appstate.edu

Abstract: Unmanned aerial vehicle (UAV)-based snow depth is mapped as the difference between snow-on and snow-off digital surface models (DSMs), which are derived using the structure from motion (SfM) technique with ground control points (GCPs). In this study, we evaluated the impacts of the quality and deployment of GCPs on the accuracy of snow depth estimates. For 15 GCPs in our study area, we surveyed each of their coordinates using an ordinary global positioning system (GPS) and a differential GPS, producing two sets of GCP measurements (hereinafter, the low-accuracy and high-accuracy sets). The two sets of GCP measurements were then incorporated into SfM processing of UAV images by following two deployment strategies to create snow-off and snow-on DSMs and then to retrieve snow depth. In Strategy A, the same GCP measurements in each set were used to create both the snow-on and snow-off DSMs. In Strategy B, each set of GCP measurements was divided into two sub-groups, one sub-group for creating snow-on DSMs and the other sub-group for snow-off DSMs. The accuracy of snow depth estimates was evaluated in comparison to concurrent in-situ snow depth measurements. The results showed that Strategy A, using both the low-accuracy and high-accuracy sets, generated accurate snow depth estimates, while in Strategy B, only the high-accuracy set could generate reliable snow depth estimates. The results demonstrated that the deployment of GCPs had a significant influence on UAV-based SfM snow depth retrieval. When accurate GCP measurements cannot be guaranteed (e.g., in mountainous regions), Strategy A is the optimal option for producing reliable snow depth estimates. When highly accurate GCP measurements are available (e.g., collected by differential GPS in open space), both deployment strategies can produce accurate snow depth estimates.

Keywords: UAV; structure from motion (SfM); snow depth; ground control points (GCP)

1. Introduction

Over one billion of people around the world rely on snowmelt water for domestic, agricultural, and industrial activities [1]. Snow accumulation in the winter season shows a rapidly decreasing trend in recent decades for North America in terms of snow depth [2], snow cover extent [3], and snow mass [4]. In particular, the continental-scale snowmelt water originating from mountainous regions is in steep decline [5]. To better understand its impact on the environment and human activities, it is necessary to accurately monitor winter snow accumulation and its dynamics in mountain ranges.

The amount of water contained in snowpack (snow water equivalent, or SWE) is jointly determined by snow density and snow depth. It has been documented in previous studies that snow depth has much stronger spatial variation than snow density [6–8]. The accurate mapping of snow depth is therefore critical for the estimation of SWE. Traditionally, snow depth is monitored by automated sensors at weather stations or measured manually using a snow probe in field surveys. These methods are typically limited to point-based locations or cover very sparse areas. Remote sensing technologies have also been used to derive snow depth information, such as satellite laser altimetry [9], passive microwave remote sensing systems [10–13], and snow radar systems [14–16]. Although these technologies have made positive contributions to snow depth mapping in relatively open space with gentle topography, their performances are mostly unsatisfactory in mountain ranges where snow accumulation varies drastically along with rapid changes in topography and vegetation within a short distance [17–19]. It has been proven that accurate estimation of snow accumulation in mountain ranges is hampered by the complex topography and varying vegetation cover [20–22]. The airborne LiDAR system has also been used to retrieve snow depth [23] and is a promising technique for mapping snow accumulation in mountainous regions. This technique, because of its high cost, is inapplicable for monitoring snow accumulation dynamics since this requires frequent and repeated observations. A more nuanced, efficient, flexible, and economic method that can accommodate the mountain environment and simultaneously provide accurate snow depth mapping is needed, particularly for a relatively small area and for ephemeral snowpack.

In recent years, the possibility of retrieving snow depth with unmanned aerial vehicle (UAV)-based structure from motion (SfM) photogrammetry has been discussed in a number of studies [24–34] due to the advancement of UAV technology, the development of low-cost lightweight cameras, and the availability of open-source and commercial SfM software packages. Compared to airborne and spaceborne remote sensing platforms, the prominent advantages of UAV-based SfM photogrammetry are its lower demand of personnel training, lower cost for purchasing and maintaining the equipment, and much higher flexibility for inaccessible areas (e.g., cliffs and valleys) and undesirable weather conditions (e.g., rain, strong wind, and heavy cloud). The basic idea of this method is to collect overlapping photos of a study area (e.g., open or vegetated, flat or rugged) with a UAV-based optical camera on snow-on and snow-off dates. Then, digital surface models (DSMs) of the study area are generated for each of the two dates by stitching the overlapping photos with an SfM workflow that has been implemented in many software packages [35]. The surface elevation increase due to snow accumulation over the study area is then calculated by differencing the two DSMs and treated as snow depth estimates. In general, the snow depth accuracy varies from several centimeters to several decimeters depending on the surface roughness, ground topography, and underlying vegetation [26,28,32,34].

In addition to the environmental characteristics of the snow accumulation field (i.e., topography, vegetation, surface roughness, etc.), the quality of snow-on and snow-off DSMs generated by SfM is a more dominant factor that determines the accuracy of UAV-based snow depth estimates. Two options are mostly adopted in existing studies to constrain the errors in DSMs and to improve snow depth accuracy. The first one uses an expensive real-time kinematic (RTK) UAV to collect optical photos of which the coordinates can be georeferenced to an accuracy of 2–4 cm by on-board RTK correction [26,27,33]. The accurate photo locations are then used directly to constrain the DSM uncertainties. In the second option, a regular customer-grade UAV is used to collect photos. Then, ground control points (GCP) evenly distributed across the study area are incorporated in SfM processing to reduce DSM errors [24,25,28,29,32,34]. The GCPs are often accurately surveyed with an RTK or post-processing kinematic (PPK) differential global positioning system (DGPS). Vander Jagt, Lucieer and Wallace et al. [27] compared the two options and reported that incorporating GCPs in the generation of DSM can produce a lower root mean square error (RMSE) between UAV-based snow depth estimates and in-situ measurements than simply using RTK UAV. Additionally, if a minimal number of GCPs is jointly used with RTK UAV, the RMSE can be further reduced. Harder, Schirmer and Pomeroy et al. [26] also suggested that inclusion of GCPs for RTK UAV would result in a great reduction of DSM bias, thus increasing the accuracy of snow depth estimates. Both studies indicated the necessity of incorporating GCPs for accurate snow depth retrieval with UAV-based SfM photogrammetry.

RTK UAV is a promising solution for snow depth mapping over inaccessible areas (e.g., areas susceptible to snow avalanche) [33]. The much higher cost of an RTK UAV than a regular customer-grade UAV limits its wide application in snow depth mapping. Most existing studies used a regular UAV to collect photos and incorporated GCPs to constrain DSM errors [24,28,31,32,34] and then retrieve snow depth. Lee, Park, Choi and Kim et al. [24] evaluated the influence of the number of GCPs on the accuracy of UAV-based snow depth and concluded that areas with higher densities of GCPs tended to have more accurate snow depth estimates. In other words, the more GCPs included, the better the snow depth accuracy. In these studies, GCPs over the study area were surveyed twice, one for snow-on date and one for snow-off date. The snow-on GCPs were usually differently from the snow-off GCPs. The two different sets of GCPs were used to produce the snow-on and snow-off DSMs, respectively. With this strategy, the errors in snow-on GCP coordinates (x, y, z) were completely independent from the errors in snow-off GCP coordinates. These errors could be propagated independently into the corresponding snow-on and snow-off DSMs, thus affecting the retrieved snow depth accuracy. However, no study has investigated how the errors in GCP coordinates influence UAV-based snow depth retrieval and how to improve the accuracy of snow depth estimates when DGPS performance is hampered by the surrounding environment.

The overall objective of this paper was to assess in detail the influence of GCP errors on the accuracy of snow depth measurements estimated by UAV-based SfM photogrammetry. Specifically, this paper (1) assesses how the errors in GCP horizontal coordinates (x, y) and vertical height (z) influence the accuracy of snow depth estimates; and (2) discusses the optimal GCP deployment strategy that can accommodate the errors in GCP coordinates (x, y, z) to help produce better snow depth estimates. The remainder of this paper is organized as follows: Section 2 describes the study area and the collection of UAV-based photos and in-situ snow depth data. Section 3 introduces SfM concepts and GCP deployment strategies. Section 4 presents the performance of UAV-based snow depth retrieval using different GCP deployment strategies. Section 5 analyzes the results and discusses the strategy of improving snow depth estimates.

2. Study Area and Data Collection

Our study area was located in the southern Appalachian Mountains at 36.215°N and 81.693°W, close to the Blue Ridge Parkway, which is the longest linear national park way in the U.S. The elevation was about 1070 m above mean sea level. The study area was beside a wind turbine near the campus of Appalachian State University in Boone, western North Carolina. It was a small open space surrounded by trees and shrubs with different heights. As shown in Figure 1, the extent of the study area was approximately 120 m × 150 m. The surface elevation decreased gradually from the west to the east (as shown in Figure 2a,b), while the middle part of the study area was relatively flat with a slope less than 3°. This study area was chosen as an ideal place for experimenting UAV-based SfM snow depth retrieval for the following reason. In the winter, mountainous areas often experience frequent wind gusts, which have a strong redistributing effect on snow accumulation. Our study area was close to a wind turbine. It represented a typical small open space in the Appalachian Mountains for snow accumulation that is heavily influenced by winter wind. In this mountainous region, the mean annual snowfall can vary from several decimeters to two meters [36–38]. On 8 January 2021, a snow storm struck Boone and yielded an average snow accumulation of 10.16 cm (4 inches) on that day and another 5.08 cm (2 inches) in the morning of January 9 (National Weather Service, https://www.weather.gov/wrh/Climate?wfo=rnk, accessed on 23 April 2023).

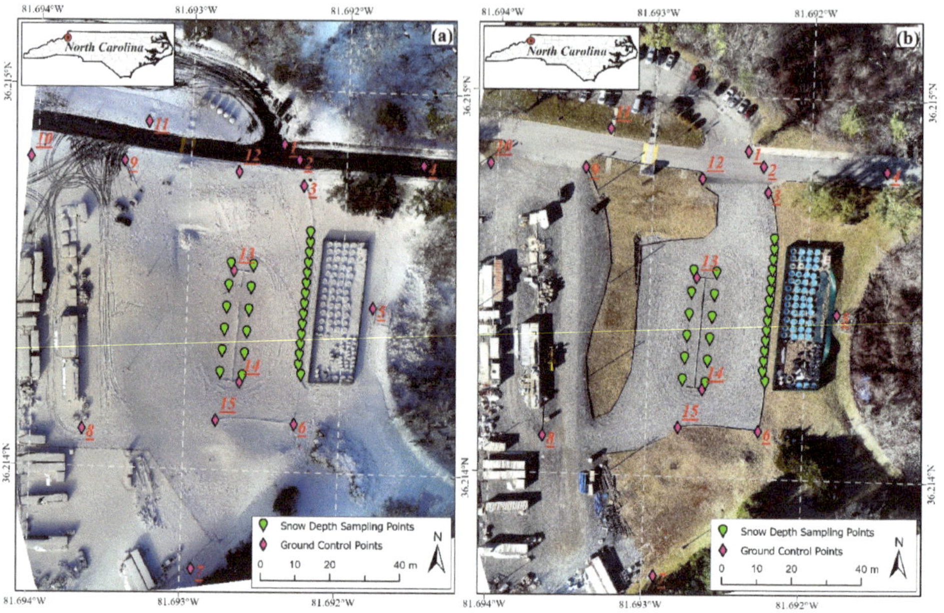

Figure 1. SfM-based orthophoto of the study area: (**a**) with snow cover on 9 January 2021, and (**b**) without snow cover on 24 February 2021. The ID number of each GCP is displayed in red color.

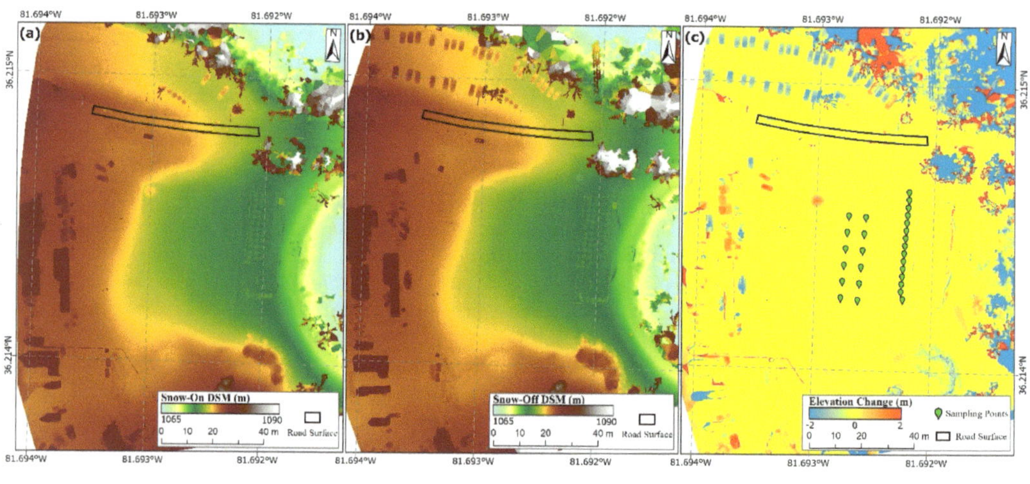

Figure 2. (**a**) DSM for snow-on date (9 January 2021); (**b**) DSM for snow-off date (24 February 2021); and (**c**) snow depth based on the difference between the two DSMs.

We used a DJI Phantom 4 quadcopter UAV to collect photos of the study area right after snowfall at 3:30 p.m. on 9 January 2021 and after snowmelt at 3:40 p.m. on 24 February 2021. The DJI Phantom 4 is a customer-grade UAV that carries a low-cost CMOS camera, which well-suited the purpose of this study. The camera has a field of view (FOV) of 84° and focal length of 8.8 mm (comparable to 24 mm focal length on a 35 mm film camera). The dimension of each photo is 5472 × 3648. The photos were acquired at an above-ground height of 45.72 m (150 feet). The front and side overlapping rates are 70%

and 80%, respectively. The ground sampling distance (spatial resolution) of the generated orthophoto and DSM is about 1.5 cm.

Concurrent in-situ snow depths accurate to one tenth of a centimeter were measured using a snow probe at 27 sampling points in the study area, as shown by the green points in Figure 1. These in-situ snow depths ranged from 5.0 to 29.3 cm. Compared to the 12 points on the left side, the 15 points on the right side had relatively higher depths (ranging from 15.0 to 29.3 cm) due to the blocking effect of the fence on the east of the 15 sampling points. We also surveyed 15 ground control points (red diamonds in Figure 1) across the entire study area using two GPS instruments, including the handheld Trimble GeoXH and the Trimble DA2 GNSS receiver. All 15 GCPs were free from snow cover on the two dates. The Trimble DA2 uses the real-time kinetic (RTK) technique and is able to achieve horizontal and vertical accuracies of 1 and 2 cm, respectively, under optimal conditions. The coordinates (x, y, z) measured by the Trimble GeoXH were not processed with any differential correction. The survey generated two sets of GCP measurements, one set with highly accurate coordinates and the other with low accuracy coordinates (hereinafter referred to as the high-accuracy and low-accuracy sets). To verify the accuracies of the two sets of GCPs, we also measured the coordinates of the 15 GCPs using a Nikon NPL-322 Series Total Stations, which provides measurements of the GCP coordinates at a millimeter-level accuracy for the spatial scale of our study area. Compared to the coordinates (x, y, z) measured by the total stations, the root mean square error (*RMSE*) values of the high-accuracy set coordinates were 3.08, 3.02, and 7.0 cm, respectively, and the *RMSE* values of the low-accuracy set coordinates were 30.25, 40.84, and 78.34 cm, respectively.

3. Methodology

In this study, the optical photos collected on the snow-on and snow-off dates were separately processed using ERSI Drone2Map software, which implemented the general SfM workflow described in Section 3.1 to generate orthophotos and DSMs for the two dates. The high-accuracy and low-accuracy sets of GCP measurements were incorporated into the SfM workflow following the strategies described in Section 3.2 in order to reduce the errors in DSMs and orthophotos and to georeference them to a real-world coordinate system. Figure 1a,b shows the orthophotos of the two dates. The SfM-based snow depth over the study area were then derived by subtracting snow-off DSM from snow-on DSM. Figure 2a,b shows the DSMs on the two dates and Figure 2c shows the derived elevation change. The accuracy of SfM-based snow depth (elevation change) was evaluated in comparison to the concurrent in-situ snow depths at the sampling points in Figure 2c (green points) with the metrics described in Section 3.3.

3.1. Generation of DSM and Orthophoto with SfM

The structure from motion (SfM) technique originated from the field of computer vision and has been implemented by many software packages, for example, ESRI Drone2Map used in this study. In general, the SfM workflow consists of the following five steps [39,40]: (1) identifying keypoints (e.g., corners, line ends, intersections, etc.) from each photo using the scale invariant feature transform (SIFT) algorithm [41,42]; (2) matching the correspondence keypoints in the overlapping areas of two adjacent photos using the approximate nearest neighbor k-dimensional trees approach [43]; (3) removing the outlier keypoints using the random sample consensus (RANSAC) approach [44]; (4) determining the extrinsic camera parameters (referred to as "motion") for each photo, such as camera location and orientation, and simultaneously estimating the 3D position of each keypoint (referred to as "structure" or sparse point cloud) by solving and optimizing a set of collinearity equations [27]; (5) generating a dense 3D point cloud for each pixel on each photo based on the sparse point cloud and the camera parameters obtained in Step 4 by applying a patch-based multi-view stereo image matching [45].

In Step 4, initial values of the extrinsic camera parameters are first obtained from the EXIF tags of UAV-collected photos, and they are used to estimate the 3D positions of keypoints through collinearity equations [39]. The camera parameters and keypoint positions are then further refined by optimizing a non-linear cost function that represents the errors between the observed and estimated keypoint positions on UAV-collected photos. The optimization process is called bundle adjustment and is the step at which GCPs are incorporated. The objectives of incorporating GCPs are twofold. First, it can help constrain the non-linear cost function and refine the camera parameters and keypoint positions. Second, it helps georeference the original sparse point cloud from the arbitrary coordinate system to a real-world geographic or projected coordinate system. A seven-parameter linear similarity transformation (i.e., three global translation parameters, three rotation parameters, and one scaling parameter) is performed to achieve the georeferencing [39]. The coordinate system used for this study was NAD 1983 StatePlane North Carolina (2011). DSMs and orthophotos are then generated based on the refined camera parameters and dense 3D point cloud.

3.2. Strategy of Incorporating GCPs in SfM

Two sets of GCP measurements, the high-accuracy and low-accuracy sets, were collected for our study area. To assess the influences of GCP measurement accuracy and GCP deployment strategy on snow depth retrieval, each set of GCP measurements was incorporated into SfM by following two strategies to produce DSMs.

In Strategy A, the same GCP measurements in each set (hereinafter referred to as the Same-GCP strategy) were used to generate both the snow-on and snow-off DSMs and then to estimate snow depth, as shown in Figure 3. We incorporated all 15 GCP measurements at the beginning (Step 1 in Table 1) to generate the two DSMs and estimate snow depth. Then, we removed one GCP measurement at each of the following steps. For example, in Step 2 in Table 1, GCP 4 was removed. The coordinate measurements of the other 14 GCPs (1–3, 5–15) were used to produce the snow-on and snow-off DSMs.

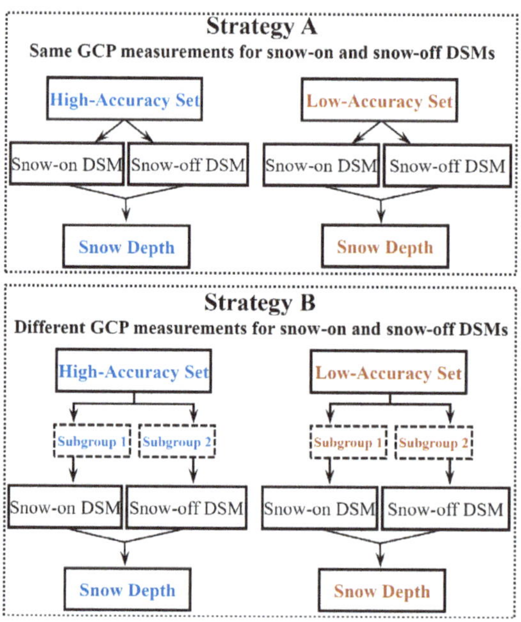

Figure 3. Strategies A and B of incorporating GCP measurements into SfM to estimate snow depth.

Table 1. GCPs incorporated at each step in Strategies A and B.

Step	Number of GCPs Incorporated	GCPs Incorporated in Strategy A	GCPs Incorporated in Strategy B	
		Snow-On/Snow-Off	Subgroup1: Snow-On	Subgroup2: Snow-Off
1	15	1–15		
2	14	1–3, 5–15		
3	13	1–3, 6–15		
4	12	1–3, 6–9, 11–15		
5	11	1–3, 6, 8, 9, 11–15		
6	10	1, 3, 6, 8, 9, 11–15		
7	9	1, 3, 6, 8, 9, 12–15		
8	8	3, 6, 8, 9, 12–15	1, 3, 4, 6, 7, 9, 13, 15	2, 5, 6, 8, 10, 11, 12, 14
9	7	3, 6, 8, 9, 12, 13, 15	1, 4, 6, 7, 9, 13, 15	2, 5, 6, 8, 11, 12, 14
10	6	3, 6, 8, 9, 12, 15	1, 4, 6, 9, 13, 15	2, 6, 8, 11, 12, 14
11	5	3, 6, 8, 9, 15	1, 6, 9, 13, 15	2, 6, 8, 11, 14
12	4	3, 6, 8, 9	1, 6, 9, 15	2, 6, 8, 11
13	3	3, 6, 8	1, 9, 15	2, 6, 8

In Strategy B, different GCP measurements in each set (hereinafter referred to as the Different-GCP strategy) were used to generate the snow-on and snow-off DSMs. As shown in Figure 3, the 15 GCP measurements in each set were divided into two subgroups, one subgroup for generating the snow-on DSMs and the other subgroup for the snow-off DSMs. At the beginning (Step 8 in Table 1), each subgroup had eight GCP measurements (with GCP 6 in Figure 1 shared by the two subgroups). Then, one GCP measurement was removed from each of the two subgroups at each step to generate the two DSMs. For example, in Step 9 in Table 1, GCP 3 was removed from subgroup 1 and GCP 10 was removed from subgroup 2. The remaining GCP measurements in the two subgroups were then used to produce the snow-on and snow-off DSMs, respectively. The GCPs in each sub-group were evenly distributed across the study area to ensure that the in-situ snow depth sampling points were encompassed by the GCPs.

Strategy B in this study, the Different-GCP strategy, replicated the approach used by most previous studies to incorporate GCPs and estimate snow depth [24,31,32,34]. Strategy A served as a reference for Strategy B. By comparing the results of the high-accuracy and low-accuracy sets under the two strategies, we were able to assess how SfM-based snow depth retrieval is influenced by the errors in GCP coordinate measurements and how GCP deployment strategies can help to mitigate the influence of the errors.

To further separate and evaluate the influences of horizontal (x, y) and vertical (z) coordinate errors on SfM-based snow depth retrieval, we replaced the GCP heights (z) in the low-accuracy set with the GCP heights in the high-accuracy set. This created a new set of GCPs with relatively inaccurate horizontal coordinates and accurate vertical coordinates, which was referred to as the accurate-Z set. We also created an accurate-XY set by replacing the GCP horizontal coordinates (x, y) in the low-accuracy set with the horizontal coordinates in the high-accuracy set. These two sets of GCPs were then incorporated into SfM following Strategy B in Table 1 to generate DSMs and estimate snow depth.

3.3. Assessment of SfM-Based Snow Depth Accuracy

For surfaces not covered by snow (e.g., the road surface in Figure 1, which had been cleared after snowfall), the elevation was supposed to remain the same in both snow-off and snow-on DSMs and the derived snow depth (i.e., elevation change) should be close to zero. When the snow depth estimates on the road surface have a systematic (positive or negative) deviation from zero, a bias between the two DSMs is observed. This bias could be wrongly interpreted as snow accumulation or depletion. This study used the road surface to estimate and remove bias in the derived snow depth. The bias was estimated as the mean value of snow depth within the black polygon on the road surface in Figure 1.

The accuracy of SfM-based snow depth was assessed in comparison to the in-situ snow depths measured at the 27 sampling points in Figure 1. At each sampling point, the SfM-based snow depth was estimated by an inverse-distance interpolation of the elevation changes in the surrounding 9 pixels (in a 3 × 3 neighborhood). The accuracy was measured by the Pearson correlation coefficient r [46], the root mean square error (*RMSE*), and the slope of regression between in-situ snow depth and SfM-based snow depth at the 27 sampling points. The *RMSE* represents the absolute accuracy of SfM-based snow depth. The smaller the *RMSE*, the higher the accuracy. The correlation coefficient r represents the capability of the SfM-based technique to capture the variation in snow depth across the study area. The closer the r value is to one, the stronger the capability. The slope of regression represents the tendency to over-estimate or under-estimate snow depth. Ideally, the slope value should be equal to one. If the intercept of regression equals zero, a slope of regression larger than one usually indicates a tendency to overestimate snow depth, whereas a slope less than one means a tendency to underestimate snow depth.

4. Results

4.1. Strategy A: High-Accuracy Set Verses Low-Accuracy Set

The high-accuracy set of GCP measurements was first incorporated into SfM following Strategy A (the Same-GCP strategy) to generate DSMs and orthophotos for the two dates and then to derive snow depth estimates. The low-accuracy set was also incorporated in the same way to derive snow depth estimates. Figure 4 shows the scatterplots of SfM-based snow depth estimates against in-situ snow depths for the two sets of GCP measurements. Figure 5 shows the variations in *RMSE*, correlation coefficient r and linear regression slope along with the number of incorporated GCPs.

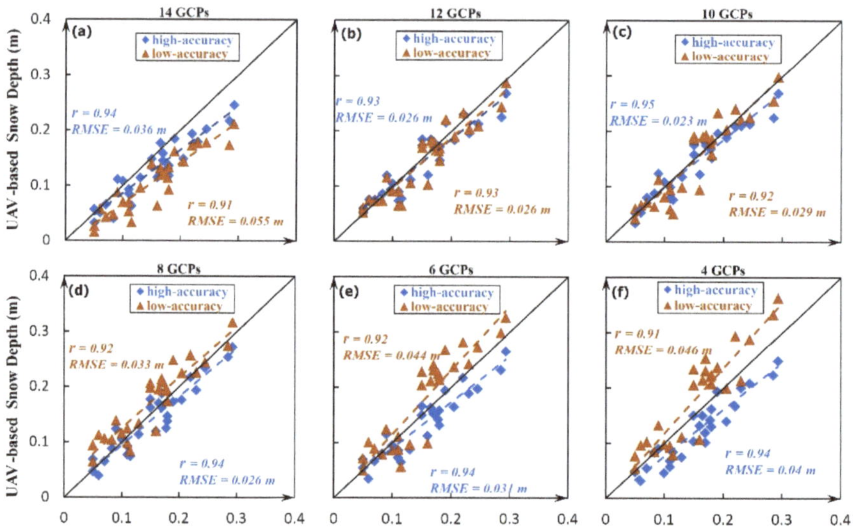

Figure 4. Scatterplots of SfM-based snow depth estimates against in-situ snow depth when using Strategy A to incorporate the high-accuracy and low-accuracy sets of GCPs. Only the scatterplots of an even number of GCPs are displayed. The number of GCPS decreases from (**a**) to (**f**).

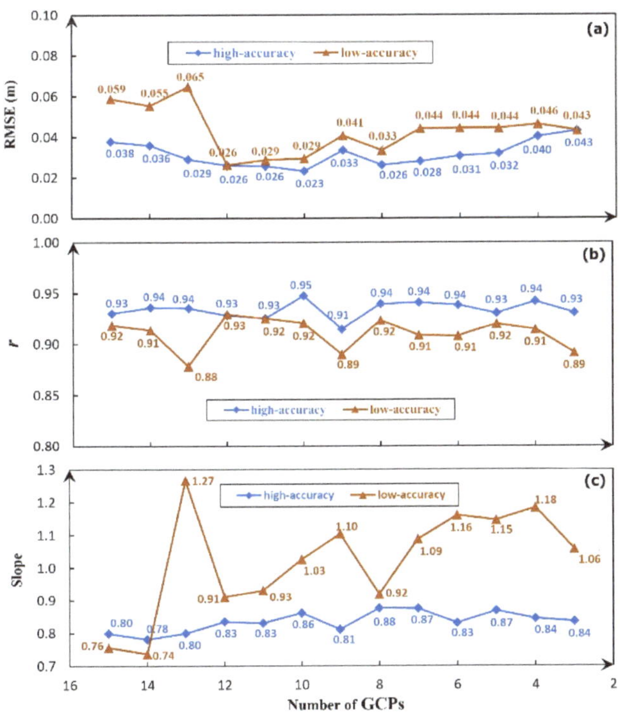

Figure 5. The performance of SfM-based snow depth retrieval when using Strategy A to incorporate the high-accuracy and low-accuracy sets of GCPs: (**a**) *RMSE*, (**b**) correlation coefficient *r*, and (**c**) linear regression slope.

As expected, the high-accuracy set of GCP measurements produced better results than the low-accuracy set in terms of the three metrics. The *RMSE* values of the high-accuracy set ranged from 2.3 to 4.3 cm, slightly better than the *RMSE* values of the low-accuracy set that ranged from 2.6 to 6.5 cm. The correlation coefficient *r* and regression slope of the high-accuracy set remained quite stable along with the number of incorporated GCPs, as shown in Figure 5, whereas clear variations were observed for the low-accuracy set. The correlation coefficients were mostly higher than 0.9 for both sets.

The regression slope of the high-accuracy set was less than one (around 0.85) all the time, indicating an underestimation of snow depth. For the low-accuracy set, the regression slope was less than one when all GCPs were incorporated at the beginning. It increased with the removal of GCPs and remained mostly larger than one (around 1.1) when the number of GCPs was less than and equal to 10. This suggested that when using the low-accuracy set of GCPs, the snow depth was underestimated at the beginning, then became slightly overestimated as the number of GCP decreased. This trend can be clearly observed in Figure 4 when the number of GCPs decreased from 14 to 4.

Another interesting observation is that as shown in Figure 5a for both sets, the *RMSE* values gradually decreased as GCPs were removed at each step and then increased when GCPs were further removed. The highest accuracies occurred when the number of GCPs was 12 for the low-accuracy set and 10 for the high-accuracy set. This suggested that when using the Same-GCP strategy, more GCPs did not necessarily produce more accurate snow depth estimates. In general, the Same-GCP strategy helped generate highly accurate snow depth estimates. With this strategy, the low-accuracy set of GCPs could generate snow depth estimates with an accuracy comparable to that of the high-accuracy set. One thing to note is that in Figure 5 there was a relatively big jump in *RMSE*, *r*, and regression slope when the number of GCPs decreased to 13. This was possibly caused by the large errors in coordinate measurements of GCP 5 in Figure 1. After the removal of this GCP, the three indicators became more stable.

4.2. Strategy B: High-Accuracy Set Verses Low-Accuracy Set

The high-accuracy and low-accuracy sets of GCP measurements were also incorporated following Strategy B (the Different-GCP strategy) to estimate snow depth for our study area. Figure 6 shows the scatterplots of SfM-based snow depth estimates against in-situ snow depths. For the high-accuracy set of GCP measurements, the accuracy of SfM-based snow depth estimates was still quite high, even though its performance in terms of the three metrics was slightly inferior to that of Strategy A (the Same-GCP strategy). The correlation between snow depth estimates and in-situ snow depths was strong, with *r* varying from 0.87 to 0.92. The regression slopes remained stably less than one (around 0.85), which was consistent with Strategy A and indicated an underestimation of snow depth. The *RMSE* values were slightly higher than those in Strategy A and increased from 4.0 to 5.4 cm when the number of GCPs decreased from 8 to 3, indicating a degradation in accuracy when fewer GCPs were incorporated. This echoed the results observed by Lee, Park, Choi and Kim [24] that more GCPs produced better accuracy. In general, when using the high-accuracy set of GCP measurements, the influence of Strategy B on the accuracy of SfM-based snow depth estimates was limited. The snow depth accuracy was slightly inferior, but still comparable to that of Strategy A.

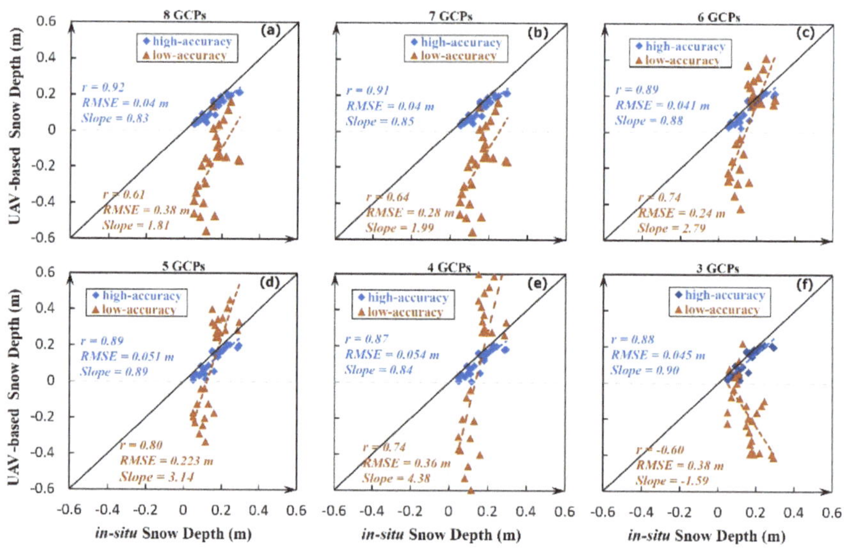

Figure 6. Scatterplots of SfM-based snow depth estimates against in-situ snow depth when using Strategy B to incorporate the high-accuracy and low-accuracy sets of GCPs. (**a**–**f**) the number of GCPs decreased from 8 to 3.

In contrast, when the low-accuracy set of GCP measurements was incorporated following Strategy B to retrieve snow depth, the accuracy was much lower than that of Strategy A. As shown in Figure 6, the *RMSE* values were high all the time and varied from 22.3 to 38 cm. Even though a correlation between SfM-based and in-situ snow depths could still be observed, the regression slopes were mostly much higher than one. The SfM technique heavily underestimated the snow depths for thin snow cover and produced negative snow depths, while it largely overestimated the snow depth for relatively thick snow. The fewer GCPs incorporated, the higher the slope. When the number of GCPs decreased to 3, a negative correlation between SfM-based and in-situ snow depths was observed, which was not reasonable for accurate snow depth mapping. By comparing the performances of the two sets of GCP measurements in Strategies A and B, it was reasonable to deduce that the errors in GCP horizontal and vertical coordinates (x, y, z) were propagated into SfM-based snow depths when using Strategy B to incorporate GCP measurements. It was also clear that Strategy A could effectively prevent the propagation of GCP coordinate errors and produce reliable SfM-based snow depths.

4.3. Strategy B: Accurate-Z Set Verses Accurate-XY Set

To further evaluate the individual influence of GCP horizontal coordinates (x, y) and vertical height (z) on SfM-based snow depth retrieval, the accurate-Z and accurate-XY sets of GCP measurements were incorporated separately following Strategy B to derive snow depth for our study area. Figure 7 shows the scatterplots of SfM-based and in-situ snow depths. Figure 8 summarizes and compares the results generated by incorporating the high-accuracy, low-accuracy, accurate-Z, and accurate-XY sets following Strategy B. As shown in Figure 7, the incorporation of the accurate-Z set produced evidently better snow depth estimates than the incorporation of the accurate-XY set. For the accurate-Z set, the *RMSE* values ranged from 6.6 to 12.5 cm, the correlation coefficient *r* varied from 0.85 to 0.92, and the regression slopes were in the range of 0.81–1.17. For the accurate-XY set, the *RMSE* values were much higher, ranging from 17 to 39.6 cm, the *r* decreased from 0.83 to 0.48 along with the removal of GCPs, and the regression slope was higher than one all the time. This comparison suggested that in this study the accuracy of GCP height (z) was more crucial than that of GCP horizontal coordinates (x, y) for accurate snow depth mapping using the SfM technique.

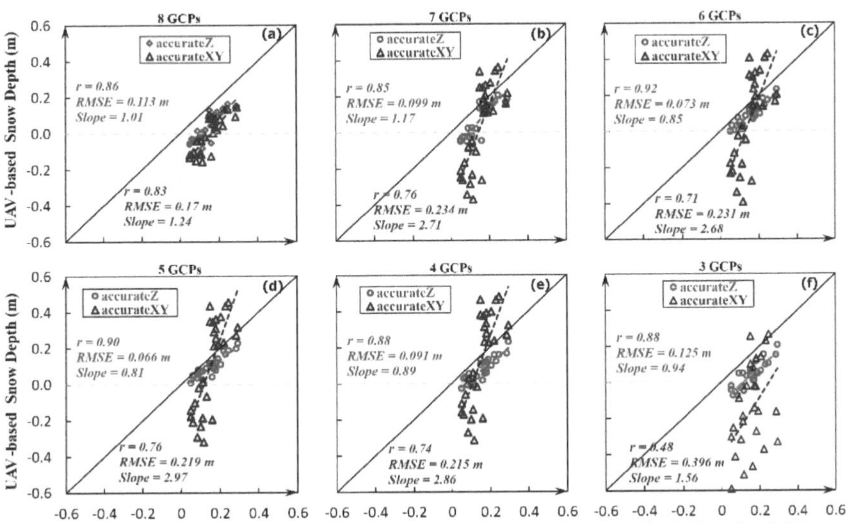

Figure 7. Scatterplots of SfM-based snow depth estimates against in-situ snow depth when using Strategy B to incorporate the accurate-Z and accurate-XY sets of GCP measurements. (**a**–**f**) the number of GCPs decreased from 8 to 3.

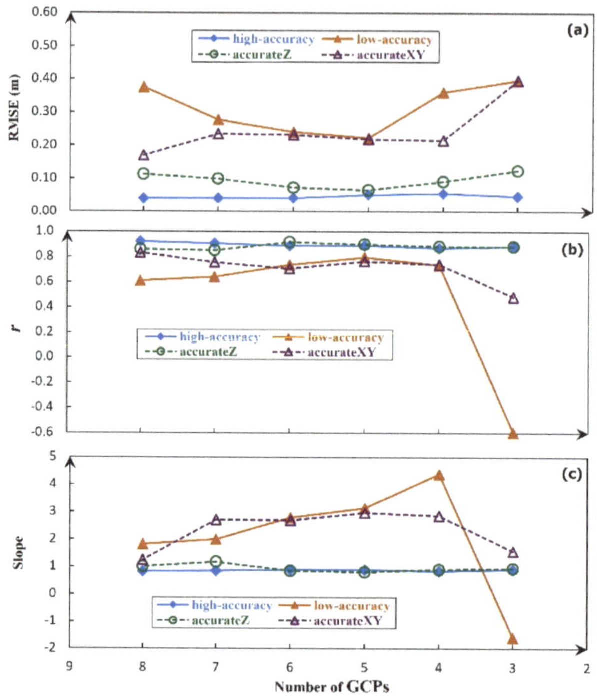

Figure 8. The performance of SfM-based snow depth retrieval when using Strategy B to incorporate the four sets of GCPs: (**a**) *RMSE*, (**b**) correlation coefficient *r*, and (**c**) linear regression slope.

On the other hand, compared to the high-accuracy set in Strategy B, the result produced by incorporating the accurate-Z set was relatively inferior in terms of all three metrics, as shown in Figure 8. In particular, the incorporation of the accurate-Z set produced negative snow depth estimates for thin snow cover, as shown in Figure 7a,b,e,f. In addition, it can be observed from Figure 8 that the results produced by the accurate-XY set were generally better than the results produced by the low-accuracy set in Strategy B. These two observations indicated that even though GCP height (z) was more important for SfM-based snow depth retrieval, reducing the errors in GCP horizontal coordinates was still a necessary step to further improve SfM-based mapping of snow depth. In particular, this was more important for mountainous regions where acute topographic change can occur within a short distance.

5. Discussion

As observed in Sections 4.1 and 4.2, the results generated by the high-accuracy and low-accuracy sets of GCP measurements in Strategies A (Same-GCP strategy) and B (Different-GCP strategy) were markedly different. The high-accuracy set produced reliable snow depth estimates in both strategies, whereas for the low-accuracy set, reliable snow depth estimates were only derived in Strategy A. The contrast highlights the effectiveness of Strategy A in mitigating the influence of GCP coordinates errors on SfM-based snow depth retrieval. It also entails a thorough discussion on how the errors in GCP coordinates are propagated in SfM-based snow depth estimates and how we should survey and incorporate GCPs in order to produce more accurate snow depth estimates.

5.1. Influence of GCP Coordinate Errors under Different GCP Deployment Strategies

As mentioned in Section 3.1, the role of GCPs in SfM workflow is twofold. It helps to optimize camera parameters and the 3D locations of keypoints (sparse point cloud) and then georeference the sparse point cloud. The DSMs and orthophotos are then generated based on the optimized camera parameters and the georeferenced sparse point cloud. Therefore, the optimizing and georeferencing processes are the two ways though which the errors in GCP coordinates (x, y, z) are propagated to the two DSMs and then to snow depth estimates. The optimizing process determines the 3D position of each keypoint relative to the surrounding keypoints. In other words, it determines the relative topographic undulation of the surface in our study area, on both the snow-off and snow-on dates. Georeferencing is achieved by using a linear similarity transformation [39]. It lines up the sparse point cloud (therefore, DSMs and orthophotos) with a real-world coordinate system through linear global translation, rotation, and scaling. These operations are applied in the same way to all points in the cloud. They are not involved in the determination of keypoints' relative locations nor change the surface's relative topographic undulation. When GCP measurements are incorporated, both the optimizing and georeferencing processes treat the GCP measurements as ground truth and try to minimize the overall discrepancy between the GCP measurements and the corresponding keypoints.

Figure 9 shows a schematic of the influence of GCP coordinate errors on snow depth estimates. Figure 9a–c illustrates the influence of coordinate errors on the snow-on and snow-off surface profiles when using Strategy A, while Figure 9d–f illustrates the influence of coordinate errors when using Strategy B. Figure 9a,d shows the true snow-on and snow-off surface profiles (snow and ground surfaces). As shown in the two figures, when the GCP coordinates are accurate and have no errors, the snow depth estimates are accurate for both strategies. These were the results observed in Sections 4.1 and 4.2 when the high-accuracy set of GCP measurements was used.

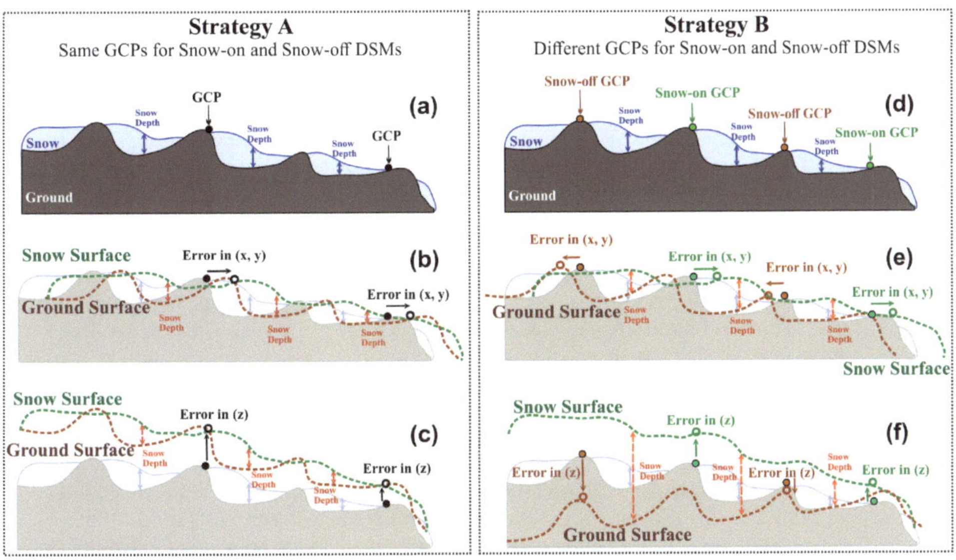

Figure 9. Influence of Strategies A and B on SfM-based snow depth estimates. (**a–c**) Strategy A. (**d–f**) Strategy B.

If errors exist in GCP horizontal coordinates (x, y) and/or vertical height (z), when incorporating GCPs using the Same-GCP strategy (Strategy A), the snow-on and snow-off surface profiles will both be modified or transformed in the same way to best fit the GCPs. For example, as shown in Figure 9b, the snow-on and snow-off profiles were shifted to the right with the same distance. Similarly, in Figure 9c, both profiles were changed in the same way. Even though the errors in GCP (x, y) and (z) were propagated to the snow-on and snow-off profiles, they were canceled out when the two profiles were subtracted, thus generating quite reliable snow depth estimates. This was the result observed in Section 4.1 when the low-accuracy set of GCP measurements was used.

However, if the Different-GCP strategy (Strategy B) is used, the errors in GCP (x, y) and (z) are propagated separately and independently to the snow-off and snow-on profiles. As shown in Figure 9e, the snow-off profile was shifted to left whereas the snow-on profile was shifted to right due to the errors in GCP (x, y). In Figure 9f, the snow-off profile was shifted to a lower location than the true snow-off surface profile whereas the snow-on profile was dragged to a higher location than the true snow-on profile due to the errors in GCP (z). These errors were largely amplified when the snow-off profile was subtracted from snow-on profile, thus producing highly inaccurate snow depth estimates. This was the result observed in Section 4.2 when the low-accuracy set of GCP measurements was used.

Note that for legibility, the errors in horizontal coordinates (x, y) and in vertical height (z) are displayed separately in Figure 9b,e and Figure 9c,f. In addition, the snow-on and the snow-off surface profiles in Figure 9b,c,e,f maintain the same shapes of the true snow-on and snow-off surface profiles in Figure 9a,d, for the purpose of legibility. This may deliver a false message that the accuracy of snow depth estimates could be improved by manually co-registering the snow-on and snow-off DSMs, particularly for Figure 9e,f. In real situations, however, the GCP horizontal and vertical errors are entangled and can hardly be separated from each other. Therefore, simply reducing the horizontal or vertical errors cannot significantly improve the accuracy of snow depth estimates. This was the result observed in Section 4.3. Additionally, when inaccurate GCPs are incorporated into the SfM optimizing process, the errors in GCP coordinates would result in inaccurate camera parameters and sparse point clouds, which could largely change the relative locations of surface features on either the snow-on or snow-off DSMs. In other words, the shape of the snow-on and snow-off surface profiles in Figure 9b,c,e,f could be quite different from the shape of the true snow-on and snow-off profiles in Figure 9a,d. The errors caused by inaccurate camera parameters and sparse point clouds cannot be easily removed by co-registering the snow-on and snow-off DSMs. In fact, we co-registered the snow-on and snow-off DSMs generated by low-accuracy set in Section 4.2 and by the accurate-Z set in Section 4.3 to derive snow depth estimates. The results showed no big improvement in the accuracy of snow depth estimates.

5.2. Optimal GCP Deployment Strategy for UAV-Based SfM Snow Depth Retrieval

Existing studies have rarely investigated whether the GCP deployment strategy could help improve snow depth retrieval using the UAV-based SfM technique. Rigorous comparisons of the results reported by previous studies were not possible due to the huge differences in their study areas and snow cover conditions. However, such comparisons could provide a general sense of how the GCP deployment strategy may influence UAV-based SfM snow depth retrieval. For example, Harder et al. [26] used the same GCPs (the Same-GCP strategy in this study) to generate snow-on and snow-off DSMs and reported *RMSE* values of 8.5, 8.8, and 13.7 cm for areas with alpine, short, and tall vegetation, respectively. Fernandes et al. [25] used the Same-GCP strategy and reported RMSE values varying from 1.58 to 10.56 cm. Goetz et al. [32] and De Michele et al. [31] used different GCPs (the Different-GCP strategy in this study) to produce snow-on and snow-off DSMs and reported overall *RMSE* values of 15.2 and 14.3 cm, respectively. Lendzioch et al. [47] used the Different-GCP strategy and reported *RMSE* values of 16, 32, and 31 cm for open area (snow ablation), forest (snow accumulation), and forest (snow ablation), respectively.

In this study, we conducted a rigorous investigation of how different GCP deployment strategies could mitigate errors in GCP measurements and improve snow depth estimates. It is reasonable to conclude that the Same-GCP strategy is a better option than the Different-GCP strategy for accurately mapping snow depth with SfM-based photogrammetry technology. Even though DGPS (either PPK or RTK) is usually used to survey GCPs for SfM-based snow depth mapping, incorporating GCPs using the Same-GCP strategy can further improve the accuracy of snow depth estimates, as observed in Sections 4.1 and 4.2. Furthermore, DGPS performs best in open space. In mountainous areas, its accuracy is often heavily hampered by the surrounding topography and vegetation. For example, the horizontal and vertical uncertainties of the DGPS used in this study were observed to be two times larger in the valley than in open space. The Same-GCP strategy was less susceptible to errors in GCP horizontal and vertical coordinates, as observed in Section 4. It is a more suitable option for mapping snow depth in mountainous regions.

The Same-GCP strategy requires the same set of GCP measurements for both snow-on and snow-off DSMs. In this study, the average snow depth was about 15 cm. We managed to identify 15 GCPs that were not covered by snow in our study area. In some areas, the snow can be thick enough to cover most of the surface features in a natural setting. In this case, it would be hard to identify GCPs that are free from snow cover on both snow-on and snow-off dates. A viable solution for those areas is to manually set up a series of GCPs that are higher than the snowpack and are maintained during the whole period between the snow-on and snow-off dates. For example, Fernandes et al. [25] used 30 cm square plywood and 15 cm diameter plastic disks as GCPs, which were suspended 1 and 1.3 m above ground surface. As observed in Section 4.1, when using the Same-GCP strategy, more GCPs did not necessarily improve snow depth accuracy. Actually, three GCPs were sufficient to produce snow depth estimates with quite high accuracy (4.3 cm for the high-accuracy and low-accuracy sets). This does not require significant extra labor effort in the field and is worthwhile for accurate snow depth mapping.

One thing to note is that the snow depth in our study area varied from 5 to 30 cm, with an average of about 15 cm. According to the experiments in this study, Strategy A performed best to retrieve snow depth in our study area. In the future, this strategy will be further tested in areas with relatively thick snow using the method discussed above and in areas with complex underlying surface conditions (e.g., grass, short shrubs), different topographic undulations, and various surrounding environments. In addition, when collecting drone optical photos for snow depth mapping, the drone should be flown following several communication protocols to avoid potential security issues, as denoted in [48,49].

6. Conclusions

This study investigated the influence of errors in GCP coordinates on the accuracy of snow depth estimates. Two sets of GCP measurements with different accuracy levels (high-accuracy and low-accuracy) were incorporated following two strategies (Same-GCP and Different-GCP) into SfM processing to generate snow-on and snow-off DSMs and derive snow depths for our study area. The results showed that the Different-GCP strategy tended to amplify and propagate GCP coordinate errors into snow depth estimates, thus heavily decreasing the accuracy. Conversely, if the Same-GCP strategy was applied, the GCP coordinate errors were propagated in the same way to the snow-on and snow-off DSMs and were canceled out when the two DSMs were subtracted to derive the snow depth estimates. These results therefore demonstrated that the Same-GCP strategy effectively mitigated the influence of GCP coordinate errors on SfM-based snow depth mapping and increased the accuracy of derived snow depth estimates. When using the Different-GCP strategy, increasing the number of GCPs improved the accuracy of snow depth estimates, which was consistent with the observations in a previous study [24]. When using the Same-GCP strategy, increasing the number of GCPs did not help produce better snow depth accuracy. In fact, three GCPs that enclosed the area of interest helped to produce highly accurate snow depth estimates.

After comprehensively evaluating the effectiveness of different sets of GCP measurements using the two strategies, the Same-GCP strategy—using the same GCP measurements to generate snow-on and snow-off DSMs—is highly recommended for UAV-based SfM snow depth retrieval. For areas in which thick snow covers most of the natural surface features and identifying the same GCPs on the two dates is difficult, man-made targets higher than the snow cover, e.g., the square plywood and plastic disks used by Fernandes et al. [25], are recommended to be set up and maintained for UAV flights collecting data throughout the whole snow season.

Author Contributions: S.S. and H.L. conceived the study; S.S., O.-Y.Y. and C.S. designed and performed the field surveys; S.S. performed formal analysis and drafted the original manuscript; S.S., H.L., O.-Y.Y., C.S. and B.Y. contributed to the review and editing of the manuscript. All authors have read and agreed to the published version of the manuscript.

Funding: This research received no external funding.

Data Availability Statement: The data in this study will be shared to readers upon request to the corresponding author.

Conflicts of Interest: The authors declare no conflict of interest.

References

1. Barnett, T.P.; Adam, J.C.; Lettenmaier, D.P. Potential impacts of a warming climate on water availability in snow-dominated regions. *Nature* **2005**, *438*, 303–309. [CrossRef] [PubMed]
2. Kunkel, K.E.; Robinson, D.A.; Champion, S.; Yin, X.G.; Estilow, T.; Frankson, R.M. Trends and Extremes in Northern Hemisphere Snow Characteristics. *Curr. Clim. Change Rep.* **2016**, *2*, 65–73. [CrossRef]
3. Dery, S.J.; Brown, R.D. Recent Northern Hemisphere snow cover extent trends and implications for the snow-albedo feedback. *Geophys. Res. Lett.* **2007**, *34*. [CrossRef]
4. Pulliainen, J.; Luojus, K.; Derksen, C.; Mudryk, L.; Lemmetyinen, J.; Salminen, M.; Ikonen, J.; Takala, M.; Cohen, J.; Smolander, T.; et al. Patterns and trends of Northern Hemisphere snow mass from 1980 to 2018. *Nature* **2020**, *581*, 294. [CrossRef]
5. Musselman, K.N.; Addor, N.; Vano, J.A.; Molotch, N.P. Winter melt trends portend widespread declines in snow water resources. *Nat. Clim. Change* **2021**, *11*, 418. [CrossRef]
6. Shook, K.; Gray, D. Small-scale spatial structure of shallow snowcovers. *Hydrol. Process.* **1996**, *10*, 1283–1292. [CrossRef]
7. Jonas, T.; Marty, C.; Magnusson, J. Estimating the snow water equivalent from snow depth measurements in the Swiss Alps. *J. Hydrol.* **2009**, *378*, 161–167. [CrossRef]
8. López-Moreno, J.I.; Fassnacht, S.; Heath, J.; Musselman, K.; Revuelto, J.; Latron, J.; Morán-Tejeda, E.; Jonas, T. Small scale spatial variability of snow density and depth over complex alpine terrain: Implications for estimating snow water equivalent. *Adv. Water Resour.* **2013**, *55*, 40–52. [CrossRef]
9. Shu, S.; Liu, H.; Frappart, F.; Huang, Y.; Wang, S.; Hinkel, K.M.; Beck, R.A.; Yu, B.; Jones, B.M.; Arp, C.D.; et al. Estimation of snow accumulation over frozen Arctic lakes using repeat ICESat laser altimetry observations—A case study in northern Alaska. *Remote Sens. Environ.* **2018**, *216*, 529–543. [CrossRef]
10. Chang, A.; Foster, J.; Hall, D. Nimbus-7 SMMR derived global snow cover parameters. *Ann. Glaciol.* **1987**, *9*, 39–44. [CrossRef]
11. Derksen, C.; Walker, A.; Goodison, B. Evaluation of passive microwave snow water equivalent retrievals across the boreal forest/tundra transition of western Canada. *Remote Sens. Environ.* **2005**, *96*, 315–327. [CrossRef]
12. Green, J.; Kongoli, C.; Prakash, A.; Sturm, M.; Duguay, C.; Li, S. Quantifying the relationships between lake fraction, snow water equivalent and snow depth, and microwave brightness temperatures in an arctic tundra landscape. *Remote Sens. Environ.* **2012**, *127*, 329–340. [CrossRef]
13. Kim, R.S.; Durand, M.; Liu, D. Spectral analysis of airborne passive microwave measurements of alpine snowpack: Colorado, USA. *Remote Sens. Environ.* **2018**, *205*, 469–484. [CrossRef]
14. Brucker, L.; Markus, T. Arctic-scale assessment of satellite passive microwave-derived snow depth on sea ice using Operation IceBridge airborne data. *J. Geophys. Res. C Ocean.* **2013**, *118*, 2892–2905. [CrossRef]
15. Kanagaratnam, P.; Markus, T.; Lytie, V.; Heavey, B.; Jansen, P.; Prescott, G.; Gogineni, S.P. Ultrawideband radar measurements of thickness of snow over sea ice. *IEEE Trans. Geosci. Remote Sens.* **2007**, *45*, 2715–2724. [CrossRef]
16. Kurtz, N.T.; Farrell, S.L. Large-scale surveys of snow depth on Arctic sea ice from Operation IceBridge. *Geophys. Res. Lett.* **2011**, *38*, L20505. [CrossRef]
17. Liston, G.E.; Sturm, M. A snow-transport model for complex terrain. *J. Glaciol.* **1998**, *44*, 498–516. [CrossRef]
18. Liston, G.E.; Elder, K. A distributed snow-evolution modeling system (snowmodel). *J. Hydrometeorol.* **2006**, *7*, 1259–1276. [CrossRef]

19. Essery, R.; Pomeroy, J. Vegetation and topographic control of wind-blown snow distributions in distributed and aggregated simulations for an arctic tundra basin. *J. Hydrometeorol.* **2004**, *5*, 735–744. [CrossRef]
20. Snauffer, A.M.; Hsieh, W.W.; Cannon, A.J. Comparison of gridded snow water equivalent products with in situ measurements in British Columbia, Canada. *J. Hydrol.* **2016**, *541*, 714–726. [CrossRef]
21. Broxton, P.D.; Zeng, X.; Dawson, N. Why do global reanalyses and land data assimilation products underestimate snow water equivalent. *J. Hydrometeorol.* **2016**, *17*, 2743–2761. [CrossRef]
22. Wrzesien, M.L.; Durand, M.T.; Pavelsky, T.M.; Howat, I.M.; Margulis, S.A.; Huning, L.S. Comparison of methods to estimate snow water equivalent at the mountain range scale: A case study of the California Sierra Nevada. *J. Hydrometeorol.* **2017**, *18*, 1101–1119. [CrossRef]
23. Deems, J.S.; Painter, T.H.; Finnegan, D.C. Lidar measurement of snow depth: A review. *J. Glaciol.* **2013**, *59*, 467–479. [CrossRef]
24. Lee, S.; Park, J.; Choi, E.; Kim, D. Factors Influencing the Accuracy of Shallow Snow Depth Measured Using UAV-Based Photogrammetry. *Remote Sens.* **2021**, *13*, 828. [CrossRef]
25. Fernandes, R.; Prevost, C.; Canisius, F.; Leblanc, S.G.; Maloley, M.; Oakes, S.; Holman, K.; Knudby, A. Monitoring snow depth change across a range of landscapes with ephemeral snowpacks using structure from motion applied to lightweight unmanned aerial vehicle videos. *Cryosphere* **2018**, *12*, 3535–3550. [CrossRef]
26. Harder, P.; Schirmer, M.; Pomeroy, J.; Helgason, W. Accuracy of snow depth estimation in mountain and prairie environments by an unmanned aerial vehicle. *Cryosphere* **2016**, *10*, 2559–2571. [CrossRef]
27. Vander Jagt, B.; Lucieer, A.; Wallace, L.; Turner, D.; Durand, M. Snow Depth Retrieval with UAS Using Photogrammetric Techniques. *Geosciences* **2015**, *5*, 264. [CrossRef]
28. Cimoli, E.; Marcer, M.; Vandecrux, B.; Bøggild, C.E.; Williams, G.; Simonsen, S.B. Application of low-cost uass and digital photogrammetry for high-resolution snow depth mapping in the Arctic. *Remote Sens.* **2017**, *9*, 1144. [CrossRef]
29. Avanzi, F.; Bianchi, A.; Cina, A.; De Michele, C.; Maschio, P.; Pagliari, D.; Passoni, D.; Pinto, L.; Piras, M.; Rossi, L. Centimetric accuracy in snow depth using unmanned aerial system photogrammetry and a multistation. *Remote Sens.* **2018**, *10*, 765. [CrossRef]
30. Miziński, B.; Niedzielski, T. Fully-automated estimation of snow depth in near real time with the use of unmanned aerial vehicles without utilizing ground control points. *Cold Reg. Sci. Technol.* **2017**, *138*, 63–72. [CrossRef]
31. De Michele, C.; Avanzi, F.; Passoni, D.; Barzaghi, R.; Pinto, L.; Dosso, P.; Ghezzi, A.; Gianatti, R.; Della Vedova, G. Using a fixed-wing UAS to map snow depth distribution: An evaluation at peak accumulation. *Cryosphere* **2016**, *10*, 511–522. [CrossRef]
32. Goetz, J.; Brenning, A. Quantifying Uncertainties in Snow Depth Mapping From Structure From Motion Photogrammetry in an Alpine Area. *Water Resour. Res.* **2019**, *55*, 7772–7783. [CrossRef]
33. Masný, M.; Weis, K.; Biskupič, M. Application of Fixed-Wing UAV-Based Photogrammetry Data for Snow Depth Mapping in Alpine Conditions. *Drones* **2021**, *5*, 114. [CrossRef]
34. Bühler, Y.; Adams, M.S.; Bösch, R.; Stoffel, A. Mapping snow depth in alpine terrain with unmanned aerial systems (UASs): Potential and limitations. *Cryosphere* **2016**, *10*, 1075–1088. [CrossRef]
35. Jiang, S.; Jiang, C.; Jiang, W. Efficient structure from motion for large-scale UAV images: A review and a comparison of SfM tools. *ISPRS J. Photogramm. Remote Sens.* **2020**, *167*, 230–251. [CrossRef]
36. Perry, L.B.; Konrad, C.E. Relationships between NW flow snowfall and topography in the Southern Appalachians, USA. *Clim. Res.* **2006**, *32*, 35–47. [CrossRef]
37. Perry, L.B.; Konrad, C.E.; Hotz, D.G.; Lee, L.G. Synoptic Classification of Snowfall Events in the Great Smoky Mountains, USA. *Phys. Geogr.* **2010**, *31*, 156–171. [CrossRef]
38. Sugg, J.W.; Fuhrmann, C.M.; Perry, L.B.; Hall, D.K.; Konradll, C.E. Sub-regional snow cover distribution across the southern Appalachian Mountains. *Phys. Geogr.* **2017**, *38*, 105–123. [CrossRef]
39. Smith, M.W.; Carrivick, J.L.; Quincey, D.J. Structure from motion photogrammetry in physical geography. *Prog. Phys. Geogr. Earth Environ.* **2015**, *40*, 247–275. [CrossRef]
40. Carrivick, J.L.; Smith, M.W.; Quincey, D.J. *Structure from Motion in the Geosciences*; John Wiley & Sons: Hoboken, NJ, USA, 2016.
41. Lowe, D.G. Distinctive image features from scale-invariant keypoints. *Int. J. Comput. Vis.* **2004**, *60*, 91–110. [CrossRef]
42. Lowe, D.G. Object recognition from local scale-invariant features. In Proceedings of the Seventh IEEE International Conference on Computer Vision, Kerkyra, Greece, 20–27 September 1999; Volume 1152, pp. 1150–1157.
43. Muja, M.; Lowe, D.G. Fast approximate nearest neighbors with automatic algorithm configuration. *VISAPP* **2009**, *2*, 2.
44. Fischler, M.A.; Bolles, R.C. Random sample consensus: A paradigm for model fitting with applications to image analysis and automated cartography. *Commun. ACM* **1981**, *24*, 381–395. [CrossRef]
45. Furukawa, Y.; Ponce, J. Accurate, dense, and robust multiview stereopsis. *IEEE Trans. Pattern Anal. Mach. Intell.* **2009**, *32*, 1362–1376. [CrossRef] [PubMed]
46. Cohen, I.; Huang, Y.; Chen, J.; Benesty, J.; Benesty, J.; Chen, J.; Huang, Y.; Cohen, I. Pearson correlation coefficient. In *Noise Reduction in Speech Processing*; Springer: Berlin/Heidelberg, Germany, 2009; pp. 1–4.
47. Lendzioch, T.; Langhammer, J.; Jenicek, M. Estimating Snow Depth and Leaf Area Index Based on UAV Digital Photogrammetry. *Sensors* **2019**, *19*, 1027. [CrossRef]

48. Krichen, M.; Adoni, W.Y.H.; Mihoub, A.; Alzahrani, M.Y.; Nahhal, T. Security Challenges for Drone Communications: Possible Threats, Attacks and Countermeasures. In Proceedings of the 2022 2nd International Conference of Smart Systems and Emerging Technologies (SMARTTECH), Riyadh, Saudi Arabia, 9–11 May 2022; pp. 184–189.
49. Ko, Y.; Kim, J.; Duguma, D.G.; Astillo, P.V.; You, I.; Pau, G. Drone Secure Communication Protocol for Future Sensitive Applications in Military Zone. *Sensors* **2021**, *21*, 2057. [CrossRef]

remote sensing

Technical Note

Drone-Based Ground-Penetrating Radar with Manual Transects for Improved Field Surveys of Buried Ice

Adam R. Tjoelker [1,2,*], Michel Baraër [3], Eole Valence [4], Bastien Charonnat [3], Janie Masse-Dufresne [3], Bryan G. Mark [1,2] and Jeffrey M. McKenzie [4]

1 Department of Geography, The Ohio State University, Columbus, OH 43210, USA; mark.9@osu.edu
2 Byrd Polar and Climate Research Center, The Ohio State University, Columbus, OH 43210, USA
3 École de Technologie Supérieure (ÉTS), Montreal, QC H3C 1K3, Canada; michel.baraer@etsmtl.ca (M.B.); bastien.charonnat.1@ens.etsmtl.ca (B.C.); janie.masse-dufresne@etsmtl.ca (J.M.-D.)
4 Department of Earth and Planetary Sciences, McGill University, Montreal, QC H3A 0E8, Canada; eole.valence@mail.mcgill.ca (E.V.); jeffrey.mckenzie@mcgill.ca (J.M.M.)
* Correspondence: tjoelker.4@osu.edu

Abstract: The steep and unstable terrain found on debris-covered glaciers, rock glaciers, talus slopes, moraines and other proglacial features often make terrestrial ground-penetrating radar (GPR) surveys unsafe or cost-prohibitive. To address these challenges, this research introduces a novel approach for studying buried ice using multi-low-frequency drone-based GPR. Monostatic antennas of 50, 100, and 200 MHz were flown along a transect spanning a debris-covered glacier and an ice–debris complex at Shár Shaw Tagà (Grizzly Creek) in southwest Yukon, Canada. The drone-based results were compared to manual GPR at two locations along the transect. The two manual segments were conducted using the same radar system in a bi-static mode and included common mid-point (CMP) surveys. Overall, the drone-based radar successfully identified buried ice and enabled estimation of ice body thickness. Notably, CMP results confirmed layer characteristics and enabled depths to be measured across the entire drone-based transect. Discrimination of detail across a range of depths was made possible by comparing the three low frequencies, highlighting the possibility of using this method for future investigations of debris thickness in addition to quantifying buried ice. This study confirms the effectiveness of drone-based GPR combined with manual CMP for surveying ice beneath previously inaccessible terrain.

Keywords: drones; ground-penetrating radar; buried ice; debris-covered glacier

1. Introduction

1.1. Ground-Penetrating Radar for Surveying Buried Ice

Debris-covered ice is present on an estimated 44% of Earth's glaciers (excluding Antarctica) and may account for 7.3% of the global mountain glacier area [1]. In addition to debris-covered glaciers, buried glacial ice is commonly found in deglaciating environments where retreating glaciers have left residual ice that is covered by sediments through paraglacial and gravitational processes [2]. Methods have been developed to improve the detection of buried ice, including the application of terrain models and deep learning methods to satellite remote sensing data [3,4]. Despite these advancements, high-quality field data are needed on the ground truth of the extent of buried ice. As warming continues to impact glaciers globally, heavily debris-covered features such as debris-covered glaciers and rock glaciers will become more important to quantify as hydrologic resources [5]. Additionally, the impacts of debris cover feedback on ice mass balance are inherently difficult to study and remain a poorly quantified part of glacier modeling efforts due to difficulties measuring buried ice melt [6,7].

Ground-penetrating radar (GPR) is a standard geophysical approach that has been used extensively over the last two and a half decades to detect and characterize buried massive ice in debris-covered glaciers, rock glaciers and permafrost [6,8–11]. However, past studies investigating the ice content of steep terrain using ground-based manual GPR surveys often feature a limited number of transects with paths determined by the landscape topography [12,13]. Airborne GPR, whether mounted on helicopters or fixed-wing aircraft, is a potential solution to overcome these challenges. For instance, a helicopter-mounted GPR has been successfully used to create a quasi-three-dimensional map of a rock glacier in the Swiss Alps using multiple hatched passes, a feat only achievable with an airborne system [14]. However, airborne options have most often been limited to mapping the depths of large glaciers and ice sheets due to high costs [15–18]. These surveys use lower-frequency radars for measuring glacier thickness and have a lower resolution as a result. In addition, most airborne GPR surveys conducted using aircraft or helicopters sacrifice spatial resolution as measurements are made at several tens of meters above the surface.

Altitude separation between airborne GPR and the surface can be reduced with recent advancements in relatively low-cost drone-based GPR [19]. Drone-based GPR systems have been used in recent years for snow hydrology surveys, where GPR enables undisturbed detection of strata within snowpack [20]. Fixed-wing drone-based ice-penetrating radar systems have also been developed for surveys of larger, mostly clean ice glaciers [21,22]. These systems use lower-energy radars which work well for mapping glacier thickness where there is limited debris coverage. More recently, rotary wing drone-based GPR has been developed for 3D and 4D coverage of mostly clean ice glacial tongues [23].

The development of drone-based GPR methods for the detection of buried glacial ice over complex terrain is a logical next step, building on the capability of drone-based radar technology. To the best of our knowledge there are no other publications that document the application of drone-based GPR for studying buried ice. The objective of this study is to provide the first test of the effectiveness of a drone-based radar survey to survey buried ice content in terrain that would be effectively inaccessible to ground-based GPR.

1.2. Shár Shaw Tagà (Grizzly Creek) Field Site

This study was conducted within a deglaciating sub-catchment of Shár Shaw Tagà valley in southwest Yukon, Canada. The study area is located on the Kluane and White River First Nation territory within Kluane National Park and Reserve. In the literature, Shár Shaw Tagà is referred to as Grizzly Creek, its non-indigenous toponym. The sub-catchment exhibits a glacier to debris-covered glacier to rock glacier continuum typical of the St. Elias Mountains greenbelt [24]. Moving down the valley, the clean glacier ice above 2100 m a.s.l. transitions to a debris-covered glacier and an ice–debris complex (as described by Bolch et al. [25]) between 2100 and 1900 m a.s.l, and then to a rock glacier from 1900 to 1700 m a.s.l., where the sub-catchment meets the main valley (Figure 1). This sub-catchment at Shár Shaw Tagà remains unnamed by the Kluane First Nation community to the best of our knowledge. It was identified as the "Ice-cored moraine and rock glacier" in initial fieldwork in the area by Johnson [24] and as the "glacier debris system rock glacier" with a "tributary glacier" by Evin et al. [26].

The drone-based GPR transect for this study is located along the transition from the debris-covered glacier to the ice–debris complex (Figure 2). Talus slopes intersect with the debris-covered glacier, and morainic accumulations near the study transect, obscuring clear distinctions between these features. At the upper end of the transect, exposed ice cliffs indicate the presence of buried glacial ice. The transect location was chosen to evaluate the continuity of buried ice between the exposed ice cliffs and the ice beneath the unstable talus, downhill from the end of the debris-covered glacier.

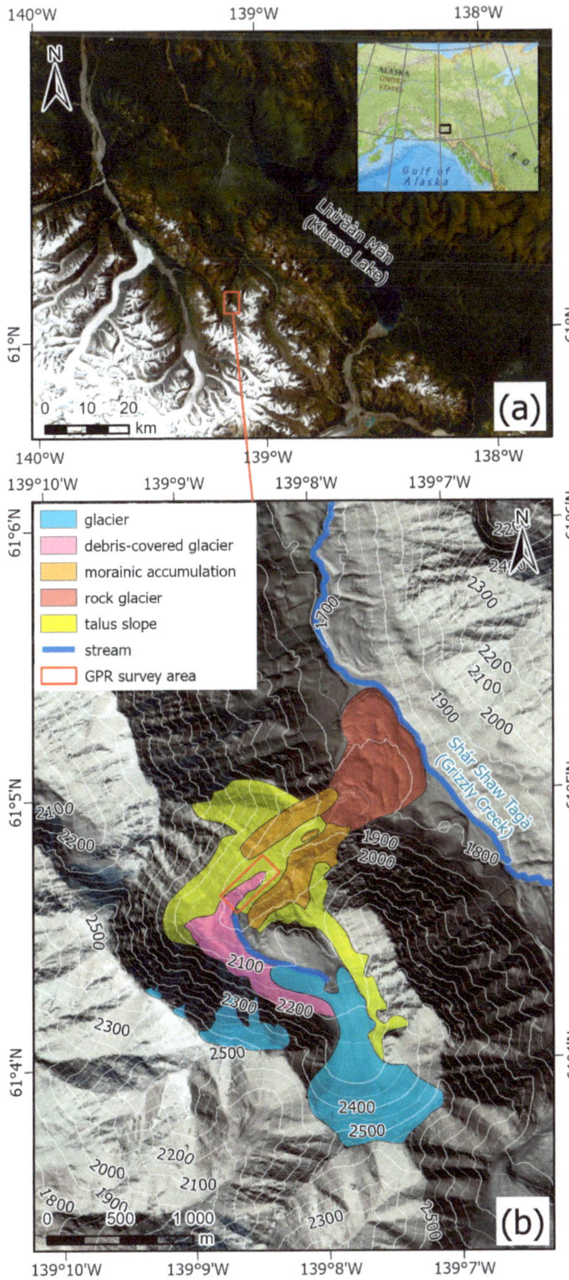

Figure 1. (**a**) Location of Shár Shaw Tagà (Grizzly Creek) valley within southwestern Yukon, Canada, with the study area identified by a red rectangle. (**b**) Geomorphological interpretation of the study area. Terrain data are from ArcticDEM by the Polar Geospatial Center under NSF-OPP awards 1043681, 1559691, 1542736, 1810976, and 2129685 [27]. Basemap credits: Esri, Maxar, Earthstar Geographics, National Geographic, and the GIS User Community. Place names are given in Southern Tutchone and English.

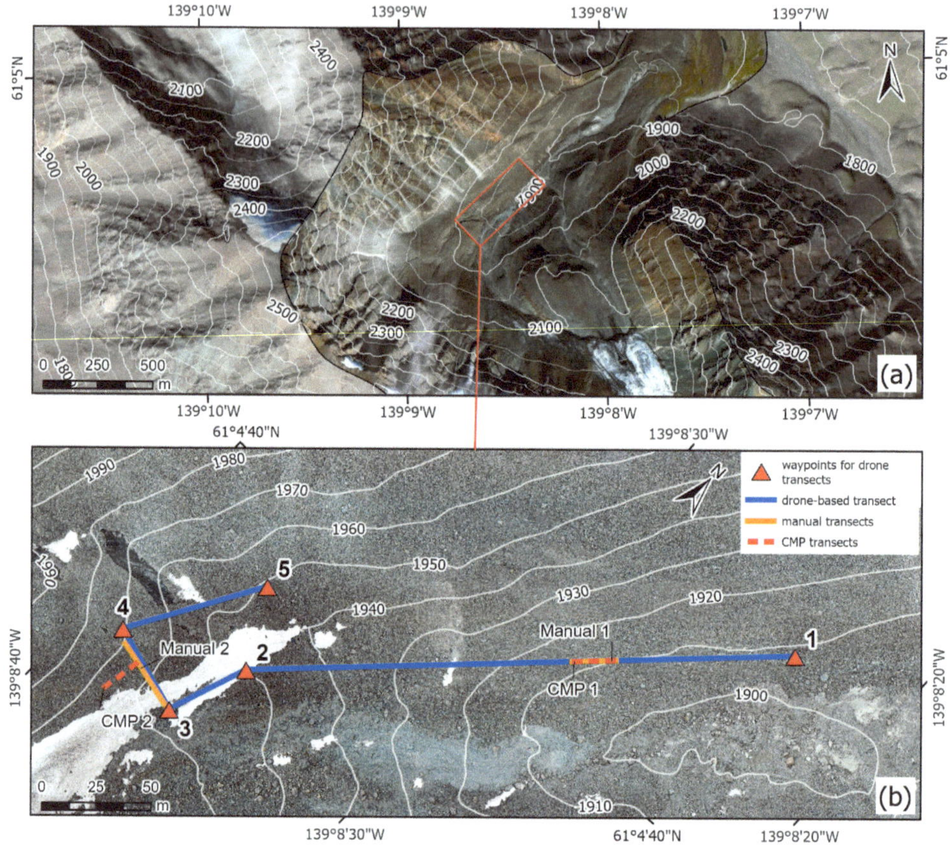

Figure 2. Map showing the location of the drone-based, manual, and CMP transects within the study site (**b**) and larger sub-catchment valley (**a**).

2. Materials and Methods

2.1. Drone-Based GPR

During June 2023, a Radar Systems Zond Aero LF GPR was flown in a monostatic configuration onboard a DJI M600 Pro drone along a 430 m transect (Figure 2). Three complete passes of the transect were flown, each with one of the 50, 100, and 200 MHz unshielded dipole antennas. A laser altimeter was used to set a target altitude of 5 m above the ground, to prevent collision with boulders on the undulating terrain. The Zond Aero LF radar was integrated with the DJI M600 using an SPH Engineering SkyHub 3 onboard computer with firmware version 2.13.1 and UgCS version 4.15 flight planning software. Drone-borne radargrams were georeferenced using the post-processed kinematic (PPK) method, with signals collected by an onboard Emlid Reach M2 RTK module and an Emlid Reach RS2+ base station. The pre-programmed flight paths of the drone followed the path indicated in Figure 2 for the three antennas. The drone transect paths after post-processing were shown to remain within 3 m of each other.

2.2. Manual Ground-Penetrating Radar

The Zond Aero LF system was operated in a bi-static configuration for ground-based manual surveys, serving as a reference for the drone-based configuration. The detachable nature of the antenna and receiver in the bi-static operation allowed for a common mid-point (CMP) survey at each of two manual transects to aid in layer identification and thickness estimation. The first manual and CMP transects, "Manual 1" and "CMP 1," were conducted along the same 20 m span of the drone-based transect on the ice-cored moraine and talus slope. The second manual and CMP transects, "Manual 2" and "CMP 2," were conducted on the debris-covered glacier in the upper portion of the transect near the exposed ice cliffs. The antenna configurations for both drone-based CMP and manual GPR transects are shown in the Supplementary Information in Table S1.

The Manual 2 GPR transect was aligned as close as possible with the path of the drone but did not correspond perfectly to the drone-based transect due to the steep slope on the debris-covered glacier (Figure 2b). The Manual 2 GPR transect spanned a 40 m path upslope onto the debris-covered glacier. Unlike the CMP 1 transect, the 20 m CMP 2 transect did not align with the Manual 2 transect. CMP 2 was conducted perpendicular to the manual transect, roughly parallel to the slope of the debris-covered glacier for safety and practical reasons. Conducting the CMP perpendicular to the slope at this location avoided the issue of varying interface depth, allowing for a more consistent substrate beneath the CMP 2 transect.

2.3. Drone Photgrammetry and DEM

Aerial photogrammetry with 620 images of the study area was conducted using a DJI Mavic 2 Pro within two days following the drone-based GPR data collection. Ground control points (n = 9) and the ends of the CMP and manual transects were surveyed using an Emlid Reach M2 RTK module connected to an Emlid Reach RS2+ base station. The orthomosaic and digital elevation model (DEM) used for the survey map and terrain rectification of the radargrams were produced using Pix4D Mapper and are shown in Figure 2b.

2.4. Radar Data Processing

Subsequent processing of drone-based and manual radargrams was conducted in the Radar Systems Prism2 software version 2.70.05. All radargrams underwent the same processing sequence: (i) background removal; (ii) Ormsby bandpass filter targeting the central frequency +/− 50%; and (iii) linear gain adjustment for optimal readability. Reflections at layer interfaces were picked manually. In addition, terrain rectification was applied to both the drone-based transects based on PPK outputs and to the manual transect using elevations from the DEM produced for the study area. Calculation of velocities from the CMP transects and depth correction followed methods used in comparable ice–debris environments [13,28,29]. The layer velocities were applied across the drone-based transect, enabling depths to be compared across the two methods. Interpretation of GPR facies and debris thickness followed standard interpretation methods for debris-covered glaciers (e.g., [6,30]). Processed radargrams without annotations can be found in the Supplementary Materials of this article.

2.5. Measuring Buried Ice Heights and Depths

Velocities determined through the CMP analysis enabled buried ice heights and depths of the ice base reflector to be compared between the drone-based and Manual 1 transects. These heights and depths were calculated from the GPR traces by multiplying one-half the two-way travel time between the top and bottom of ice and debris layers by the respective velocities derived from the CMP analysis. The calculation of velocities allowed the depths to the base reflector and the height of the buried ice to be calculated. The interfaces between each layer for calculating the two-way travel time were manually picked from the data. Depth measurements were taken at points 0 m, 10 m and 20 m along the CMP 1 transect and compared to the same locations in the drone-based radargrams. These locations were chosen to compare the results at three (3) points evenly spaced along the CMP transect.

3. Results

3.1. Drone-Based Radargrams

The processed and annotated drone-based radargrams for all three frequencies are shown in Figure 3. Three major areas of buried ice, B.I.$_1$, B.I.$_2$ and B.I.$_3$, are identified along the drone-based transects as voids of low reflection observed with all three antennas. The detection of a base reflector beneath the ice bodies was possible using only the 50 MHz and 100 MHz antennas. The delineated shape of the ice bodies varies slightly from antenna to antenna, with the 50 MHz antenna yielding the thickest debris estimates due to its inherent low resolution. Surface multiples can be seen in the 100 MHz and 50 MHz radargrams caused by repeated reflections from the near surface. This noise in the 100 MHz and 50 MHz radargrams makes determination of the debris layer thickness, T, difficult to discern. On the top of the debris-covered glacier, at B.I.$_2$, the apparent thickness shown in the radargram can be compared across the results from the three frequencies. In the 200 MHz radargram, the debris layer appears thinner than that shown in the two lower frequencies, showing a thickness closer to the field observation of approximately 30 cm depth at the top of the debris-covered glacier.

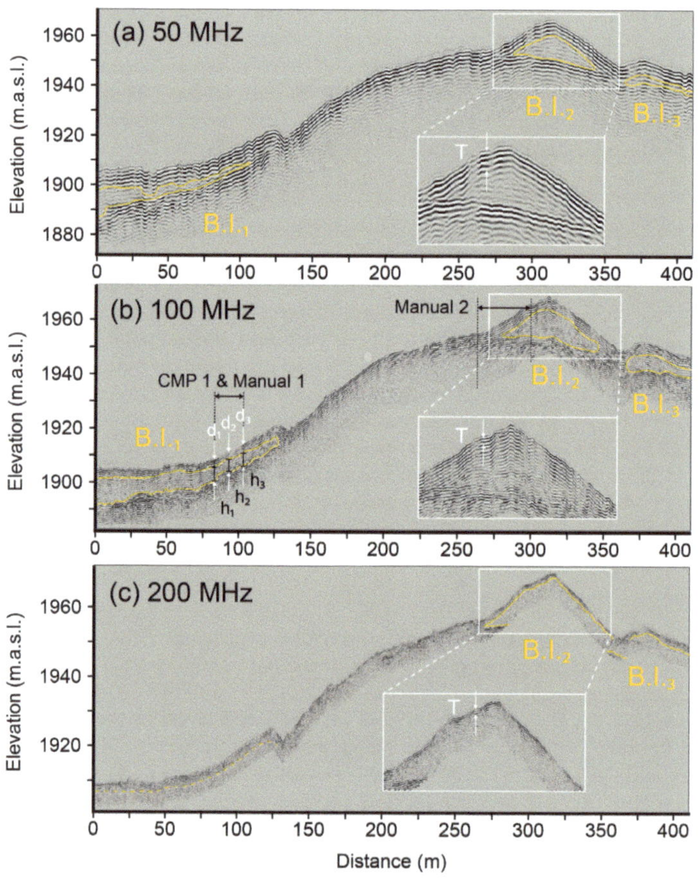

Figure 3. Processed and annotated radargrams of the (**a**) 50 MHz, (**b**) 100 MHz and (**c**) 200 MHz drone-based GPR transects. Three areas of buried ice, B.I.$_1$, B.I.$_2$ and B.I.$_3$, are outlined for emphasis across the three transect radargrams. Areas of interest at the top of the debris-covered glacier are

zoomed in by a factor of 1.5× to emphasize the differing levels of detail of debris thickness, T, discernable across the three antenna frequencies. In (**b**), h_1, h_2 and h_3 indicate buried ice layer heights, and d_1, d_2 and d_3 show the depths from the surface to the base reflector measured at the locations corresponding to the beginning, middle and end of the CMP and Manual 1 transect. The location of the Manual 2 transect is also indicated in (**b**).

3.2. Manual Radargrams

Each of the manual GPR transects surveyed show similar locations and thicknesses of the ice masses (Figure 4). Comparing the results from the drone-based radargrams in Figure 3, the manual GPR radargrams from the Manual 2 transect (Figure 4a–c), agree with the shape and location of B.I.$_2$. The improved quality of the manual transect radargrams is likely due to the coupling between the antenna and the surface, reducing the signal loss and enabling the 200 MHz antenna to penetrate deeper into the debris-covered glacier. Radargrams from the Manual 1 transect along the ice-cored moraine also capture the same shape and location of the B.I.$_1$ feature identified in the drone-based transects. The B.I.$_1$ at this location is covered by a thicker debris layer than what was found on the debris-covered glacier. Overall, the 200 MHz and 100 MHz radargrams show a 5–7 m thick layer of ice, while the 50 MHz radargram is particularly noisy, confounding the identification of a clear ice layer.

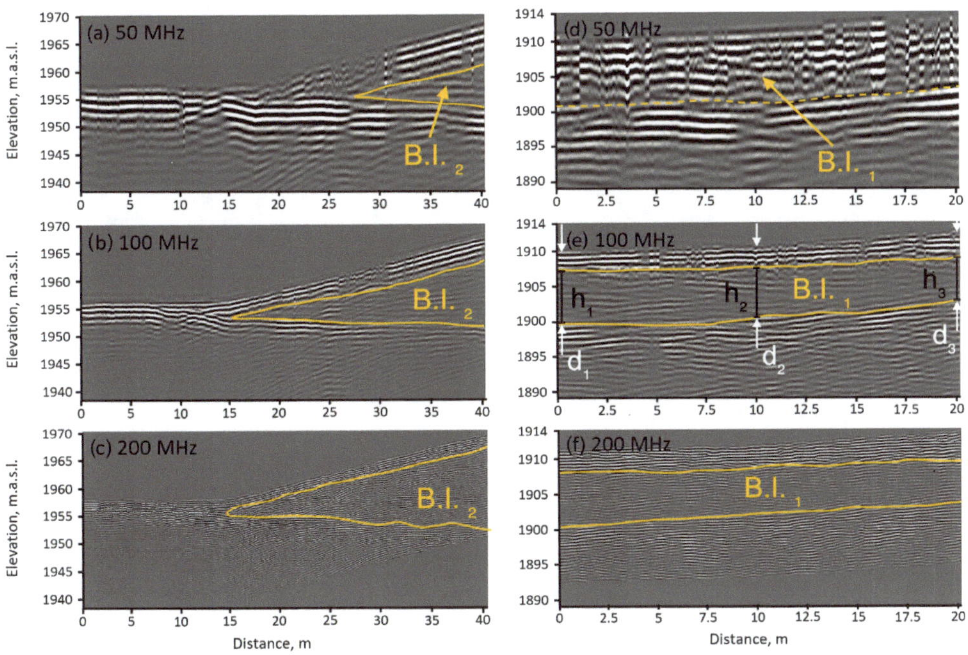

Figure 4. Manual GPR radargrams from the debris-covered glacier, Manual 2 (**a–c**), and ice–debris complex, Manual 1 (**d–f**). In (**e**) h_1, h_2 and h_3 indicate ice layer height measurements and d_1, d_2 and d_3 show depths measured from the surface to the base reflector.

The radargrams collected at the two CMP transects were used to compute the relative permittivity and the wave velocities and estimate the thickness of both debris and ice layers (Figure 5). The 50 and 100 MHz CMP transects showed clear signal trajectories. The 200 MHz CMP transects did not provide clear separations between the different layers, preventing the estimation of layer characteristics with this antenna. The values for the thickness, velocity, and relative permittivity of the ice and debris layers at each CMP transect are presented in Table 1. The values of permittivity for ice range from 2.9 to 4.1 and debris primitivities range from 3.3 to 4.7.

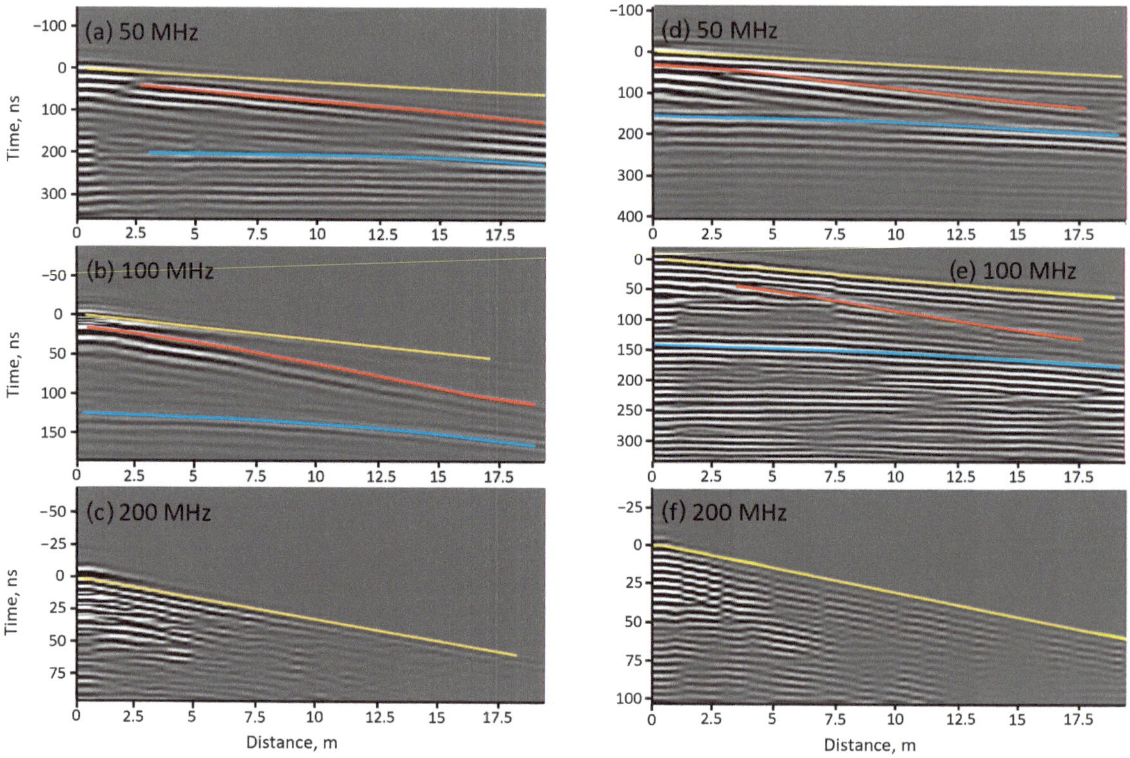

Figure 5. CMP results showing the trajectory signals of the air wave (yellow), the surface debris/ice interface (red), and the ice/bottom layer (blue) for each of the three antenna frequencies at the CMP 2 (**a–c**) and CMP 1 (**d–f**) transects.

Table 1. Results of CMP analysis.

		CMP 1			CMP 2		
		50 MHz	**100 MHz**	**200 MHz**	**50 MHz**	**100 MHz**	**200 MHz**
Debris Layer	Thickness (m)	2.9	3	N/A	2.6	1.6	N/A
	Velocity (cm/ns)	13.8	14.1	N/A	15.6	16.7	N/A
	Permittivity (dim) *	4.7	4.5	N/A	3.7	3.3	N/A
Ice Layer	Thickness (m)	9.4	7.6	N/A	13.6	9.8	N/A
	Velocity (cm/ns)	14.8	17.0	N/A	16.6	17.8	N/A
	Permittivity (dim) *	4.1	3.2	N/A	3.2	2.9	N/A

* Relative permittivity values are reported as dimensionless values, indicated by units of "dim".

3.3. Height and Depth Measurements

At the 0 m, 10 m and 20 m distances along the 100 MHz Manual 1 transect, ice body heights of $h_1 = 8.5$ m, $h_2 = 7$ m and $h_3 = 6.8$ m were estimated. The 100 MHz frequency was chosen for making comparisons between the airborne and ground observations since it was the best compromise between depth and resolution. The 200 MHz radar did not have sufficient penetration in the drone-based configuration and the 50 MHz radar sacrificed resolution.

The corresponding 100 MHz drone-based GPR heights sampled resulted in $h_1 =$ 10.3 m, $h_2 = 9.7$ m and $h_3 = 7.7$ m. Comparing the ice layer height measurements at each location, the mean absolute difference between the measurements is 1.8 m. Measurements of the depth of the ice base reflector at the same locations were calculated by adding the thickness of the surficial debris layer to the buried ice layer height. Total depths from the Manual 1 transect were found to be d_1 = 12.9 m, d_2 = 10 m and d_3 = 9.8 m. Drone-based depths were calculated as d_1 = 12.1 m, d_2 = 11.8 m and d_3 = 10.5 m. The mean absolute difference between the depth measurements was found to be 1.1 m. Differences between heights and depths measured with the drone and manual GPR are discussed further in the discussion in Section 4.2.

4. Discussion

4.1. CMP Results

The relative permittivity values for ice across both transects range between 2.9 and 4.1, aligning with commonly reported values in the literature. For example, Thomson et al. [31] place relative permittivity for ice under debris between three and five, while Wu and Liu [29] report values between three and four. The results indicate that CMP1 (4.7 to 4.5) and CMP2 (3.7 to 3.3) exhibit different relative permittivity for the debris layers, highlighting the variability reported in the literature for such media [32]. Overall, the addition of CMP to the aerial GPR acquired in this study enable high confidence in the layers of ice and debris identified in the radargrams across the drone-based transect and allow ice thickness measurements.

4.2. Differences between Drone and Manual GPR Measurements

Discrepancies between the depths and heights measured in the drone and manual radargrams can be linked to (i) limitations inherent to the airborne radar method used and (ii) the uncertainty introduced by the surface and debris media which impact both GPR methods. The mean absolute error is 1.8 m between the ice heights measured and 1.1 m between the base reflector depths. It is important to note that there is still uncertainty associated with the manual observations, so the manual measurements cannot be considered ground truth. The uncertainty in georeferencing uncovered after post-processing the exact position of the drone indicate that the measurements were not always taken at exactly the same spot between the manual and drone-based surveys. Future work could be carried out to improve transect positioning through RTK. Digging debris pits for ground truthing would also allow for accurate validation of debris thickness measurements.

The sources of uncertainty associated with the drone-based operation alone limit the performance of the drone radar compared to manual surveys, but these considerations must be acknowledged when deploying drone-based radar beyond where manual surveys would be possible. Drone-based GPR also enables much larger areas to be covered at the same time, improving the efficiency of field surveys. In this case, it is important to consider the tradeoffs in the data quality associated with an aerial GPR survey. The following two sections address the possible uncertainties associated with the airborne method and the field conditions.

4.2.1. Uncertainty Related to Aerial GPR Operation

The first noted source of uncertainty in the drone-based radargrams is the user identification of the air/ground interface. When the antenna is decoupled from the surface during an aerial GPR survey, the exact position of the ground is more difficult to determine due to the separation of the antenna from the surface [33]. This error impacts the measurements used in our study to locate the surface of the debris layer, which was used for calculating the debris thickness and total depth of the ice base reflector.

The height of the drone above the surface also presents a challenge for the quality of the results. Increasing the separation of the GPR antenna from the surface is known to result in a lower signal to noise ratio (SNR) because of the increase in the effective antenna beam pattern footprint [34]. This altitude effect reduces the precision that can be used to identify small targets and causes subsurface features to have locational error since reflections will not always come from directly beneath the antenna. The altitude of the drone varied due to the altimeter compensation lag over the rough debris surface, which contained large boulders. This effect complicates the interpretation of the GPR through a variable layer of error.

Additionally, aerial GPR antennas that are decoupled from the surface present the additional challenge of high power loss caused during the first reflection at the air/ground interface [35]. The change in permittivity between the air and ground combined with the debris surface roughness both contribute to power loss, which limits the penetration of the radar. This factor contributes to the issue seen with the 200 MHz antenna used in the drone-based configuration in the present study, where penetration depth is dramatically limited when compared to the corresponding manual surface GPR transect. For example, in the drone-based GPR in Figure 3, the location of the debris/ice interface is not clear and indicated by a dotted line, but it is clearly visible along the Manual 1 transect in Figure 4.

Reflections from outside the survey plane are an additional consideration for airborne radar surveys over steep terrain, such as the terrain at the study site at Shár Shaw Tagà. The pitch and roll of the drone also contribute to additional reflections originating outside the mapped path of the drone. Migration techniques have been described in the literature to address these concerns [36], but these methods were not applied to the standard GPR processing conducted to produce the radargrams used for this study. Migration was not applicable for our study since it would involve the development of new techniques to account for the chaotic surface media comprising various sized rocks and boulders in addition to the surface topography. Future research could address these concerns to improve radargrams obtained from an airborne GPR over steep and rough terrain.

The effect of the target altitude above the surface was not investigated in this study due to the limited flight time permitted by field conditions and battery life. Additionally, the high risk of collision with large boulders scattered on the steep slopes prevented testing the GPR at lower altitudes. It can be concluded that the target altitude separation of 5 m used for the drone-based transects contributed to these described errors of lower SNR and diminished capacity to identify the ice–debris interface.

4.2.2. Other Sources of Uncertainty

Beyond the error originating from airborne GPR operation, conditions in the field contributed to potential uncertainties. The first of these sources originates from the potential misinterpretation of interfaces [30]. This misinterpretation could be caused by a potential integration of small debris into the top layer of the ice, limiting the dielectric contrast between the ice and debris in the radargrams and causing a weaker reflection. This effect would be present with the data acquired using both methods but would have an increased effect with the drone-based GPR which already experiences higher energy loss due to antenna decoupling from the surface.

The manual GPR measurements presented additional potential uncertainty due to difficulty maintaining a consistent antenna position over the surface [37]. Changes in angle and height varied as the antenna was moved over the irregular rocky surface. The noisy data from the 50 MHz radargram in Figure 4 can be explained by the technical challenge of maintaining a consistent antenna position over the ground when handling larger antennas exceeding 2 m in length over steep, unstable terrain. Accidental collisions of the antenna with the surface were difficult to avoid when conducting the 50 MHz survey and, as a result, caused the drone-based GPR to outperform the manual radar in this case.

The thickness of the surficial debris layer varied from 30 cm to a few meters across the study area. Depending on the antenna frequency used, the minimum depth of investigation varies based on the wavelength of the radar. The thinner areas of debris, especially at the top of the debris-covered glacier where debris is as thin as 30 cm, would prevent the two lower-frequency antennas from being able to resolve the debris–ice interface. Wave theory indicates that the best vertical resolution that can be achieved is one-quarter of the dominant wavelength, which would be around 0.75 m for the 100 MHz antenna [38]. This detection limit impacts the measurements of the debris thickness estimations and the calculations of total ice body height reported in the results.

If the 200 MHz radar had been able to identify the ice–debris interface, it could be able to be used to measure debris thickness more accurately than the 100 MHz antenna. However, the 200 MHz radar did not have sufficient penetration in the drone-based transect to facilitate a direct comparison between the drone and ground-based radar transects.

4.3. Methodological Recommendations

The results of this study demonstrate that the detection of large continuous buried ice bodies is particularly effective using drone-based GPR. The dielectric contrast of the ice and the surrounding media at this site permits effective detection of the ice base reflector. Improvements in processing and field collection could likely further improve the measurement of the debris layer thickness given the difficulties seen with surface multiples and complex reflections from the air/ground interface.

For geophysical investigations using drone-based GPR, ice detection is most effective when operating the drone as close to the surface as possible, at slow speeds, with flat terrain that has a high dielectric contrast between the media and ice. Thinner debris layers permit easier detection of ice compared to thicker debris layers.

For thinner debris layers, higher-frequency radars should have sufficient penetration to resolve the base reflector. For example, in this study, the 200 MHz drone-based radar was able to detect the base reflector below the thinner debris near the Manual 2 transect and was not able to detect the base reflector to the deeper debris along Manual 1 transect. Specific frequencies for drone-based GPR investigations should be chosen specifically for the field conditions present at a study site.

As discussed in the methods and results, the thickness of the debris layer was not directly measured in this study and was not the focus of the investigation. However, given the results seen with the surface multiples and difficulty discerning the debris layer thickness using the drone-based GPR, it could be suggested that finer debris would likely improve the determination of the surface reflector and could contribute to more effective determination of debris layer thickness when compared to manual GPR.

5. Conclusions

This study demonstrates the effectiveness of drone-based GPR for detecting and measuring buried ice where traditional manual GPR methods would not be possible. The inclusion of the two manual transects with a CMP survey using the same radar system provide a valuable comparison of the performance of the radar in airborne and ground-based operations. Comparing the two methods, the drone-based radargrams agree with the location and shape of buried ice detected in the manual GPR performed at the same frequencies and locations. The addition of CMP along the two manual transects enabled depths to be calculated accurately using the drone-based radar.

Across three measurements of the total height and depth of $B.I._1$, the drone-based GPR is within 1.8 m of the height and 1.1 m of the depths determined using the manual GPR on average. Despite these discrepancies, which can likely be improved upon with additional processing and methodological improvements, the ability to detect buried ice over complex and dangerous terrain provides significant opportunities for future investigations of processes and changes in debris-covered glacier systems.

Acknowledging the inherent limitations in data resolution when performing GPR with an elevated antenna that is decoupled from the surface, the drone-based methods matched the detection ability of the manual radar. Notably, the 50 MHz drone-based radar outperformed the manual antenna of the same frequency, a testament to the difficulty of performing GPR surveys in challenging terrain. However, noise and signal loss caused by the lack of antenna coupling with the surface when using the drone-based radar complicated efforts to determine debris layer thickness in the radargrams. Future research directions building upon this work point towards the application of more advanced algorithms for processing radargrams to improve the near-surface signal quality, enabling more precise measurements of debris layer thickness.

Supplementary Materials: The following supporting information can be downloaded at https://www.mdpi.com/article/10.3390/rs16132461/s1: Table S1: Configuration parameters for the Zond Aero LF GPR system; Figure S1: The 50 MHz raw processed radargram; Figure S2: The 100 MHz raw processed radargram; Figure S3: The 200 MHz raw processed radargram. Figure S4: Manual Transect 1 50 MHz Raw. Figure S5: Manual Transect 1 100 MHz Raw. Figure S6: Manual Transect 1 200 MHz Raw. Figure S7: Manual Transect 2 50 MHz Raw. Figure S8: Manual Transect 2 100 MHz Raw. Figure S9: Manual Transect 2 200 MHz Raw.

Author Contributions: Conceptualization, A.R.T., M.B., E.V. and B.C.; data curation, E.V.; formal analysis, A.R.T., M.B. and E.V.; funding acquisition, M.B., J.M.M. and B.G.M.; investigation, A.R.T., M.B., E.V. and B.C; methodology, A.R.T., M.B., E.V. and B.C; project administration, M.B. and J.M.-D.; supervision, M.B., B.G.M., J.M.M. and J.M.-D.; visualization, A.R.T., M.B., E.V. and B.C; writing—original draft, A.R.T.; writing—review and editing, A.R.T., M.B., E.V., B.C., J.M.-D., B.G.M. and J.M.M. All authors have read and agreed to the published version of the manuscript.

Funding: This research was funded by the Natural Sciences and Engineering Research Council of Canada, grant numbers RGPIN-2020-05612 and RGPNS-2020-05612, and by Natural Resources Canada—Polar Continental Shelf Program, grant number 62723. Funding for the expedition was also provided by the Geochemistry and Geodynamics Research Center (GEOTOP) Quebec and The Ohio State University Department of Geography.

Data Availability Statement: The raw GPR data from this survey are available for download at https://zenodo.org/records/10689923 [39].

Acknowledgments: We would like to thank Kluane First Nation and White River First Nation for the opportunity to conduct research on their land and acknowledge their long history and relationship with the land. We would also like to thank the Arctic Institute of North America and the University of Calgary for providing us service and access to their facilities at Kluane Lake Research Station. Permission to conduct research in Kluane National Park and Reserve was provided by Parks Canada.

Conflicts of Interest: The authors declare no conflicts of interest. The funders had no role in the design of the study; in the collection, analyses, or interpretation of data; in the writing of the manuscript; or in the decision to publish the results.

References

1. Herreid, S.; Pellicciotti, F. The State of Rock Debris Covering Earth's Glaciers. *Nat. Geosci.* **2020**, *13*, 621–627. [CrossRef]
2. Everest, J.; Bradwell, T. Buried Glacier Ice in Southern Iceland and Its Wider Significance. *Geomorphology* **2003**, *52*, 347–358. [CrossRef]
3. Paul, F.; Huggel, C.; Kääb, A. Combining Satellite Multispectral Image Data and a Digital Elevation Model for Mapping Debris-Covered Glaciers. *Remote Sens. Environ.* **2004**, *89*, 510–518. [CrossRef]
4. Xie, Z.; Haritashya, U.K.; Asari, V.K.; Young, B.W.; Bishop, M.P.; Kargel, J.S. GlacierNet: A Deep-Learning Approach for Debris-Covered Glacier Mapping. *IEEE Access* **2020**, *8*, 83495–83510. [CrossRef]
5. Jones, D.B.; Harrison, S.; Anderson, K.; Whalley, W.B. Rock Glaciers and Mountain Hydrology: A Review. *Earth-Sci. Rev.* **2019**, *193*, 66–90. [CrossRef]
6. Mertes, J.R.; Thompson, S.S.; Booth, A.D.; Gulley, J.D.; Benn, D.I. A Conceptual Model of Supra-Glacial Lake Formation on Debris-Covered Glaciers Based on GPR Facies Analysis: GPR Facies Analysis of Spillway Lake, Ngozumpa Glacier, Nepal. *Earth Surf. Process. Landf.* **2017**, *42*, 903–914. [CrossRef]
7. Strickland, R.M.; Covington, M.D.; Gulley, J.D.; Kayastha, R.B.; Blackstock, J.M. Englacial Drainage Drives Positive Feedback Depression Growth on the Debris-Covered Ngozumpa Glacier, Nepal. *Geophys. Res. Lett.* **2023**, *50*, e2023GL104389. [CrossRef]
8. Berthling, I.; Etzelmüller, B.; Isaksen, K.; Sollid, J.L. Rock Glaciers on Prins Karls Forland. II: GPR Soundings and the Development of Internal Structures. *Permafr. Periglac. Process.* **2000**, *11*, 357–369. [CrossRef]
9. Brandt, O.; Langley, K.; Kohler, J.; Hamran, S.-E. Detection of Buried Ice and Sediment Layers in Permafrost Using Multi-Frequency Ground Penetrating Radar: A Case Examination on Svalbard. *Remote Sens. Environ.* **2007**, *111*, 212–227. [CrossRef]
10. De Pascale, G.P.; Pollard, W.H.; Williams, K.K. Geophysical Mapping of Ground Ice Using a Combination of Capacitive Coupled Resistivity and Ground-Penetrating Radar, Northwest Territories, Canada. *J. Geophys. Res.* **2008**, *113*, F02S90. [CrossRef]
11. Degenhardt, J.J. Development of Tongue-Shaped and Multilobate Rock Glaciers in Alpine Environments—Interpretations from Ground Penetrating Radar Surveys. *Geomorphology* **2009**, *109*, 94–107. [CrossRef]
12. Capt, M.; Bosson, J.-B.; Fischer, M.; Micheletti, N.; Lambiel, C. Decadal Evolution of a Very Small Heavily Debris-Covered Glacier in an Alpine Permafrost Environment. *J. Glaciol.* **2016**, *62*, 535–551. [CrossRef]
13. Petersen, E.I.; Levy, J.S.; Holt, J.W.; Stuurman, C.M. New Insights into Ice Accumulation at Galena Creek Rock Glacier from Radar Imaging of Its Internal Structure. *J. Glaciol.* **2020**, *66*, 1–10. [CrossRef]

14. Merz, K.; Green, A.G.; Buchli, T.; Springman, S.M.; Maurer, H. A New 3-D Thin-skinned Rock Glacier Model Based on Helicopter GPR Results from the Swiss Alps. *Geophys. Res. Lett.* **2015**, *42*, 4464–4472. [CrossRef]
15. Blindow, N.; Salat, C.; Casassa, G. Airborne GPR Sounding of Deep Temperate Glaciers — Examples from the Northern Patagonian Icefield. In Proceedings of the 2012 14th International Conference on Ground Penetrating Radar (GPR), Shanghai, China, 4–8 June 2012; pp. 664–669.
16. Casassa, G.; Rodríguez, J.L.; Blindow, N. Airborne GPR on High Andean Glaciers—First Results from 6000 m Altitude. In Proceedings of the 15th International Conference on Ground Penetrating Radar, Brussels, Belgium, 30 June–4 July 2014; pp. 728–733.
17. Langhammer, L.; Rabenstein, L.; Schmid, L.; Bauder, A.; Grab, M.; Schaer, P.; Maurer, H. Glacier Bed Surveying with Helicopter-Borne Dual-Polarization Ground-Penetrating Radar. *J. Glaciol.* **2019**, *65*, 123–135. [CrossRef]
18. Zamora, R.; Ulloa, D.; Garcia, G.; Mella, R.; Uribe, J.; Wendt, J.; Rivera, A.; Gacitúa, G.; Casassa, G. Airborne Radar Sounder for Temperate Ice: Initial Results from Patagonia. *J. Glaciol.* **2009**, *55*, 507–512. [CrossRef]
19. Francke, J.; Dobrovolskiy, A. Challenges and Opportunities with Drone-Mounted GPR. In *First International Meeting for Applied Geoscience & Energy Expanded Abstracts; Society of Exploration Geophysicists, Denver, CO, USA and Virtual, 26 September–1 October 2021;* Society of Exploration Geophysicists: Houston, TX, USA, 2021; pp. 3043–3047.
20. Valence, E.; Baraer, M.; Rosa, E.; Barbecot, F.; Monty, C. Drone-Based Ground-Penetrating Radar (GPR) Application to Snow Hydrology. *Cryosphere* **2022**, *16*, 3843–3860. [CrossRef]
21. Leuschen, C.; Hale, R.; Keshmiri, S.; Yan, J.B.; Rodriguez-Morales, F.; Mahmood, A.; Gogineni, S. UAS-Based Radar Sounding of the Polar Ice Sheets. *IEEE Geosci. Remote Sens. Mag.* **2014**, *2*, 8–17. [CrossRef]
22. Teisberg, T.O.; Schroeder, D.M.; Broome, A.L.; Lurie, F.; Woo, D. Development of a Uav-Borne Pulsed ICE-Penetrating Radar System. In Proceedings of the IGARSS 2022—2022 IEEE International Geoscience and Remote Sensing Symposium, Kuala Lumpur, Malaysia, 17–22 July 2022; pp. 7405–7408.
23. Ruols, B.; Baron, L.; Irving, J. Development of a Drone-Based Ground-Penetrating Radar System for Efficient and Safe 3D and 4D Surveying of Alpine Glaciers. *J. Glaciol.* **2023**, *volume*, 1–12. [CrossRef]
24. Johnson, P.G. Glacier-Rock Glacier Transition in the Southwest Yukon Territory, Canada. *Arct. Alp. Res.* **1980**, *12*, 195. [CrossRef]
25. Bolch, T.; Rohrbach, N.; Kutuzov, S.; Robson, B.A.; Osmonov, A. Occurrence, Evolution and Ice Content of Ice-debris Complexes in the Ak-Shiirak, Central Tien Shan Revealed by Geophysical and Remotely-sensed Investigations. *Earth Surf. Process. Landf.* **2019**, *44*, 129–143. [CrossRef]
26. Evin, M.; Fabre, D.; Johnson, P.G. Electrical Resistivity Measurements on the Rock Glaciers of Grizzly Creek, St Elias Mountains, Yukon. *Permafr. Periglac. Process.* **1997**, *8*, 179–189. [CrossRef]
27. Porter, C.; Howat, I.; Noh, M.-J.; Husby, E.; Khuvis, S.; Danish, E.; Tomko, K.; Gardiner, J.; Negrete, A.; Yadav, B.; et al. *ArcticDEM—Mosaics, Version 4.1 2023*; Polar Geospatial Center: Saint Paul, MN, USA, 2023.
28. McCarthy, M.; Pritchard, H.; Willis, I.; King, E. Ground-Penetrating Radar Measurements of Debris Thickness on Lirung Glacier, Nepal. *J. Glaciol.* **2017**, *63*, 543–555. [CrossRef]
29. Wu, Z.; Liu, S. Imaging the Debris Internal Structure and Estimating the Effect of Debris Layer on Ablation of Glacier Ice. *J. Geol. Soc. India* **2012**, *80*, 825–835. [CrossRef]
30. Giese, A.; Arcone, S.; Hawley, R.; Lewis, G.; Wagnon, P. Detecting Supraglacial Debris Thickness with GPR under Suboptimal Conditions. *J. Glaciol.* **2021**, *67*, 1108–1120. [CrossRef]
31. Thomson, L.I.; Osinski, G.R.; Pollard, W.H. The Dielectric Permittivity of Terrestrial Ground Ice Formations: Considerations for Planetary Exploration Using Ground-penetrating Radar. *J. Geophys. Res.* **2012**, *117*, 2012JE004053. [CrossRef]
32. Meng, T.M.; Petersen, E.I.; Holt, J.W. Rock Glacier Composition and Structure from Radio Wave Speed Analysis with Dipping Reflector Correction. *J. Glaciol.* **2023**, *69*, 639–657. [CrossRef]
33. Noviello, C.; Gennarelli, G.; Esposito, G.; Ludeno, G.; Fasano, G.; Capozzoli, L.; Soldovieri, F.; Catapano, I. An Overview on Down-Looking UAV-Based GPR Systems. *Remote Sens.* **2022**, *14*, 3245. [CrossRef]
34. Booth, A.D.; Koylass, T.M. Drone-Mounted Ground-Penetrating Radar Surveying: Flight-Height Considerations for Diffraction-Based Velocity Analysis. *Geophysics* **2022**, *87*, WB69–WB79. [CrossRef]
35. Lambot, S.; Antoine, M.; Vanclooster, M.; Slob, E.C. Effect of Soil Roughness on the Inversion of Off-ground Monostatic GPR Signal for Noninvasive Quantification of Soil Properties. *Water Resour. Res.* **2006**, *42*, 2005WR004416. [CrossRef]
36. Dujardin, J.-R.; Bano, M. Topographic Migration of GPR Data: Examples from Chad and Mongolia. *Comptes Rendus. Géoscience* **2013**, *345*, 73–80. [CrossRef]
37. Conyers, L.B.; Goodman, D. *Ground-Penetrating Radar: An Introduction for Archaeologists*; Altamira Press: London, UK, 1997.
38. Sheriff, R.E. Limitations on Resolution of Seismic Reflections and Geologic Detail Derivable from Them[1]. In *Seismic Stratigraphy—Applications to Hydrocarbon Exploration*; American Association of Petroleum Geologists: Tulsa, OK, USA, 1977; ISBN 978-1-62981-205-2.
39. Baraer, M. *GPR Data Collected at Shar Shaw Tagà in June 2023*; Zenodo: Geneva, Switzerland, 2024.

MDPI AG
Grosspeteranlage 5
4052 Basel
Switzerland
Tel.: +41 61 683 77 34

Remote Sensing Editorial Office
E-mail: remotesensing@mdpi.com
www.mdpi.com/journal/remotesensing